Solid and Hazardous Waste Disposal and Resource Utilization

Solid and Hazardous Waste Disposal and Resource Utilization

Guest Editors

Yang Chen
Qinzhong Feng
Liyuan Liu

MDPI

Basel • Beijing • Wuhan • Barcelona • Belgrade • Novi Sad • Cluj • Manchester

Guest Editors

Yang Chen
College of Resources and Environment
University of Chinese Academy of Sciences
Beijing
China

Qinzhong Feng
College of Resources and Environment
University of Chinese Academy of Sciences
Beijing
China

Liyuan Liu
College of Resources and Environment
University of Chinese Academy of Sciences
Beijing
China

Editorial Office
MDPI AG
Grosspeteranlage 5
4052 Basel, Switzerland

This is a reprint of the Special Issue, published open access by the journal *Processes* (ISSN 2227-9717), freely accessible at: https://www.mdpi.com/journal/processes/special_issues/72F4KQTZJY.

For citation purposes, cite each article independently as indicated on the article page online and as indicated below:

Lastname, A.A.; Lastname, B.B. Article Title. *Journal Name* **Year**, *Volume Number*, Page Range.

ISBN 978-3-7258-5047-1 (Hbk)
ISBN 978-3-7258-5048-8 (PDF)
https://doi.org/10.3390/books978-3-7258-5048-8

Contents

About the Editors

Yang Chen

Yang Chen received his Ph.D. in Environmental Science from the Institute of Applied Ecology, Chinese Academy of Sciences, Shenyang, China, in 2007. From 2015 to 2020, he was a Research Professor and concurrently the Executive Deputy Director of the National Environmental Protection Engineering Center for Mercury Pollution Control and Resource Recovery, Beijing Comprehensive Research Center, Chinese Academy of Sciences. Since May 2020, he has been a Professor with the College of Resources and Environment, University of Chinese Academy of Sciences, Beijing, China. His research interests include solid and hazardous waste treatment, resource recovery, mercury and dioxin pollution control, non-thermal plasma technologies, and environmental standard and policy development. He has led or participated in more than 40 national and international projects, including the National Key R&D Program, National High-Tech R&D Program (863 Program), and Global Environment Facility initiatives. He has been granted over 40 national and international patents, authored or co-authored 15 monographs, and published more than 100 peer-reviewed papers. It has won more than 20 provincial and ministerial science and technology and talent awards; among his honors are the United Nations Industrial Development Organization Project Merit Award, the National Leading Talent in Environmental Science and Technology of China award, and the Provincial Leading Talent in Innovation and Entrepreneurship award.

Qinzhong Feng

Qinzhong Feng received his Ph.D. in Environmental Science from the Research Center for Eco-Environmental Sciences, Chinese Academy of Sciences, Beijing, China, in 2009. Since 2020, he has been an Associate Professor with the College of Resources and Environment, University of Chinese Academy of Sciences, Beijing, China, and concurrently serves as Director of the Technology R&D Department at the National Environmental Protection Engineering Center for Mercury Pollution Control and Resource Recovery. His research interests include mercury and other heavy-metal pollution control, non-thermal plasma technologies, solid-waste and hazardous-waste treatment, resource recovery, and environmental policy and standards. He has led or participated in more than ten national and provincial projects, including the National High-Tech R&D Program (863 Program), National Key R&D Program, and National Natural Science Foundation of China projects. He has published over 80 peer-reviewed papers, applied for more than 20 national invention patents, edited or co-edited 7 monographs, and contributed to the formulation of more than 10 national environmental standards and technical guidelines.

Liyuan Liu

Liyuan Liu received her Ph.D. in Environmental Engineering from China University of Mining and Technology (Beijing), Beijing, China, in 2012 . She was an Assistant Researcher with the Institute of High Energy Physics, Chinese Academy of Sciences, Beijing, China, from 2012 to 2015. From 2015 to 2020, she was an Associate Researcher with the Beijing Comprehensive Research Center, Chinese Academy of Sciences. Since May 2020, she has been an Associate Professor with the College of Resources and Environment, University of Chinese Academy of Sciences, Beijing, China. Her research interests include solid waste and chemical management, disposal and resource utilization of solid waste, emission inventories of pollutants, and coordinated pollution and carbon reduction.

processes

Article

New Trends in Pollution Prevention and Control Technology for Healthcare and Medical Waste Disposal in China

Liyuan Liu [1], Yue Gong [2,*], Yanrong Miao [3], Jianbo Guo [1], Hongfei Long [1], Qinzhong Feng [1] and Yang Chen [1,*]

[1] College of Resources and Environment, University of Chinese Academy of Sciences, Beijing 101408, China; liuliyuan@ucas.ac.cn (L.L.); guojianbo@ucas.ac.cn (J.G.); longhongfei22@mails.ucas.ac.cn (H.L.); fengqinzhong@ucas.ac.cn (Q.F.)

[2] National Science Library, Chinese Academy of Sciences, Beijing 100190, China

[3] National Library of China, Beijing 100081, China; myr@nlc.cn

* Correspondence: gongy@mail.las.ac.cn (Y.G.); chenyang@ucas.ac.cn (Y.C.)

Abstract: This study explores the progression of global healthcare and medical waste (HMW) disposal technologies and emerging practices in China including the COVID-19 pandemic period through patent technology innovation analysis. Trends were identified through both the Derwent Innovation Index database and bibliometric methods. Based on the bibliometric analysis of 4128 patents issued from 2002 to 2021, the development status and research trends of HMW disposal technology were revealed. Regarding patents, China significantly advanced post-2011. However, a large number of applications are filed only in China and are more focused on domestic rather than overseas markets. As the pandemic remains a threat, and increasing amounts of medical waste are generated, new technologies are being sought in China that will be safer for humans and the environment, and will also be in line with the zero waste technology trend. Incineration and waste crushing are core methodologies in medical waste disposal. Future directions pivot towards innovations in large-scale and distributed processing equipment, automation and unmanned systems and high-temperature steam disinfection collaborative disposal methods—including the "High temperature steam–municipal solid waste incineration collaborative technology" and the "High temperature steam–thermal magnetic gasification collaborative technology". This patent analysis enhances our understanding of the impact of COVID-19 on HMW disposal practices, guiding improved policymaking and research in the HMW sector.

Keywords: healthcare and medical waste (HMW); COVID-19; waste disposal; technology innovation evolution

1. Introduction

Recent pandemics, including H1N1 influenza A, Ebola, and Middle East respiratory syndrome, underscored global healthcare challenges associated with the management of healthcare and medical waste (HMW). The emergence of COVID-19 in 2020 catalyzed an unparalleled strain on global medical infrastructure, accentuating the intricacies of HMW management [1–7]. Such medical waste, characterized by its infectious, toxic, and environmentally deleterious nature, poses an augmented environmental threat during pandemics [8].

Although esteemed bodies, including the WHO, UNICEF, and UNEP, have disseminated guidelines for healthcare waste management during pandemics [9–15], a comprehensive understanding of the dynamic landscape of HMW disposal and associated research trends remains nascent.

China, post-SARS, has emerged as one of the focal points in pandemic response, developing an intricate HMW management system. A series of HMW disposal technologies, e.g., rotary kilns [16], pyrolysis [16], high-temperature hot air treatment [17], high-temperature steaming [18], chemical disinfection [19], microwave [20], and technical documents that permit joint management among relevant management agencies, formed a complete closed-loop medical waste management system in China. Yet, the unprecedented surge in medical waste due to COVID-19 underscored the need for agility and resilience in waste management systems. Between January and June 2020, China addressed approximately 447,000 metric tons of medical waste, ameliorating their disposal strategies and capacities [21]. Factors such as the amount of waste, costs, maintenance and types of waste, etc. should be taken into consideration when selecting appropriate disposal technologies [22]. Incineration technology could be adopted when the amount of waste is large and the investment is sufficient when disinfecting pathological or pharmaceutical waste. If the scale of the hospital is smaller and the investment is limited, chemical disinfection and high-temperature steam disinfection—which are easily maintained—are preferred. At the same time, the emergence of some combination technologies such as "high temperature steam–municipal solid waste incineration collaborative technology [23]" has also contributed to the diversification of medical waste disposal technologies due to the emergence of needs such as emergency disposal and local disposal.

Patents, pivotal for the dissemination and safeguarding of technical innovations, present a tangible metric for delineating technological trajectories in specific domains [24–27]. Harnessing repositories such as the Derwent Innovations Index (DII) and Patsnap, this study undertook a bibliometric excursion. (Patsnap is a worldwide patent search, analytics and collaboration platform (PatSnap Information Technology (Suzhou) Ltd., Suzhou, China), including patent data from 162 countries (regions) around the world, including China, the United States, Japan, South Korea, Germany, the United Kingdom, France, Australia, the European Patent Office and WIPO etc. More than 180 million patent documents are collected, covering 170 countries/regions are collected, with a time range from 1790-now. The update frequency of Patsnap is more than once a week. https://www.zhihuiya.com/ (accessed on 18 December 2023)). A total of 4128 patents were identified. This study was geared towards unearthing market evolution and technological categorizations in patent filings pertinent to HMW disposal. A pronounced emphasis was allocated to discerning the technological pathways in China, with attention concurrently cast to the burgeoning trends during the COVID-19 pandemic.

2. Materials and Methods

2.1. Data Collection

Data were obtained from searches of DII and Patsnap databases. Details were as follows: (1) ALL = ((Medical incineration Flue gas) OR (Medical incineration exhaust gas) OR "Hospital waste liquid" OR "Hospital liquid waste" OR "Hospital Sewage" OR "Hospital wastewater" OR "Medical waste liquid" OR "Medical liquid waste" OR "Medical Sewage" OR "Medical wastewater" "medical waste" OR "clinical waste" OR "Hospital waste" OR "Infectious Medical waste" OR "Infectious waste" OR "pathological waste" OR "Pathogenic waste" OR "Pathogenic Medical waste"). (2) The application year was set from 2002 to 2021, as the first patent related to HMW was collected in the database in 2002. (3) The language of the patent was set to Chinese, English, Japanese, Korean and French. (4) Numbers of patent families related to HMW disposal technology were grouped by the earliest priority year. (5) The relevant data are exported in CSV format from the DII and Patsnap databases, integrated according to the patent publication number, and a total of 5580 patents were collected. (6) In order to obtain accurate articles related to HMW disposal technology, a second round comprising manual screening of 4128 patents based on title, abstract, and content was performed (Figure 1), and the patents were imported into the Patsnap system in Excel data format for online analysis.

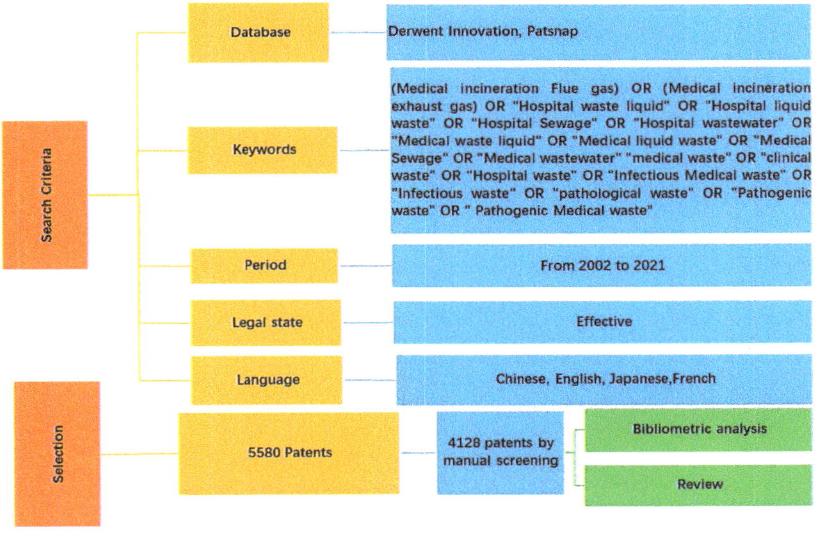

Figure 1. Document selection and flow diagram of data collection on HMW disposal technology.

2.2. Patent Analysis and Visualization

Bibliometric searches were conducted using the Patsnap system, i.e., application trends, patent competition layout, followed by the research hotspot analysis, i.e., keywords cluster. Here, we employed the following criteria for the bibliometric analysis:

Application trends: This was based on the year of patent application to reveal the trend of patent applications over time.

Patent competition layout: This analysis combined the dimensions of application year and priority countries. We observed the target market by counting the priority countries where all members of the patent family applied. By comparing the "earliest priority" country and the "priority countries" of family members, we performed an analysis of patent flow.

Microsoft Excel 2020 and Apache ECharts 5.3 software were used for the visualization of the above results.

Keywords cluster: A visual analysis of co-occurrence of keywords was prepared based on the International Patent Classification (IPC). Keywords for HMW disposal were extracted from abstracts and divided into nine clusters distributed on a 3D map using Patsnap. The white peak represents hotpots of current research, and the valley indicates blind spots. Categories are displayed in different colors.

The 3D patent map serves as a visual representation of the primary technological layout based on IPC classification. In this context, Patsnap utilizes text clustering to construct the topography, followed by an in-depth analysis of all patent texts within each cluster. From this analysis, nine groups of key terms were extracted, as illustrated in Supplementary Materials. Within this depiction, the "terrain" of the patent map reflects the quantity of patents within different technology clusters: white peaks signify areas of concentrated technological focus, while valleys signify technological blind spots. To enhance our comprehension of these technologies, the nine clusters were manually consolidated into three major categories, guided by the technical labels and IPC classification codes. These categories are visually represented on the map in distinct colors: blue, red, and yellow. Each colored point on the map corresponds to an individual patent.

The blue category, focused on pollution prevention, includes pretreatment (IPC: B02C18), incineration/non-incineration, and landfilling (IPC: F23G5, A61L11). The red category includes pollution-control technologies of wastewater disposal (IPC: C02F9), liquid waste disposal (IPC: C02F1), and flue gas disposal (IPC: B01D53). The yellow category introduces waste management technologies, such as management in collection and transportation (IPC: A41D13), management systems (IPC: G05B19), and special evaluation methods (IPC: G06Q50).

3. Results and Discussion

3.1. Patent Application Activities of HMW Disposal (2002 to 2021)

An examination of China's patent application trends over the past two decades reveals that the patent application activity experienced a "slow growth period" in 2002–2010 and a "rapid development period" in 2011–2019, mirroring global trends (Figure 2). Several technical guidelines [28–35] were issued in 2003 for regulation of SARS medical waste generated by medical and health institutions in Figure 1. Between 2009 and 2010, the Response Plan on the Management of Medical Waste for Influenza A (H1N1) [36,37] was issued by the Ministry of Environmental Protection of China to ensure that medical waste was disposed of in a timely, orderly, efficient, and harmless manner. A pronounced increase in applications in China was observed with the AGR reaching 20.4% after 2011. Meanwhile, the AGR remained high at 25.0% (2015), 32.7% (2016), and 37.2% (2018), which aligns notably with the emergence of significant public health crises, such as Zika in 2015–2016 and Ebola in 2018. With an average AGR of 26.8%, there is a discernible uptick in Chinese research and development efforts. Post-2011, China rolled out a plethora of pertinent technical standards [38,39] to support the development of HMW disposal technology, including promotion of innovation and technological transformation, disposal technology selection and performance testing and further cultivating a conducive ecosystem for the evolution of HMW disposal technologies. The strengthening of government regulation in this period also affected the growth in patent submissions.

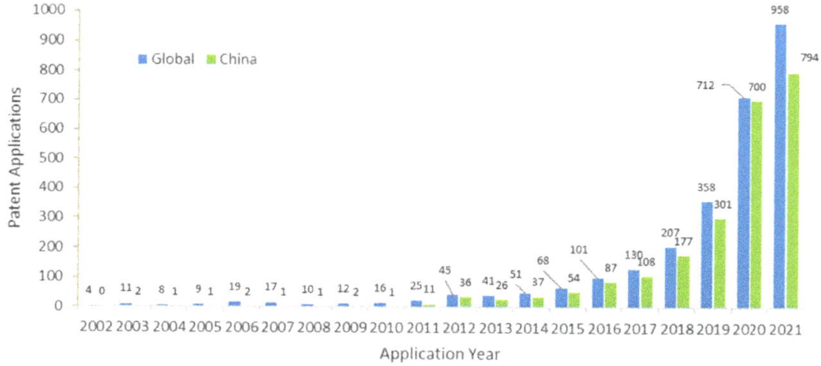

Figure 2. Chronological application activities of medical waste disposal technology patents. The statistics are based on publicly available documents—Patent applications are normally published 3–18 months after filing; therefore, rendering the 2021 data as provisional.

The emergence of the COVID-19 pandemic and ensuing lockdown measures profoundly impacted HMW management, thereby garnering intensified research scrutiny during this period due to the infectious nature of the SARS-CoV-2 virus. In February 2020, the Chinese government issued its Work Plan for Comprehensive Treatment of Waste in Medical Institutions, which requires all cities above the prefecture-level to build at least one centralized medical waste disposal facility by the end of 2020 and each county to build a medical waste system that consists of the collection, transfer, and disposal parts by 2022 [40]. Notwithstanding the proposal of an ambitious large-scale blueprint for medical waste disposal infrastructure, the intricacies of their execution remain paramount. By the close of 2018, HMW incineration facilities comprised 37% of the total facility count in

China. Incineration facilities are primarily pyrolysis furnaces; non-incineration facilities such as high-temperature steam treatment, microwave, chemical disinfection and other technologies are also used [41]. Prior to the advent of COVID-19, centralized disposal facilities predominantly employed unwieldy equipment, which were challenging both in terms of mobility and installation. Furthermore, these facilities exhibited rigidity, falling short of the specialized requirements for managing voluminous amounts of highly contagious materials. Movable waste disposal facilities, hazardous and municipal solid waste incineration facilities, industrial furnaces, cement rotary kilns, and other facilities can be used for emergency disposal in China [40]. Operational parameters necessitate calibration to accommodate medical waste processing [42].

3.2. Geography of Patent Filings in HMW Disposal (2002 to 2021)

Assessing the geographical distribution of patent filings helps to understand the activity of relevant technologies in China, as well as to identify the main countries producing technologies and their markets. Historically, key patent offices, which indicate the perceived target markets for HMW disposal technology, included Korea, the United States, Japan, Ireland, and Spain. Nevertheless, these jurisdictions have witnessed a decline in patenting activities since 2011. Concurrently, China has experienced a burgeoning trend in patent submissions (see Supplementary Materials), eclipsing those of the United States and constituting over 72.2% of the global aggregate since 2012 (Figure 3).

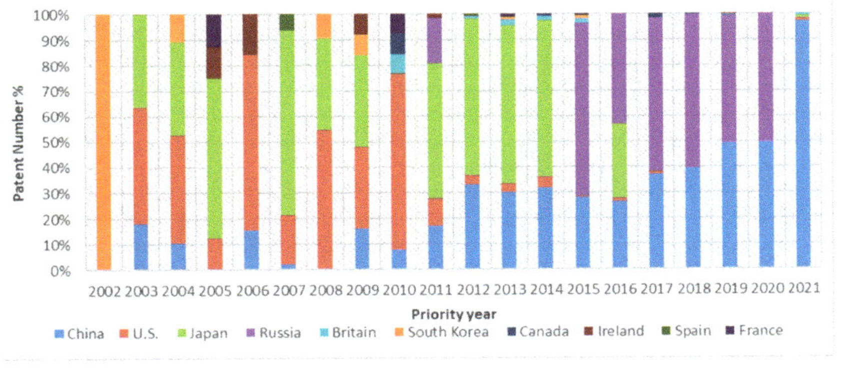

Figure 3. Patent filing trends related to HMW disposal technologies across primary market countries/regions by year.

In examining target markets, overall, >85.5% of the patent families in the dataset were safeguarded within a single jurisdiction, reflecting a predominant interest in a single (mostly local) market, especially in China. A mere 1% of initial patents filed in China are subsequently filed in another jurisdiction, compared with 6.7–23.1% percent being re-filed from all other offices. China trails among the five offices (China, Korean, Japan, EPO, U.S.) containing families with at least one granted patent. While myriad factors may underpin these statistics, the pronounced volume of applications exclusively filed in China suggests a Chinese proclivity for the domestic over the international market in the realm of HMW disposal. The substantial proportion of applications localized to China, juxtaposed against a relatively augmented refiling rate in the USPTO and the other three offices, may indicate that HMW disposal technologies are innately tethered to local development and operational contexts (Figure 4).

China's HMW disposal technologies, such as steam-based treatment, microwave treatment, frictional heat treatment, and other non-incineration processes, have been locally developed and refined. These technologies are notably tailored to suit China's unique environmental conditions, offering cost-effective, operationally streamlined, and space-efficient solutions.

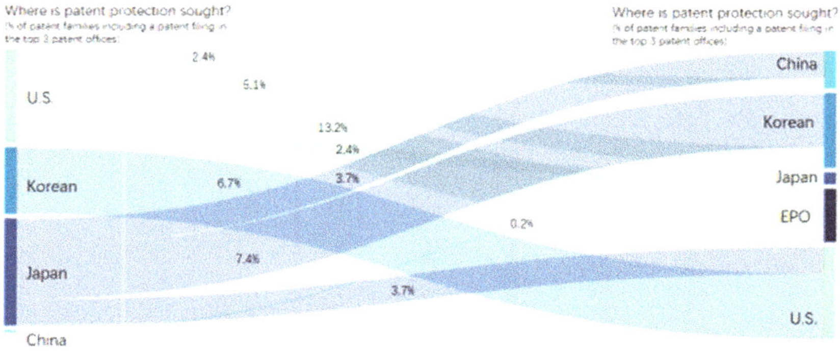

Figure 4. Patent applications in the main patent offices in terms of number of patent applications filed in 2002–2021 for HMW disposal technology.

In contrast the United States, Japan, and European nations have traditionally shared more convergent process systems distinct from those prevalent in China. Historically, these regions have represented crucial markets for one another, a trend evident in their corresponding patent activities. For example, incineration technology was extensively embraced for HMW disposal in the United States throughout the last century, driven by active promotion from the Environmental Protection Agency (USEPA). However, the ubiquity of such technologies has gradually dwindled since the late 1990s. In Japan and European countries, incineration and vacuum autoclave technologies have predominated as primary options for HMW disposal.

In just over a decade, China's medical waste disposal technology has gone through a process of self-disposal by hospitals, simple incineration during the SARS period, centralized incineration under national planning, and separation of planned incineration and non-incineration technologies. With the application and promotion of BAT/BEP, after nearly a decade of development, the overall level of medical waste disposal technology in China has improved, and incineration is developing towards large-scale. The scale of non-incineration treatment is constantly expanding, and the technology tends to be regional synergy and complementarity.

A noteworthy development since 2020 is the heightened international reach of Chinese patents related to HMW disposal technologies. This expansion can be attributed to adaptive enhancements in HMW disposal technologies during the COVID-19 pandemic, coupled with the burgeoning overseas application experience of Chinese patentees. This trend is anticipated to persist, with Chinese patents extending to an even wider array of overseas patent offices in the future.

3.3. Technical Status of HMW Disposal Process in China

Nine clusters were manually merged into three major categories created using technical labels and IPC classification codes (Figure 5). Active technical areas of waste disposal are provided with labels on the three prominent peaks: (1) Crushing equipment, waste plastics, grinding, crushing; (2) Disinfection devices, ultraviolet light, composite light, visible light, antibiotics; and (3) Plastics, recycling, mixing, incinerators. The two peaks in wastewater treatment reflect: (1) Contact oxidation, regulating, deacidification, and sedimentation tanks, and filtration; and (2) Wastewater purification, drainage pipe, sewage collection. It is worth noting that during the COVID-19 pandemic, movable disposal technology (IPC: F23G7) was also an active area of development (marked in orange). At the intersection of these categories, a conspicuous low-patent-density valley emerges. Technologies on either side of this valley encompass HMW pollution risk assessment methods (CN106096268B), devices for the treatment of radioactive medical waste (CN110526245A), and methods for treating carcinogens produced in medical wastewater (CN107758924A). These technologies potentially signify prospects for further advancements in these domains.

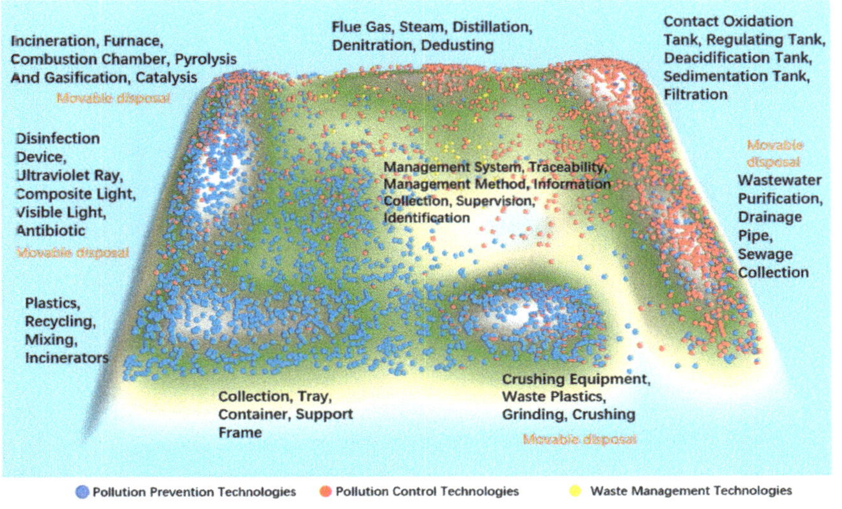

Figure 5. 3D patent map of HWM keyword co-occurrence.

Managing wastes associated with the COVID-19 epidemic requires special attention due to viral infectivity [12]. Proper waste management procedures are crucial to maintain the safety of health workers and communities [9]. Proper pollution prevention and selection of control technology are also critical for healthcare facilities and COVID-19 treatment units [14]. Different from previous conventional technologies, the main innovations of HMW disposal technology in China during the COVID-19 epidemic are discussed below (Figure 6). WHO guidelines [43] categorize healthcare waste into eight major groups, including both hazardous and nonhazardous components. While medical waste in China is normally divided into five categories –infectious, pathological, sharps, pharmaceutical, and chemical [44]. Much waste generated by healthcare facilities during the COVID-19 pandemic was general noninfectious waste, e.g., packing, food, and disposable towels. General waste was typically segregated from infectious waste in clearly marked bins, and then bagged and disposed with general municipal waste. Infectious waste produced during patient care, including those patients with confirmed COVID-19 infections (e.g., sharps, bandages, and pathological waste) was intended for collection in clearly marked lined containers and sharps boxes [41]. Therefore, the collection of COVID-19 medical waste needs to use negative pressure isolation, hands-free opening, intelligent classification, convenient packaging, automatic packaging, and compression technology (Figure 6I) to avoid excessive contact with medical staff and waste-collection personnel. Intelligent and automatic technologies such as RFID, Scan QR code, tracking systems, etc. are widely used for the transfer, tracking, loading/unloading, and cleanup for waste transfer and transport processes (Figure 6II). Intelligent and automatic technologies are also widely used for sterilization, disinfection, crushing, solid–liquid separation, feeding, and packing waste during pretreatment (Figure 6III). Patent applications were active in waste crushing which accounts for 13.2% of the total. Disposal technologies of HMW were the dominant direction in the whole field which accounts for 39.4%, with incineration technology (Figure 6IV) accounting for 22.9%. Movable and collaborative disposal of medical wastes applications greatly reduce risks of multiple contacts between infectious waste and workers during collection, loading, unloading, and other processes.

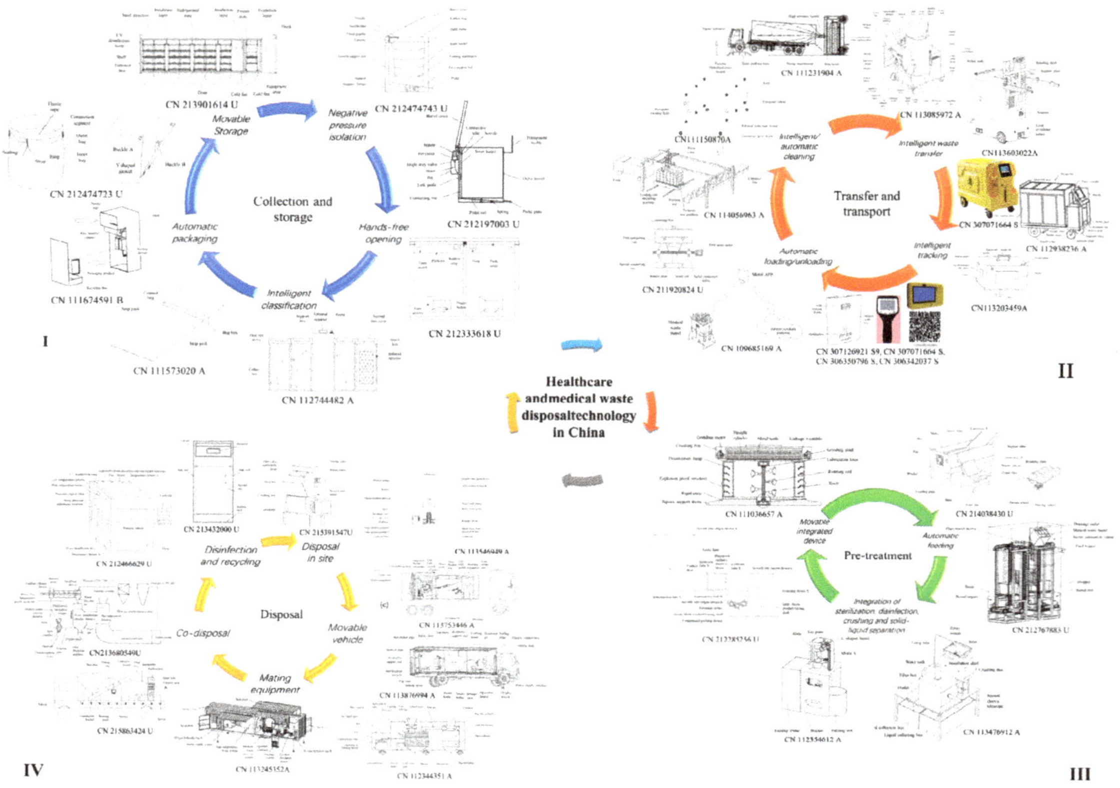

Figure 6. Active technology development for healthcare and medical waste disposal in China. (**I** indicate collection and storage process, **II** indicate transfer and transport process, **III** indicate pretreatment process, **IV** indicate disposal process).

Prior to the COVID-19 outbreak in China, centralized medical waste disposal facilities used large-scale equipment that was difficult to move and install, and lacked the flexibility to adapt to the special needs of disposing of large quantities of highly infectious medical wastes unexpectedly during an outbreak. After the outbreak of the epidemic—and in addition to research and development into centralized medical waste disposal facilities—there was more research into new methods. Mobile medical waste disposal facilities and the co-processing of medical wastes by co-disposal facilities such as hazardous waste incineration facilities (CN219346508U, incineration temperature higher than 1100 °C), municipal solid waste incineration facilities (CN110486727A, CN217732843U, incineration temperature higher than 850 °C) and industrial kilns (CN112207115B, kiln temperature higher than 850 °C)—which meet the higher demand for medical waste disposal, and also improve the non-incineration technologies—have been developed, to more rapidly eliminate or reduce the intensity of infection of medical waste during epidemics. However, incineration technology dominates in the centralized disposal of medical waste. Due to the high operating costs and the high requirements in controlling dioxin pollution of centralized incineration facilities, in the case of decentralized or mobile emergency disposal, non-incineration technologies have also undergone considerable technological innovation. Against the background of Basel Convention implementation, Chinese medical waste disposal technology also underwent a gradual transition from incineration to non-incineration methods, such as high-temperature steam, microwave, and chemical disinfection. Among them, the high-temperature steam disinfection collaborative technology will be widely applied. (The scope of application of different medical waste treatment and disposal technologies is shown in Table 1) These collaborative technologies can fully leverage the advantages of low operating cost and low pollution in the treatment process of high-temperature steam disinfection process, achieving harmless treatment of medical waste. At the same time, the range of high-temperature steam disinfection technologies can be expanded by combining other medical waste disposal technologies. The "High temperature steam–municipal solid waste incineration collaborative technology (CN215489776U)" can send the disinfected waste residue to the municipal solid waste incineration plant for incineration, solving the final problem of reducing the volume of medical waste after disinfection. The "High temperature steam–thermal magnetic gasification collaborative technology (CN218146491U)" can disposal all five types of medical waste, among which infectious, damaging, and pathological medical waste is disinfected by high-temperature steam; chemical and pharmaceutical medical waste is disposed of through thermal magnetic gasification. The measured dioxin emission value of 0.013 ngTEQ/m^3 of medical waste disposed by the thermal magnetic gasification disposal technology is in line with the emission standard for medical waste disposal technology in China (0.5 ngTEQ/m^3) and also lower than that for medical waste disposal technology in the EU (0.1 ngTEQ/m^3).

Table 1. Scope of application of different medical waste treatment and disposal technologies.

Technology		Infectious Waste	Pathological Waste	Injurious Waste	Pharmaceutical Waste	Chemical Waste
Incineration	Incineration	√	√	√	√	√
	Pyrolysis	√	√	√	√	√
	Grate furnace	√	√	√	√	allow part
	Plasma	√	√	√	√	√
Non-incineration	High-temperature steam disinfection	√	×	√	×	×
	Microwave disinfection	√	×	√	×	×
	Chemistry disinfection	√	×	√	×	×
	High temperature dry heat disinfection	√	×	√	×	×
	Electronic irradiation disinfection	√	×	√	×	×
New technology	High temperature steam—municipal solid waste incineration collaborative	√	√	√	√	√
	High temperature steam—thermal magnetic gasification collaborative	√	√	√	√	√
	Friction heat treatment	√	√	√	×	×

"√" indicate that the category of medical wastes can be handled through this technology, "×" means indicate that the category of medical wastes cannot be handled through this technology.

Waste disposal processes also require terminal control capability to prevent the release of waste to the surrounding environment. Control technologies include flue gas purification, wastewater treatment, and solid waste disposal (Figure 7). Patent analysis in flue gas purification shows focus on high-temperature steam treatment, microwave and chemical disinfection process (30.4%), incineration process (64.5%), and wastewater treatment processes (5.1%). The main IPC classification number involved is B01D53. Flue gas purification technologies include desulfurization, denitrification, dust removal, dioxin and heavy metal removal, disinfection, and sterilization. Flue gas purification from incineration processes includes organic and inorganic waste gas treatment, odor abatement, flue gas control, and waste heat boiler-related processes. Medical wastewater mainly comprises liquid and wastewater contaminated with pathogens generated during the collection and disposal of medical waste. The main IPC classification number was C02F9 and mainly in pretreatment (60.3%), biological (0.7%) and disinfection processes (39.4%), which comprised 6.1% of patents in the HMW disposal field. Pretreatment mainly concerns physical treatment, such as grids, sedimentation, and air flotation. Biological treatment mainly focuses on degradation and transformation of pollutants in sewage into sludge via microorganism activity. Disinfection treatment examines sewage disinfection with chlorine, ultraviolet radiation, and ozone. Solid wastes generated during the HMW disposal include fly ash, slag, and sewage sludge. These materials are all managed as hazardous waste in China, and thus, technical patents primarily address slag tapping, cooling, conveying, solidification/stabilization, recycling, disinfection/drying, and incineration.

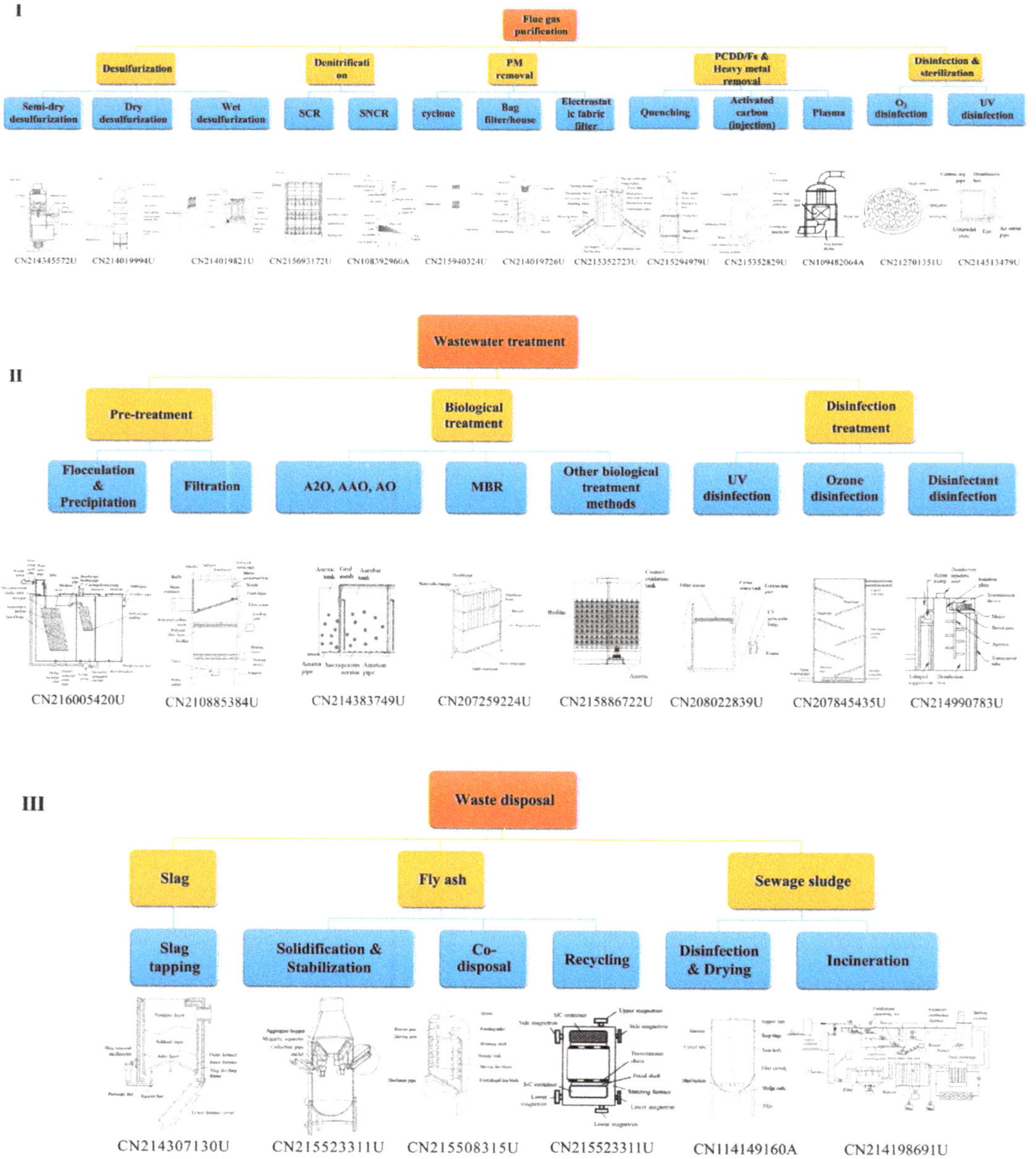

Figure 7. Active areas of technology for flue gas purification, wastewater treatment and solid waste disposal in the HMW disposal process (**I–III**).

Waste management mainly focuses on overall HMW systems. Waste management often uses decision analysis based on data obtained from sensor networks via the Internet of Things (CN104700210B), cloud platforms (CN205788259U), and intelligent networks (CN206348776U). The goal is to achieve intelligent control over production, classification, collection, storage, transportation, and disposal. Waste management often focuses on the collection and transportation processes for medical waste, including logistics of transportation positioning (CN210140185U), remote monitoring (CN203012472U), classification, screening, and collection (CN206046503U), waste weighing (CN210924622U), and collection of other data (CN207319709U). In addition, specific disposal processes may require special data collection (CN210775280U).

3.4. Challenges and Future Prospects

3.4.1. Challenges

1. As of the end of 2022, China has issued a total of 577 hazardous waste business licences for the disposal of medical waste (including 538 separate disposal units and 39 collaborative disposal units). Except for Danzhou and Sansha in Hainan, 337 cities at or above the prefecture level have built centralized disposal facilities. In 2022, the disposal capacity of medical waste management units nationwide was 2.61 million tons per year, with an actual disposal capacity of 2.09 million tons and a disposal load rate of 80%. Among them, 800,000 tons of medical waste generated in grassroots and remote areas account for about one-third of the national medical waste production, and the composition is complex and difficult to collect and cover. During the epidemic, the production volume increased sharply: Wuhan, for example, saw a daily production volume of over 200 tons of highly infectious waste—which was 3–5 times higher than usual).
2. Medical waste disposal technology has increasingly high requirements for efficiency, cost, safety, environmental protection, and occupational health. The existing incineration and non-incineration technologies have shortcomings, and the contradiction is becoming increasingly prominent. The incineration process is prone to producing dioxins and requires complex exhaust gas treatment systems, resulting in high operating and maintenance costs; non-incineration processes do not have exhaust emissions issues, but capacity reduction is limited, and waste residue still needs to be landfilled or sent to municipal solid waste incineration plants for incineration.
3. The treatment of chemical and pharmaceutical waste will inevitably become more important in future. The difficulty in disposing of these two types of waste lies in the wide variety and varying harmfulness of chemical and pharmaceutical waste. The current situation of the collection and transportation of pharmaceutical and chemical waste in Chinese hospitals is significantly different from that in Europe and America. Meanwhile, high quality and recyclable plastics account for 40–50% of the total amount of medical waste. After being landfilled, plastic degradation takes hundreds of years or even longer. At the same time, plastic is a key source of dioxin production during incineration. Due to the constraints of China's medical waste management policies, the resource utilization of medical waste still requires joint efforts from all parties.

3.4.2. Future Prospects

Based on the above research results, it can be concluded that some innovative technologies will be well applied in the future:

1. The large-scale and distributed processing equipment will develop in parallel. In North America, medical waste treatment operators generally use large high-temperature steam treatment equipment with a single processing capacity of over 15 tons/day, with a maximum processing capacity of up to 3 tons/h per unit. For large centralized processing plants: high-temperature cooking equipment with a daily processing capacity of over 15 tons, or incineration equipment with a single processing capacity

of over 30 tons, will become the mainstream configuration. For small and medium-sized medical waste treatment plants, a new generation of high-temperature cooking equipment with a daily processing capacity of less than 5 tons and ultra-low emission thermal magnetic gasification equipment will be vigorously developed.
2. Automation and unmanned systems that are more reliable and cost-effective will become increasingly popular. Due to the increasing cost of labor, automation equipment is crossing the critical point of cost-effectiveness, and the safety and occupational health risks of on-site workers are increasing. Reducing manual operations and pursuing zero contact between humans and medical waste are the future development trends. Real unmanned workshops and intelligent factories will be popularized in the near future.
3. The high-temperature steam disinfection collaborative technology will be widely applied. These collaborative technologies can fully leverage the advantages of low operating cost and low pollution in the treatment process of high-temperature steam disinfection process, achieving harmless treatment of medical waste. At the same time, the range of high-temperature steam disinfection technologies can be expanded by combining medical waste disposal technologies.

4. Conclusions

Chinese patents will be extended to overseas patent offices more frequently in the future with the adaptive improvement of HMW disposal technologies during the COVID-19 pandemic and the increasing overseas application experience of Chinese patentees. Within the realm of HMW disposal, incineration and waste crushing emerged as core methodologies. Future directions pivot towards innovations in large-scale and distributed processing equipment, automation and unmanned systems and high-temperature steam disinfection and incineration/thermal magnetic gasification disposal methods. This patent analysis enhances our understanding of the impact of COVID-19 on HMW disposal practices, guiding improved policymaking and research in the HMW sector.

Supplementary Materials: The following supporting information can be downloaded at: https://www.mdpi.com/article/10.3390/pr12010007/s1, Table S1: Top 10 patentees in the field of healthcare and medical waste disposal technology in China. Table S2: Details of major technologies of the IPC categories.

Author Contributions: Conceptualization, L.L. and Y.G.; methodology, Y.G. and Y.M.; software, Y.M.; validation, J.G. and H.L.; investigation, Q.F. and Y.C.; resources, Y.C.; data curation, L.L. and Y.G.; writing—original draft preparation, L.L. and Y.G.; writing—review and editing, L.L. and Y.G.; project administration, Y.C.; funding acquisition, Y.C. All authors have read and agreed to the published version of the manuscript.

Funding: This research was funded by the National Key Research and Development Program of China (2022YFC3902300), and the Program of National Science Library, Chinese Academy of Sciences: Intelligence Analysis of mercury and solid waste/hazardous waste polltion control (E0901154).

Data Availability Statement: Data are contained within the article and Supplementary Materials.

Acknowledgments: The authors acknowledge the invaluable comments by reviewers.

Conflicts of Interest: The authors declare no conflict of interest. The funders had no role in the design of the study; in the collection, analyses, or interpretation of data; in the writing of the manuscript; or in the decision to publish the results.

References

1. Zhou, F.; Yu, T.; Du, R.; Fan, G.; Liu, Y.; Liu, Z.; Xiang, J.; Wang, Y.; Song, B.; Gu, X.; et al. Clinical course and risk factors for mortality of adult inpatients with COVID-19 in Wuhan, China: A retrospective cohort study. *Lancet* **2020**, *395*, 1054–1062. [CrossRef] [PubMed]
2. Zhu, N.; Zhang, D.; Wang, W.; Li, X.; Yang, B.; Song, J.; Zhao, X.; Huang, B.; Shi, W.; Lu, R.; et al. A Novel Coronavirus from Patients with Pneumonia in China, 2019. *N. Engl. J. Med.* **2020**, *382*, 727–733. [CrossRef] [PubMed]

3. Yu, H.; Sun, X.; Solvang, W.D.; Zhao, X. Reverse Logistics Network Design for Effective Management of Medical Waste in Epidemic Outbreaks: Insights from the Coronavirus Disease 2019 (COVID-19) Outbreak in Wuhan (China). *Int. J. Environ. Res. Public Health* **2020**, *17*, 1770. [CrossRef] [PubMed]
4. Sadia, L.; Ranjan, S.R.; Kim, H. Disinfection technology and strategies for COVID-19 hospital and bio-medical waste management. *Sci. Total Environ.* **2020**, *749*, 141652.
5. Saadat, S.; Rawtani, D.; Hussain, C.M. Environmental perspective of COVID-19. *Sci. Total Environ.* **2020**, *728*, 138870. [CrossRef] [PubMed]
6. Zhu, Y.J.; Xie, J.G.; Huang, F.M.; Cao, L.Q. Association between short-term exposure to air pollution and COVID-19 infection: Evidence from China. *Sci. Total Environ.* **2020**, *727*, 138704. [CrossRef] [PubMed]
7. WHO. *Water, Sanitation, Hygiene and Waste Management for COVID-19*; World Health Organization: Geneva, Switzerland; United Nations Children's Fund: New York, NY, USA, 2020.
8. Mochammad, C.; Masaru, T.; Ashok, V.S. A system dynamics approach for hospital waste management. *Waste Manag.* **2008**, *28*, 442–449.
9. WHO. *Interim Guidance: Water, Sanitation, Hygiene, and Waste Management for SARS-CoV-2, the Virus that Causes COVID-19*; World Health Organization: Geneva, Switzerland, 2020. Available online: https://www.who.int/publications/i/item/WHO-2019-nCoV-IPC-WASH-2020.4 (accessed on 13 December 2023).
10. UNEP (United Nations Environment Programme). BEP Guideline. 2019. Available online: http://www.pops.int/Implementation/BATandBEP/Progressold/BATBEPGuidelines/tabid/377/Default.aspx (accessed on 13 December 2023).
11. WHO. *Management of Waste from Injection Activities at the District Level: Guidelines for District Health Managers*; World Health Organization: Geneva, Switzerland, 2006. Available online: https://apps.who.int/iris/handle/10665/43476 (accessed on 13 December 2023).
12. WHO. *Overview of Technologies for the Treatment of Infectious and Sharp Waste from Health Care Facilities*; World Health Organization: Geneva, Switzerland, 2019. Available online: https://www.who.int/publications/i/item/9789241516228 (accessed on 13 December 2023).
13. UNICEF. *Appropriate Disposal of Immunization Waste (ADIW) Platform*; United Nations Children's Fund: New York, NY, USA, 2020. Available online: https://www.technet-21.org/en/library/main/6388-appropriate-disposal-of-immunization-waste-(adiw)-platform (accessed on 13 December 2023).
14. UNICEF. *COVID-19 Emergency Preparedness and Response: WASH and Infection Prevention and Control in Health Care Facilities*; Guidance Note; United Nations Children's Fund: New York, NY, USA, 2020. Available online: https://www.unicef.org/media/66386/file/WASH-COVID-19-infection-prevention-and-control-in-health-care-facilities-2020.pdf (accessed on 13 December 2023).
15. UNEP/IGES (United Nations Environment Programme/Institute for Global Environmental Strategies). *Waste Management during the COVID-19 Pandemic: From Response to Recovery*; United Nations Environment Programme: Nairobi, Kenya; Institute for Global Environmental Strategies: Hayama, Japan, 2020. Available online: https://reliefweb.int/sites/reliefweb.int/files/resources/WMC-19.pdf (accessed on 13 December 2023).
16. MEE (Ministry of Ecology and Environment of the People's Republic of China). Technical Specifications for Centralized Incineration Facility Construction on Medical Waste (HJ/T 177). 2005. Available online: http://www.mee.gov.cn/ywgz/fgbz/bz/bzwb/other/hjbhgc/200505/t20050524_67082.shtml (accessed on 13 December 2023). (In Chinese)
17. Chen, Y.; Feng, Q.Z.; Liu, L.Y.; Chen, R.Z.; Zhang, Y.H.; Bao, Z. Reform and development of medical waste disposal technology system in the new era. *Chin. J. Environ. Eng.* **2021**, *15*, 382–388.
18. MEE (Ministry of Ecology and Environment of the People's Republic of China). Technical Specifications for Steam-Based Centralized Treatment Engineering on Medical Waste (on Trial) (HJ/T 276). 2006. Available online: http://www.mee.gov.cn/ywgz/fgbz/bz/bzwb/other/hjbhgc/200608/W020110127377094539894.pdf (accessed on 13 December 2023). (In Chinese)
19. MEE (Ministry of Ecology and Environment of the People's Republic of China). Technical Specifications for Chemical Disinfection Centralized Treatment Engineering on Medical Waste (on Trial) (HJ/T 228). 2006. Available online: http://www.mee.gov.cn/gkml/zj/gg/200910/t20091021_171615.htm (accessed on 13 December 2023). (In Chinese)
20. MEE (Ministry of Ecology and Environment of the People's Republic of China). Technical Specifications for Microwave Disinfection Centralized Treatment Engineering on Medical Waste (on Trial) (HJ/T 229). 2006. Available online: http://www.mee.gov.cn/ywgz/fgbz/bz/bzwb/other/hjbhgc/200603/W020111221394894641492.pdf (accessed on 13 December 2023). (In Chinese)
21. Liu, L.Y.; Wu, G.L.; Feng, Q.Z.; Chen, Y. Emergency disposal and management of medical waste during the COVID-19 outbreak in China. *E3S Web Conf.* **2021**, *245*, 02011. [CrossRef]
22. Wang, J.; Shen, J.; Ye, D.; Yan, X.; Zhang, Y.; Yang, W.; Li, X.; Wang, J.; Zhang, L.; Pan, L. Disinfection technology of hospital wastes and wastewater: Suggestions for disinfection strategy during coronavirus Disease 2019 (COVID-19) pandemic in China. *Environ. Pollut.* **2020**, *262*, 114665. [CrossRef] [PubMed]
23. Chen, Y.; Zhang, K.; Xu, J.Y.; Shao, Z.; Ji, R. Analysis of Cooperative disposal of Medical Waste Treatment and Municipal Solid Waste Incineration. *IOP Conf. Ser. Earth Environ. Sci.* **2019**, *295*, 012052.
24. Yoon, B.; Magee, C.L. Exploring technology opportunities by visualizing patent information based on generative topographic mapping and link prediction. *Technol. Forecast Soc. Chang.* **2018**, *132*, 105–117. [CrossRef]

25. Wang, J.; Chen, Y.J. A novelty detection patent mining approach for analyzing technological opportunities. *Adv. Eng. Inform.* **2019**, *42*, 100941. [CrossRef]
26. Park, Y.; Yoon, J. Application technology opportunity discovery from technology portfolios: Use of patent classification and collaborative filtering. *Technol. Forecast Soc. Chang.* **2017**, *118*, 170–183. [CrossRef]
27. Lukas, J.A.; Jens, L.; Chie, H.S. Analysis of technological knowledge stock and prediction of its future development potential: The case of lithium-ion batteries. *J. Clean. Prod.* **2019**, *223*, 301–311.
28. Michele, G.; Livio, C.; Martina, D.G.; Francesco, R. The patent portfolio value analysis: A new framework to leverage patent information for strategic technology planning. *Technol. Forecast Soc. Chang.* **2015**, *94*, 286–302.
29. Fischer, T.; Leidinger, J. Testing patent value indicators on directly observed patent value: An empirical analysis of Ocean Tomo patent auctions. *Res. Policy* **2014**, *43*, 519–529. [CrossRef]
30. SC (State Council of the People's Republic of China). *Regulations on the Administration of Medical Waste*; State Council of the People's Republic of China: Beijing, China, 2003; Volume 380.
31. NHC (Nation Health Commission of the People's Republic of China). *Administrative Measures of Medical Waste in Medical Institutions*; Nation Health Commission of the People's Republic of China: Beijing, China, 2003.
32. MEP (Ministry of the Environment Protection of the People's Republic of China). *Technical Specification on Centralized Disposal of Medical Waste (Trial)*; Ministry of Environment Protection of the People's Republic of China: Beijing, China, 2003.
33. MOH (Ministry of Health of the People's Republic of China). *Classification Catalogue of Medical Waste*; Ministry of Health of the People's Republic of China: Beijing, China, 2003; Volume 287.
34. *GB19217-2003*; Technical Standard for Medical Waste Transport Vehicle. SEPA (State Environmental Protection Administration): Beijing, China, 2003.
35. *GB19218-2003*; Technical Standard for Medical Waste Incinerator. SEPA (State Environmental Protection Administration): Beijing, China, 2003.
36. MEP (Ministry of the Environment protection of the People's Republic of China). *Response Plan on the Management of Medical Waste for Influenza A (H1N1)*; Ministry of Environment Protection of the People's Republic of China: Beijing, China, 2009.
37. MEP (Ministry of the Environment protection of the People's Republic of China). *Technical Specification of Performance Testing for Facilities of Hazardous Waste (Including Medical Waste) Incineration*; Ministry of the Environment protection of the People's Republic of China: Beijing, China, 2010.
38. *HJ-BAT-008*; Guidelines on Best Available Technologies of Pollution Prevention and Control for Medical Waste Treatment and Disposal (on Trial). MEP (Ministry of the Environment protection of the People's Republic of China): Beijing, China, 2012.
39. MEP (Ministry of the Environment protection of the People's Republic of China). *Environmental Protection Technical Specification for Co-Processing of Solid Wastes in Cement Kiln*; Ministry of the Environment Protection of the People's Republic of China: Beijing, China, 2013.
40. MEE (Ministry of Ecology and Environment of the People's Republic of China). *Management and Technical Guidelines for Emergency Disposal of Medical Waste during the COVID-19 Epidemic (on Trial)*; Ministry of the Environment Protection of the People's Republic of China: Beijing, China, 2020. (In Chinese)
41. *DB31/768-2013*; Emission Standard of Air Pollution for Municipal Solid Waste Incineration. SEPB (Shanghai Environmental Protection Bureau): Shanghai, China, 2013.
42. MEE (Ministry of Ecology and Environment of the People's Republic of China). Annual Report on the Prevention and Control of Environmental Pollution by Solid Wastes in Large and Medium Sized Cities [Z/OL]. 2019. Available online: https://www.mee.gov.cn/ywgz/gtfwyhxpgl/gtfw/201912/P020191231360445518365.pdf (accessed on 13 December 2023).
43. WHO. *Safe Management of Wastes from Healthcare Activities*, 2nd ed.; World Health Organization: Geneva, Switzerland, 2014.
44. Zhao, H.L.; Liu, H.Q.; Wei, G.X.; Zhang, N.; Qiao, H.Y.; Gong, Y.Y.; Yu, X.N.; Zhou, J.H.; Wu, Y.H. A review on emergency disposal and management of medical waste during the COVID-19 pandemic in China. *Sci. Total Environ.* **2022**, *810*, 152302. [CrossRef] [PubMed]

processes

MDPI

Article

Non-Thermal Plasma Technology for Further Purification of Flue Gas in the Resource Utilization Process of Waste Mercury Catalyst: A Case Study in Xinjiang, China

Qinzhong Feng [1,†], Kaiyue Wang [1,†], Shitong Yang [1], Jianbo Guo [1], Jun Chen [1,2], Tongzhe Wang [1], Liyuan Liu [1,*] and Yang Chen [1,*]

1 College of Resources and Environment, University of Chinese Academy of Sciences, Beijing 100049, China; 2021205117@cqust.edu.cn (J.C.)

2 School of Chemistry and Chemical Engineering, Chongqing University of Science and Technology, Chongqing 401331, China

* Correspondence: liuliyuan@ucas.ac.cn (L.L.); chenyang@ucas.ac.cn (Y.C.); Tel.: +86-01069672842 (Y.C.)

† These authors contributed equally to this work.

Abstract: This study aims to advance the recycling of mercury-containing waste and promote sustainable development within the polyvinyl chloride (PVC) industry. Our innovative system integrates pre-treatment technology (spraying potassium permanganate and demisting and dust removal) with efficient non-thermal plasma oxidation, resulting in excellent treatment efficiency, low cost, and simple operation. With a processing capacity of 3000 m^3/h, the concentration of mercury emissions in flue gas can achieve the target of <0.01 mg/m^3, boasting a removal efficiency exceeding 98%, which satisfies the standard "Emission standard of air pollutants for industrial kiln and furnace" (GB 9078-1996). Our results can provide technical support for the comprehensive purification of mercury-containing flue gas during the resource recovery process from mercury-containing waste. The application of our system can contribute to reducing mercury emissions in the PVC industry, lowering occupational exposure risks for workers, and promoting China's better compliance with "the Minamata Convention on Mercury".

Keywords: non-thermal plasma technology; mercury-containing flue gas; waste mercury catalyst; resource utilization process; environmentally friendly

1. Introduction

Due to China's energy structure, which is characterized by a high coal content, low oil content, and low natural gas content, the predominant method for polyvinyl chloride (PVC) production is the mercury-based calcium carbide process [1,2]. This process will continue to exist for a long time until technology that utilizes mercury-free catalysts becomes viable. Notably, each ton of PVC production in China typically consumes between 1.2 kg and 1.6 kg of mercury catalyst (mercury chloride, $HgCl_2$), resulting in an annual consumption of $HgCl_2$ ranging from 700 to 1200 tons [3]. In the PVC production process, the main mercury-containing waste consists of waste mercury catalyst and waste activated carbon. The commonly used $HgCl_2$ catalyst is susceptible to carbon poisoning and deactivation, which requires frequent disposal of the spent mercury catalyst. The waste typically contains a mass fraction of $HgCl_2$ of approximately 2% to 3% [4]. Similarly, waste activated carbon used to treat mercury-containing waste gas in PVC production by adsorbing $HgCl_2$ has a similar composition to waste mercury catalysts [5,6]. Based on statistics, the calcium carbide PVC industry in China generates an estimated 10,000 to 15,000 tons of mercury-containing waste, which requires transportation to certified facilities for resource recovery [5,6]. Within the waste stream, waste mercury catalysts and mercury-containing activated carbon account for 36% and 51% of the total $HgCl_2$ used, respectively [7].

The mercury in these waste materials has the potential for resource recovery. Mercury (Hg) is persistent, transported long-range and bioconcentrated, and is one of the most toxic heavy metal pollutants in ecosystems [8]. Improper treatment or disposal by resource recovery enterprises can lead to mercury pollution, posing significant threats to the environment and human health [9]. Exposure to $HgCl_2$ has been linked to various adverse health effects, such as neurotoxicity leading to cognitive impairment and motor dysfunction, dysfunction of the central auditory system, as well as nephrotoxicity, hepatotoxicity, and increased mortality rates [10]. Before "the Minamata Convention on Mercury" came into effect, China had relatively weak management of mercury pollution prevention and control, and specialized mercury reduction technologies for the PVC industry were immature. "The Minamata Convention on Mercury" sets clear requirements for the chlor-alkali industry, promoting various measures to reduce reliance on mercury originating from primary mercury mining and implementing measures to minimize mercury emissions and releases into the environment. In 2015, the Ministry of Ecology and Environment of China issued the "Technical Policy for Mercury Pollution Prevention and Control", which is a landmark document in China's efforts to prevent and control mercury pollution, providing comprehensive support for mercury emission reduction in the PVC industry.

Representative recovery processes mainly include distillation and controlled oxygen dry distillation [6,11]. The distillation method primarily recovers elemental mercury (Hg^0), while the controlled oxygen dry distillation method primarily recovers $HgCl_2$. The distillation–condensation method is commonly used, but it has low efficiency in recovering materials and poses difficulties with regard to pollution control. In recent years, there have been reports of new processes for treating waste mercury catalysts. Qiu et al. [12]. and Yu et al. [13]. invented a technology for treating waste mercury catalysts using hydrometallurgy. Ren et al. [14] utilized ball mills to immobilize mercury in waste mercury catalysts by grinding it with sulfur, but their widespread application has not been observed. Distillation is the central component of the entire resource utilization process, and the distillation furnace is a sealed structure. During the heating process, it is essential to maximize the thermal volatilization of mercury, as it is crucial for improving the mercury recovery rate [15]. The condensation system is a process that involves the recovery of crude mercury from mercury-containing vapor through condensation. The recovered mercury is then processed and purified to obtain the final product.

Currently, the treatment of waste gas containing mercury after distillation and condensation typically involves methods such as dust removal, potassium permanganate oxidation, and activated carbon adsorption [16]. Mercury in flue gas typically exists in three forms: elemental mercury (Hg^0), divalent mercury ions (Hg (II)), and particulate-bound mercury (Hg^p) [17]. Among them, Hg (II) is highly soluble in water and can be effectively treated through spraying or adsorption technology. Hg^p can be removed through dust removal processes. However, a significant portion of Hg^0 in flue gas is difficult to remove. Conventional flue gas purification technologies are unable to effectively remove Hg^0 [7,11–13]. From an industry-wide perspective, the overall processing technology is outdated and the process routes represented by spray oxidation and activated carbon adsorption make it difficult to achieve stable and standardized emissions. Furthermore, the remaining concentration of mercury in waste activated carbon is high at 1000 mg/kg, classifying it as hazardous waste and requiring special treatment. If an efficient technology can be used to oxidize relatively insoluble elemental mercury, it is possible to achieve a comprehensive treatment of mercury-containing waste gas. This signifies a significant advancement in technological innovation for preventing and controlling mercury pollution.

The non-thermal plasma reactor generates a large number of high-energy electrons that collide with mercury atoms, oxidizing them into divalent mercury. This process leads to the conversion of mercury from insoluble to soluble, or from uncharged to charged forms. Through the process of salt formation, Hg^0 is removed from the gas phase and transformed into other products that are easily removable [18]. Non-thermal plasma technology has a wide range of applications in the field of environmental protection, including the disposal of waste fluorescent lamps, non-ferrous metal smelting, co-disposal of fly ash in cement kilns, medical waste incineration, and other fields, and is efficient and low-cost [19,20]. This technology can also be used as a calibration approach for gaseous oxidized mercury by nonthermal plasma oxidation of elemental mercury due to its quantitative oxidation capabilities [21].

Environmental Technology Verification (ETV), developed by the United States Environmental Protection Agency, encompasses a set of procedures and methods that involve the authorization of technology developers (owners), users, governments, or other relevant parties in accordance with applicable national laws, regulations, and standards [22,23]. Since the mid-1990s, environmental protection departments in countries and regions such as Canada, Japan, South Korea, and the European Union have successively established ETV evaluation systems [24,25]. As of May 2013, a total of 1368 validation and evaluation projects have been completed in various countries worldwide. In 2015, over 30 countries carried out ETV evaluations, which had a positive impact on technological innovation and transfer [26]. However, ETV evaluations also have some shortcomings, such as requiring significant time, manpower, and financial investment, particularly in data collection, testing, and validation.

The goal of this technology verification is to confirm the efficacy of non-thermal plasma coupling treatment technology in treating mercury-containing flue gas produced during the disposal of waste mercury catalysts. Xinjiang, China, is an important base for the production and sale of $HgCl_2$ catalysts for PVC as well as the recycling of mercury-containing waste resources. Thus, in our research, we selected Xinjiang, China, as our experimental site.

Our research team conducted a study on the efficiency of Hg^0 oxidation using a self-developed direct current (DC) high-voltage narrow-pulse non-thermal plasma reaction device. The study focused on the effects of power supply parameters (voltage, frequency) and simulated concentrations of O_2, H_2O, HCl, NO, and SO_2 in flue gas. The study investigated the synergistic control mechanism of non-thermal plasma for reducing NOx, SO_2, and Hg^0, and determined the feasibility of this technology. This technology has been partially applied in related engineering.

2. Materials and Methods

2.1. Small-Scale Experiment

By simulating the composition of flue gas generated during the heat treatment process of typical mercury-containing waste, we conducted a study at a small-scale experimental level to investigate the factors influencing non-thermal plasma (NTP) treatment for mercury-containing flue gas.

2.1.1. Equipment

The experimental setup, as illustrated in Supplementary Figure S1, includes a gas distribution system, a high-voltage power supply, a plasma reactor, a water vapor generator, a U-shaped mixer, a VM-3000 Mercury Vapor Monitor (Karlsfeld, Germany), a flue gas treatment system, and a ventilation system. The gas distribution system consists of N_2, O_2, SO_2, and NO. The main gas used to generate plasma is N_2. The concentration of Hg used in our experiments is about 200 mg/m^3.

The experiment uses a non-thermal plasma power supply with a pulse width and rising edge in the range of hundreds of nanoseconds. The parameters for the non-thermal plasma power supply for the small-scale experiment can be found in Supplementary Table S1. The non-thermal plasma reactor used employs a wire-tube configuration, consisting of a 320 mm long discharge tube with an inner diameter of 32 mm. The actual volume is 257 cm^3. The discharge tube is arranged in parallel. The positive terminal of the power supply is connected to the high-voltage insulator of the reactor, while the negative terminal and ground wire are connected to the low-voltage insulator of the reactor.

2.1.2. Experimental Content and Measurement

The following is the main content and testing methods of this experiment.

(1) Effect of power supply parameters of Hg^0 removal by non-thermal plasma. This study investigates the impact of voltage, frequency, pulse width, and pulse rising edge of the non-thermal plasma power supply on the rate of Hg^0 oxidation in flue gas. The voltage and current were measured using an oscilloscope (MDO3024, Tektronix, Beaverton, OR, USA). The power of the plasma discharge was calculated according to Equation (1).

$$P = f\int_0^T U(t)I(t)dt \tag{1}$$

where P is the output power (Watt, W), f is pulse frequency of pulse power supply (Hz), U is Instantaneous voltage (V), and I is Instantaneous current (A).

(2) The influence of atmospheric conditions on the removal of mercury from flue gas using non-thermal plasma. This study examines the impact of atmospheric conditions on the efficiency of Hg^0 oxidation in the presence of pollutants like NO and SO_2. The study aims to clarify the distinct effects of each pollutant on Hg^0 oxidation efficiency. The concentration of SO_2 and NO was measured using the TESTO Flue Gas Analyzer (TESTO 350, Titisee-Neustadt, Germany).

2.2. *Environmental Technology Verification (ETV Process)*

2.2.1. Site Selection

The technology demonstration and verification took place at a company in Xinjiang. The company specializes in recycling and safely disposing of waste materials containing mercury, such as discarded mercury catalysts. During the operation of a fluidized bed furnace to treat waste mercury catalysts, flue gas containing mercury is generated and emitted, with a mercury content ranging from 15 to 20 mg/m^3. Despite the implementation of a multi-stage mercury absorption process, it is not possible to achieve stable and standardized emissions. The flue gas mercury emissions do not meet the mercury emission concentration limit of 0.01 mg/m^3 as specified in the "Emission standard of air pollutants for industrial kiln and furnace" (GB 9078-1996) [27].

2.2.2. On-Site Process

A non-thermal plasma coupling treatment device is utilized to effectively control mercury, SO_2, NOx, particulate matter, and other pollutants in the flue gas [28,29]. The flue gas treatment process first involves a four-stage washing pretreatment to remove dust particles from the flue gas and reduce the concentration of mercury vapor. During the washing treatment, the generation of a certain amount of saturated water vapor and the entrainment of liquid droplets significantly increase the moisture content of the flue gas. Then, the water vapor content in the flue gas is reduced using the mist removal system to prevent any potential impact on the subsequent system. Finally, the treated flue gas enters the non-thermal plasma integrated system, which combines non-thermal plasma efficient oxidation with environmentally functional material (modified silica) adsorption, to achieve effective degradation and thorough purification, ensuring compliance with the emission standard. The parameters for the non-thermal plasma power supply for on-site experiments are provided in Supplementary Table S2.

2.2.3. Sample Collection

According to the characteristics of the technology being evaluated and the objectives of the evaluation, the samples collected for experimental testing include gas samples and solid residues. According to "The determination of particulates and sampling methods of gaseous pollutants from exhaust gas of stationary" (GB/T 16157-1996) [30] and the "Stationary source emission-Determination of mercury-Cold atomic absorption spectrophotometry" (HJ 543-2009) [31], sampling was conducted at the following locations: outlet of the tube cooler (Point 1#), outlet of the four-stage series spray tower (Point 2#), outlet of the dust and mist removal equipment (Point 3#), outlet of the primary oxidation catalyst bed (Point 4#), and outlet of the tertiary oxidation catalyst bed (Point 5#). The sampling locations are depicted in Supplementary Figure S2. To ensure accurate data, sampling must be done simultaneously at each sampling point.

2.2.4. Sample Measurement

The methods for determining mercury concentration, SO_2 concentration, NOx concentration, and particulate matter concentration are specified in the following standards: "Stationary source emission-Determination of mercury-Cold atomic absorption spectrophotometry" (HJ 543-2009) [31], "Stationary source emission-Determination of sulfur dioxide—Fixed potential by electrolysis method" (HJ/T 57-2017) [32], "Stationary source emission-Determination of nitrogen oxides-Fixed potential by electrolysis method" (HJ 693-2014) [33], and "Stationary source emission—Determination of mass concentration of particulate matter at low concentration—Manual gravimetric method" (HJ 836-2017) [34], respectively.

2.2.5. ETV Test Parameters

The testing parameters are divided into two categories: environmental effects and operating processes. In this assessment, suitable parameters were chosen according to the characteristics and validation objectives of non-thermal plasma technology, as depicted in Table 1.

Table 1. List of Test Parameters.

Parameter Category	Object	Specific Parameters
Environmental effect parameters	inlet and outlet exhaust gas	temperature, pressure, flow rate
	air pollutant	mercury content, NOx, SO_2, particulate matter, etc. in exhaust gas
Process operating parameters	non-thermal plasma system	stable operating time
		pulse frequency
		pulse voltage
		pulse current
	processing scale	unit time processing capacity

Security measures: Our research was conducted under sealed negative pressure experimental conditions, where the exhaust gas was treated with potassium permanganate solution.

Data analysis: The data for this paper was organized using Excel 2021, while the graphs were created using Origin 2022.

3. Results for Small-Scale Experiment

3.1. Effect of Power Supply Parameters of Hg^0 Removal by Non-Thermal Plasma

3.1.1. Voltage

The effect of voltage change on the rate of Hg^0 oxidation is illustrated in Figure 1. The experiment was conducted with a pulse frequency of 600 Hz. It can be seen that as the voltage reaches the threshold level, both the single pulse energy and injection energy increase with the rise in voltage. When the voltage reaches 13.1 kV, stable plasma is generated. As the voltage increases, the concentration of Hg^0 decreases rapidly, eventually reaching an optimal level of approximately 88% oxidation rate when the voltage reaches 13.1 kV. Taking cost into consideration, we have selected 13.1 kV as the optimal input voltage.

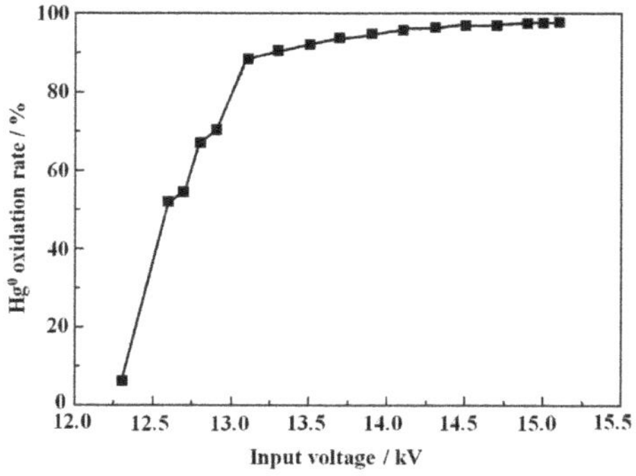

Figure 1. Effect of voltage change on Hg^0 oxidation rate.

3.1.2. Pulse Frequency

The impact of varying pulse frequency on the rate of Hg^0 oxidation is illustrated in Figure 2. The experiment was conducted at an input voltage of 13.1 kV. The oxidation rate of Hg^0 increases gradually as the power frequency increases. When the pulse frequency reaches approximately 200 Hz, the rate of mercury oxidation reaches 80%. Afterward, as the pulse frequency increased, the rate of mercury oxidation basically stabilized.

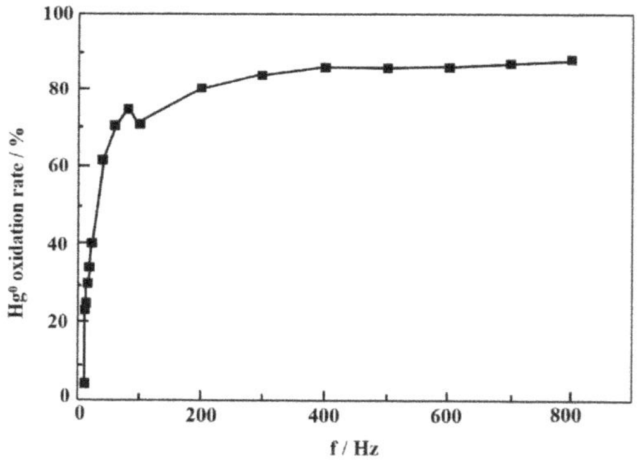

Figure 2. Effect of pulse frequency variation on Hg^0 oxidation rate.

3.1.3. Pulse Width

The effect of pulse width variation on Hg^0 oxidation rate is illustrated in Figure 3. The experiment was conducted at an input voltage of 13.1 kV. The oxidation rate of Hg^0 increases gradually with the pulse width in the range of 0–1000 ns. Upon reaching approximately 300 ns, the concentration of Hg^0 at the output ceases to decrease. With further increases in pulse width (300–1000 ns), the oxidation rate of Hg^0 remains relatively constant. This indicates that the device's oxidizing capacity for Hg^0 has reached its limit. Under these conditions, a variety of output frequencies (12 Hz, 25 Hz, 50 Hz, 100 Hz) were chosen for comparative experiments. It was observed that as the frequency increased, the oxidation rate of Hg^0 also increased to some extent. Notably, at a frequency of 100 Hz, the curve demonstrates excellent performance in oxidizing Hg^0, while at 12 Hz, the overall rate of Hg^0 oxidation is lower. The oxidation rate change curves of Hg^0 at different frequencies all exhibit a similar trend.

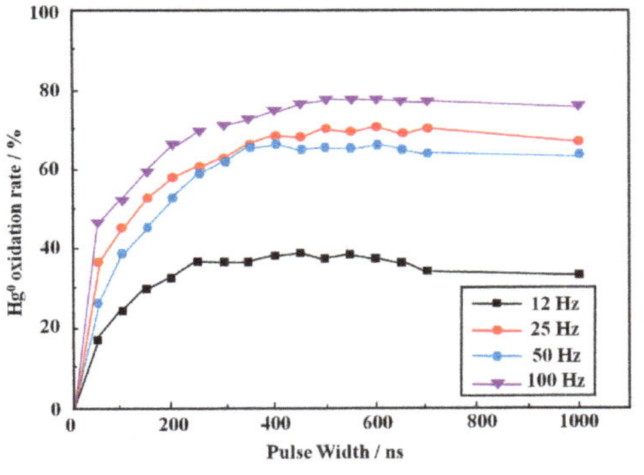

Figure 3. Effect of pulse width variation on Hg^0 oxidation rate (0–1000 ns).

3.1.4. Pulse Rising Edge

The effect of pulse rising edge variation on the Hg^0 oxidation rate is illustrated in Figure 4. It can be observed that the rising edge of the pulse has minimal influence on the oxidation of mercury. The variation in the rising edge of the pulse has little effect on both the energy of a single pulse and the rate of Hg^0 oxidation.

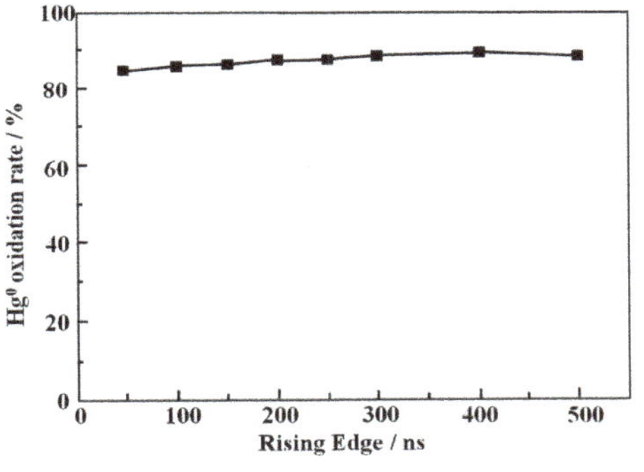

Figure 4. Effect of pulse rising edge variation on Hg^0 oxidation rate.

3.2. Effect of Atmospheric Conditions of Hg^0 Removal by Non-Thermal Plasma

3.2.1. NO Concentration

The effect of NO on non-thermal plasma oxidation of Hg^0 is illustrated in Figure 5. It can be concluded that the rate of Hg^0 oxidation gradually decreases as the NO concentration increases. This indicates that NO inhibits the oxidation of Hg^0 [35]. There is a competitive relationship between NO and Hg^0. Due to the faster reaction rate between NO and O_3, O_3 is consumed by NO, leading to a reduction in the number of active particles that react with Hg^0. As a result, the rate of Hg^0 oxidation decreases. Based on this, it can be inferred that if NO is present in the processed gas component, it is essential to elevate the pulse voltage or frequency to achieve a greater energy density and attain a specific Hg^0 oxidation rate.

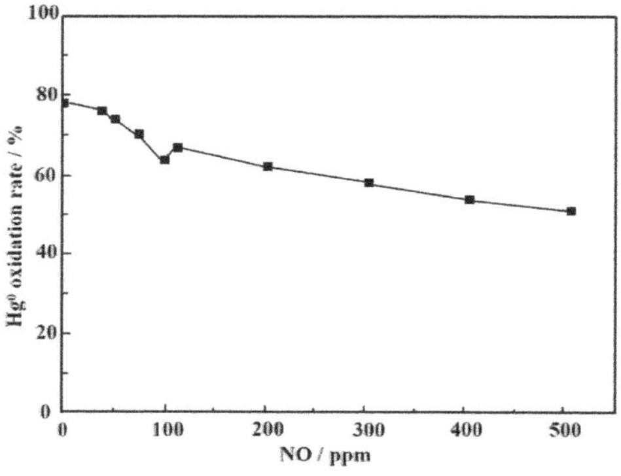

Figure 5. Effect of NO on non-thermal plasma oxidation of Hg^0.

3.2.2. SO_2 Concentration

The effect of SO_2 concentration on non-thermal plasma oxidation of Hg is illustrated in Figure 6. It can be concluded that the presence of SO_2 inhibits the oxidation of Hg^0, compared to the effect of NO on the Hg^0 oxidation rate in non-thermal plasma. However, the effect of SO_2 on the rate of Hg^0 oxidation is relatively small. This is primarily because the reaction rate between SO_2 and O_3 is relatively low, and there is also relatively little competition between SO_2 and Hg^0 [35].

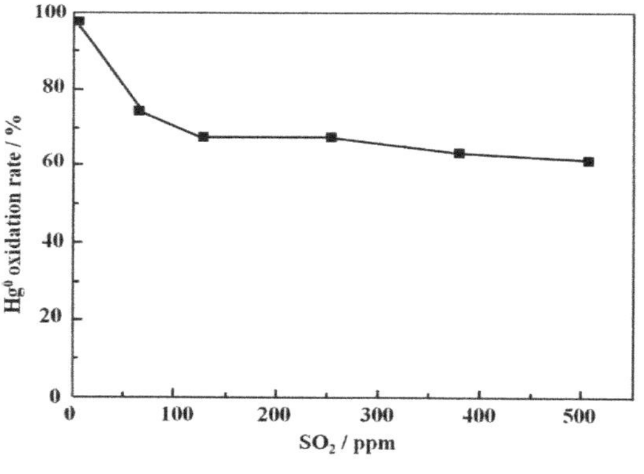

Figure 6. Effect of SO_2 concentration on non-thermal plasma oxidation of Hg^0.

The process of non-thermal plasma synergistic control of NO_x, SO_2, and Hg^0 can be delineated into three stages: ① Generation of DC high-voltage pulse discharge low-temperature plasma through the inelastic collision of high-energy electrons and gas molecules, leading to the excitation and dissociation of gas molecules, thereby producing reactive radicals such as ·N, ·O, etc. ② Reaction of pulse discharge with the excited state of gas molecules, resulting in the formation of ·O and O_3. ③ Utilization of ·O, O_3, and other reactive species to react with NO_x and Hg^0 in the gas, thereby achieving oxidative removal. The specific reaction processes have been described in the Supplementary Materials SD1.

4. ETV Process of Non-Thermal Plasma

This project is being studied based on the system and methods of ETV. ETV refers to the authorization of technology developers (owners), users, governments, or other relevant parties in accordance with applicable national laws, regulations, and standards. It is conducted in accordance with the requirements of the "Environmental management—

Environmental technology verification" (GB/T 24034-2019) [36] and "General Protocol for Environmental Technology Verification" (T/CSES 01-2015) [37].

4.1. Non-Thermal Plasma Equipment Integration

4.1.1. Pretreatment System

(1) Facilities for spraying potassium permanganate

The washing process, which uses a four-stage spray tower and a potassium permanganate solution, has the ability to reduce the mercury concentration in flue gas to 1–2 mg/m^3, or even achieve lower emission levels. This process achieves the goal of preliminary treatment for mercury in flue gas. The design parameters of the equipment are as follows: processing flue gas volume: 10,000 m^3/h; gas flow rate: less than 0.7 m/s.

(2) Demisting and dust removal facilities

By utilizing the electrons emitted from the cathode in a high-voltage electric field and the negative ions produced through electron collisions with air molecules, particulate matter and water mist particles are captured. The particles are then charged and adsorbed onto the anode, effectively achieving the goal of cleaning. The equipment requirements and technical specifications are as follows: the power supply should be high-voltage constant current DC, and the power supply power should not exceed 20 kW.

4.1.2. Non-Thermal Plasma Coupling Treatment System

(1) Non-thermal plasma power supply

The non-thermal plasma power supply primarily comprises two components: a DC charging power supply and a MARX generator module. The pulse non-thermal plasma power supply is based on the MARX generator circuit, which rapidly generates high voltage from low voltage, eliminating the need for large low-frequency and high-voltage transformers. The use of modular circuits reduces the voltage resistance requirements and minimizes the power supply volume, thereby decreasing the inductance of the output circuit. By adjusting the pulse voltage, pulse current, pulse frequency, and other parameters, the core technology of the DC high-voltage narrow-pulse plasma power supply has been developed to meet the requirements of pulse rising edges in the range of hundreds of nanoseconds. This technology enables stable and controllable power pulse discharge corona.

(2) Plasma reactor

A non-thermal plasma tube reactor is utilized, and the reactor structure is depicted in Figure S3 of the Supplementary Materials. The flue gas containing mercury is introduced into the non-thermal plasma reactor using an induced draft fan. In the reactor, Hg^0 in the flue gas is converted into Hg^{2+}, which then reacts with the negatively charged oxygen ions produced in the reactor to form electrically neutral mercury (II) oxide [38].

(3) Environmental functional material adsorption system

Utilizing ceramic nano single-molecule porous materials for targeted adsorption. The specific surface area of this material ranges from 200 to 800 m^2/g, and it demonstrates a strong adsorption effect on Hg^{2+}, Hg^+, and molecular mercury compounds [39]. The primary technical indicators of this material are as follows: ceramic nano adsorption capacity: 0.4 g/kg; ceramic nano regeneration cycle: 6 months; the flue gas flow rate inside the adsorption box/regeneration tower is 0.385 m/s. Ceramic nanomaterials, when utilized as adsorption materials for non-thermal plasma oxidation, can exhibit outstanding adsorption and capture capabilities. This provides the necessary conditions for the subsequent regeneration of ceramic nanomaterials.

4.1.3. Operating Process

The non-thermal plasma coupling treatment system ran continuously for 72 h. The spray tower's flow rate is 80 m^3/h. The concentration of potassium permanganate is 0.5%.

During the testing period, the inlet mercury concentration of the system remained stable at around 4.37 mg/m^3, while the outlet mercury concentration was below 0.012 mg/m^3. The operating parameters remained stable within the frequency range of the dust removal equipment, which was between 2.88 and 4.2 Hz. The flow rate of the spray tower is 80 mg/m^3. The electric pressure was maintained at 38 kV, and the frequency was set at 1000 Hz. Additionally, the flue gas was consistently discharged to meet the required standards. The main materials and energy consumed in this process include water, electricity, potassium permanganate powder, and the adsorbent materials. The flue gas treatment capacity is calculated based on the actual operation of 300 days per year and 24 h per day. The annual operating cost of the system and the detailed information are in the Supplementary Materials SD2.

4.2. Results

During the ETV period, the main results are as follows:

(1) The on-site experiment utilized a cascaded oxidation treatment technique that combines washing with potassium permanganate solution and coupling with non-thermal plasma technology. Elemental mercury in the flue gas is oxidized to form oxidized mercury species, which are then adsorbed by the solution and environmentally friendly functional materials, resulting in optimal treatment efficiency. This technology enables the further purification of mercury, improves treatment efficiency, reduces material usage, and lowers operating costs.

(2) The pulse voltage of the plasma power supply ranges from 10 to 35 kV, the pulse current ranges from 8 to 160 A, and the pulse frequency ranges from 100 to 1000 Hz. The facility's operating parameters are normal and meet the requirements for continuous and stable operation for 72 h.

(3) The test results indicate that this technology can meet the treatment requirements for flue gas containing mercury during the disposal of waste mercury catalysts. Furthermore, the environmental emission indicators meet the requirements of applicable national and local standards. When the system's processing capacity is 3000 m^3/h, the mercury removal efficiency can exceed 98%, and the mercury emission concentration in the flue gas can meet the mercury emission concentration limit of 0.01 mg/m^3 as specified by the "Emission standard of air pollutants for industrial kiln and furnace" (GB 9078-1996) [27]. Currently, only Guizhou in China has established emission limit requirements for mercury and its compounds in the local standard "Emission Standards for mercury and its compound industrial pollutants" (DB52-1422-2019) [40], with a limit of 0.03 mg/m^3. The concentration of mercury emissions also meets the emission standards. All other pollutants, such as particulate matter, SO_2, and NOx, also meet the standard requirements. The test results of flue gas samples are shown in Table 2.

Table 2. Test results of flue gas samples.

Pollutant	Point 1# (mg/m^3)	Point 2# (mg/m^3)	Point 3# (mg/m^3)	Point 4# (mg/m^3)	Point 5# (mg/m^3)	GB 9078-1996 (mg/m^3)
Hg	3.3 ± 0.040	0.75 ± 0.051	0.63 ± 0.026	0.024 ±0.00030	0.010 ±0.00020	0.01
Particulate matter	11 ± 0.35	11 ± 0.55	9.3 ± 0.40	8.6 ± 0.15	7.6 ± 0.25	120
SO_2	<3	<3	<3	<3	<3	550
NOx	<3	<3	<3	<3	<3	240

5. Conclusions and Perspective

The test results demonstrate the efficiency of this technology in meeting the requirements for treating mercury-containing exhaust gases during the disposal process of waste mercury catalysts. It serves as a viable alternative to traditional methods, achieving a remarkable mercury recovery rate exceeding 99%, with tail gas emissions maintained below 0.01 mg/m^3. This addresses the persistent challenge of limited adsorption capacity observed in technologies such as activated carbon.

Non-thermal plasma technology operates at ambient temperature and pressure, presenting a straightforward process with high treatment efficiency and negligible secondary pollution. It effectively circumvents the constraints of conventional mercury pollution control methods, and has become a popular research topic for scholars globally.

The oxidation of elemental mercury by non-thermal plasma is influenced by several factors including voltage, pulse width, frequency of the power supply, and waste gas composition. Various industries and process technologies exhibit disparities in gas volumes and pollutant compositions, thereby necessitating divergent parameter designs for plasma power supplies and reactors. Thus, high frequency, high voltage, and appropriate pulse width considerations are pivotal in enhancing the power factor of the power supply for optimal application in plasma reactors.

This technology finds applicability in resource recovery from waste mercury catalysts, purification of mercury-containing flue gas from waste fluorescent tube disposal, and mitigation of mercury and dioxin emissions from diverse industries including chemical engineering, non-ferrous metal smelting, and waste incineration. Looking ahead, non-thermal plasma technology holds promise in supplanting activated carbon adsorption as an energy-efficient and eco-friendly method for simultaneous control of multiple pollutants. It stands poised to aid China in fulfilling its commitments under the Minamata Convention on Mercury by providing indispensable technical support. Moreover, the recovered mercury can help alleviate China's mercury supply shortages and contribute to the sustainability of the PVC industry.

Nevertheless, it is imperative to acknowledge that the prevention and control of mercury pollution represent enduring challenges requiring sustained collaborative efforts among governmental bodies, enterprises, and research institutions. Continuous innovation aimed at meeting the evolving needs of industry development, regulatory compliance, and technological advancement remains paramount.

Supplementary Materials: The following supporting information can be downloaded at: https://www.mdpi.com/article/10.3390/pr12040691/s1, Figure S1. The experimental system for small-scale experiment; Figure S2. Schematic diagram of sampling location; Figure S3. Non-thermal plasma reactor structure; Table S1. Non-thermal plasma power supply parameters for small-scale experiments; Table S2. Non-thermal plasma power supply parameters for on-site experiments; Table S3. Power device parameters; Table S4: Annual operating cost of the system; SD1. The process of non-thermal plasma synergistic control of NOx, SO_2, and Hg^0; SD2. Detailed information for annual operating cost of the system.

Author Contributions: Conceptualization, Q.F. and L.L.; methodology, Y.C.; software, T.W.; validation, S.Y., J.C. and J.G.; formal analysis, K.W.; investigation, Q.F. and L.L.; resources, L.L.; data curation, S.Y.; writing—original draft preparation, Q.F. and L.L.; writing—review and editing, Q.F., Y.C. and L.L.; visualization, J.G.; supervision, Y.C.; project administration, L.L.; funding acquisition, J.G. All authors have read and agreed to the published version of the manuscript.

Funding: This research was funded by National Key Research and Development Plan Project "Integration and industrialization of air pollution prevention and control technologies for unconventional pollutants", grant number 2016YFC0209204.

Data Availability Statement: The data that support the findings of this study are available from the corresponding author upon reasonable request.

Acknowledgments: The authors would like to thank the funding Agencies; the Ministry of Science and Technology of the People's Republic of China.

Conflicts of Interest: The authors declare no conflicts of interest.

References

1. Sun, J.S.; Li, G.; Wang, Z.H. Optimizing China's energy consumption structure under energy and carbon constraints. *Struct. Chang. Econ. Dyn.* **2018**, *47*, 57–72. [CrossRef]
2. Liu, Y.J.; Zhou, C.B.; Li, F.; Liu, H.J.; Yang, J.X. Stocks and flows of polyvinyl chloride (PVC) in China: 1980–2050. *Resour. Conserv. Recycl.* **2020**, *47*, 104584. [CrossRef]
3. Zeng, H.X.; Hu, B.L.; Zhang, Y.L. Regeneration status of mercury-containing wastes in China. *Nonferrous Met. Sci. Eng.* **2012**, *33*, 36–38.
4. *GB/T 36382-2018*; State Administration for Market Supervision and Administration, China National Standardization Administration. Treatment and Disposal Methods of Waste Mercury Catalyst. Standards Press of China: Beijing, China, 2018.
5. Xie, Z.N.; Xu, Z.D.; Ren, F.Z.; Yang, H.; Fan, F.Z. Mercury leaching from waste mercury catalyst and the absorption property of regenerated activated carbon. *Chin. J. Environ. Eng.* **2019**, *13*, 1194–1201.
6. Wang, Y.J.; Tian, Y.; Zang, W.C.; Jian, X.D. Study on Treatment and Recycling of Mercury from Waste Mercury Catalysts in China. *Procedia Environ. Sci.* **2016**, *31*, 432–439. [CrossRef]
7. Liu, X.Y.; Zhou, X.G.; Liu, Y.B. The comprehensive analysis on the flow direction of mercury in the production of PVC by calcium carbide method and the prevention of its pollution. *Polyvinyl Chloride* **2008**, *36*, 29–31.
8. Kim, H.N.; Ren, W.X.; Kim, J.S.; Yoon, J. Fluorescent and colorimetric sensors for detection of lead, cadmium, and mercury ions. *Chem. Soc. Rev.* **2012**, *41*, 3210–3244. [CrossRef] [PubMed]
9. Pavithra, K.G.; SundarRajan, P.; Kumar, P.S.; Rangasamy, G. Mercury sources, contaminations, mercury cycle, detection and treatment techniques: A review. *Chemosphere* **2023**, *312*, 137314. [CrossRef] [PubMed]
10. Huang, C.F.; Hsu, C.J.; Liu, S.H.; Lin-Shiau, S.Y. Neurotoxicological mechanism of methylmercury induced by low- dose and long-term exposure in mice: Oxidative stress and down-regulated Na^+/K^+-ATPase involved. *Toxicol. Lett.* **2008**, *176*, 188–197. [CrossRef]
11. Li, Y.; Qiu, Y.G.; Wei, Z.H. A Method for Recovering Waste Mercury Catalysts in PVC Production by Calcium Carbide Method. CN106673055A, 17 May 2017.
12. Qiu, Y.R.; Yan, S. A Method for Recycling and Utilizing Waste Mercury Catalysts. CN103803638A, 21 May 2014.
13. Yu, G.R.; Chen, X.C. A New Technology for Wet Recovery of Mercury from Mercury Containing Catalysts. CN104451154A, 25 March 2015.
14. Ren, Y.F.; Zhu, J.X. Leaching behavior and risk control of waste mercury chloride catalyst through mechanochemical reaction using sulfur. *Environ. Eng. Sci.* **2015**, *32*, 816–823. [CrossRef]
15. Zhang, J.; Li, C.T.; Du, X.Y.; Li, S.H.; Huang, L. Recycle of waste activated coke as an efficient sorbent for Hg0 removal from coal-fired flue gas. *Fuel* **2022**, *432*, 124645. [CrossRef]
16. Liu, T.; Xiong, Z.; Ni, P.; Ma, Z.Z.; Tan, Y.; Li, Z.S.; Deng, S.N.; Li, Y.C.; Yang, Q.R.; Zhang, H.W. Review on adsorbents in elemental mercury removal in coal combustion flue gas, smelting flue gas and natural gas. *Chem. Eng. J.* **2023**, *454*, 140095. [CrossRef]
17. Qin, R.Y.; Chang, S.; Mei, J.; Hong, Q.Q.; Yang, S.J. Selective removal of Hg^{2+} from acidic wastewaters using sulfureted Fe_2TiO_5: Underlying mechanism and its application as a regenerable sorbent for recovering Hg from waste acids of smelters. *Water Res.* **2022**, *221*, 118796. [CrossRef]
18. Byun, Y.; Ko, K.B.; Cho, M.; Namkung, W.; Shin, D.N.; Lee, J.W.; Koh, D.J.; Kim, K.T. Oxidation of elemental mercury using atmospheric pressure non-thermal plasma. *Chemosphere* **2008**, *72*, 652–658. [CrossRef] [PubMed]
19. Kim, H.H. Nonthermal plasma processing for air-pollution control: A historical review, current issues, and future prospects. *Plasma Process Polym.* **2004**, *1*, 91–110. [CrossRef]
20. Barjasteh, A.; Dehghani, Z.; Lamichhane, P.; Kaushik, N.; Choi, E.H.; Kaushik, N.K. Recent Progress in Applications of Non-Thermal Plasma for Water Purification, Bio-Sterilization, and Decontamination. *Appl. Sci.* **2021**, *04*, 3372. [CrossRef]
21. Nair, S.V.; Gacnik, J.; Zivkovic, I.; Andron, T.D.; Ali, S.W.; Kotnik, J.; Horvat, M. Application of traceable calibration for gaseous oxidized mercury in air. *Anal. Chim. Acta* **2024**, *1288*, 342168.
22. Li, Y.; Wang, X.J. Introduction to environmental technology verification (ETV) program in the United States. *Chin. J. Popul. Resour. Environ.* **2001**, *11*, 121–122.
23. Cunningham, C.J.; Peshkur, T.A.; Kuyukina, M.S.; Ivshina, I.B. Environmental Technology Verification (ETV): Challenges to Verifying the Performance of Bioremediation Technologies. *Russ. J. Ecol.* **2022**, *53*, 500–506. [CrossRef]
24. Korea Environmental Industry Technology Institute. What Are the NET&ETV? [EB/OL]. Available online: https://www.koetv.or.kr/eng/home/sub_etv.jsp (accessed on 26 December 2020).
25. European Commission. EU Environmental Technology Verification (ETV) [EB/OL]. Available online: https://ec.europa.eu/environment/ecoap/etv/news/technologies-protect-environmentwin-increased-eu-support-expanded-etv-schem_en (accessed on 26 December 2020).
26. Huang, G.Z.; Tan, Y.Y.; Sun, J.R.; Kong, L.W.; Tan, L. Research on the evaluation system of ecological environment technology. *J. Electr. Power* **2020**, *35*, 162–165. (In Chinese)
27. *GB 9078-1996*; Emission Standard of Air Pollutants for Industrial Kiln and Furnace. Ministry of Ecology and Environment of the People's Republic of China: Beijing, China, 1996.
28. Mok, Y.S.; Nam, I.S. Positive pulsed corona discharge process for simultaneous removal of SO_2 and NO from iron-ore sintering flue gas. *IEEE Trans. Plasma Sci.* **1999**, *27*, 1188–1196.

29. Malik, M.A.; Kolb, J.F.; Sun, Y.; Schoenbach, K.H. Comparative study of NO removal in surface-plasma and volume-plasma reactors based on pulsed corona discharges. *J. Hazard. Mat.* **2011**, *197*, 220–228. [CrossRef] [PubMed]
30. *GB/T 16157-1996*; The Determination of Particulates and Sampling Methods of Gaseous Pollutants from Exhaust Gas of Stationary source. Ministry of Ecology and Environment of the People's Republic of China: Beijing, China, 1996.
31. *HJ 543-2009*; Stationary Source Emission—Determination of mercury-Cold Atomic Absorption Spectrophotometry. Ministry of Ecology and Environment of the People's Republic of China: Beijing, China, 2009.
32. *HJ 57-2017*; Stationary Source emission—Determination of Sulfur Dioxide—Fixed Potential by Electrolysis Method. Ministry of Ecology and Environment of the People's Republic of China: Beijing, China, 2017.
33. *HJ 693-2014*; Stationary Source Emission-Determination of Nitrogen Oxides-Fixed Potential by Electrolysis Method. Ministry of Ecology and Environment of the People's Republic of China: Beijing, China, 2014.
34. *HJ 836-2017*; Stationary Source Emission—Determination of Mass Concentration of Particulate Matter at Low Concentration—Manual Gravimetric Method. Ministry of Ecology and Environment of the People's Republic of China: Beijing, China, 2017.
35. Hao, S.S.; Chen, Y.; Yin, L.Q.; Wei, S.H.; Feng, Q.Z. Research of thermal desorption-pulse non-thermal plasma integration system to remove mercury. *Environ. Eng.* **2016**, *34*, 93–98.
36. *GB/T 24034-2019*; Environmental Management—Environmental Technology Verification. Standardization Administration of the People's Republic of China: Beijing, China, 2019.
37. *T/CSES 01-2015*; General Protocol for Environmental Technology Verification. Chinese Society for Environmental Sciences: Beijing, China, 2015.
38. Kogelschatz, U. Dielectric-barrier discharges: Their history, discharge physics, and industrial applications. *Plasma Chem. Plasma Process.* **2003**, *23*, 1–46. [CrossRef]
39. Feng, X.D.; Fryxell, G.E.; Wang, L.Q.; Kim, A.Y.; Liu, J.; Kemner, K.M. Functionalized Monolayers on Ordered Mesoporous Supports. *Science* **1997**, *276*, 923–926. [CrossRef]
40. *DB52-1422-2019*; Emission Standards for Mercury and Its Compound Industrial Pollutants. Department of Ecology and Environment of Guizhou Province: Guiyang, China, 2019.

processes

MDPI

Article

Synergetic Mechanism of Multiple Industrial Solid Waste-Based Geopolymer Binder for Soil Stabilization: Optimization Using D-Optimal Mixture Design

Xiaoli Wang [1,2], Xiancong Wang [1], Pingfeng Fu [1,*], Bolan Lei [1], Jinjin Shi [3,4] and Miao Xu [3,4]

1 School of Civil and Resources Engineering, University of Science and Technology Beijing, Beijing 100083, China; xiaoliwang@ustb.edu.cn (X.W.); runner17@163.com (X.W.); leibolan1999@163.com (B.L.)
2 State Key Laboratory of Mineral Processing, Beijing 100160, China
3 Cangzhou Municipal Engineering Company Limited, Cangzhou 061000, China; czszjsk@126.com (J.S.); czszxm1989@163.com (M.X.)
4 Road Materials and Technology Engineering Research Center of Hebei Province, Cangzhou 061000, China
* Correspondence: pffu@ces.ustb.edu.cn

Abstract: In order to improve the comprehensive utilization rate of industrial solid waste and the road quality, a novel low-carbon and environmental friendly soil stabilizer is proposed. In this study, steel slag (SS), carbide slag (CS), blast furnace slag (BFS), fly ash (FA), and desulfurized gypsum (DG) were used as raw materials to develop a multiple industrial solid waste-based soil stabilizer (MSWSS). The optimal mix ratio of the raw materials determined by D-optimal design was as follows: 5% SS, 50% CS, 15% BFS, 15% DG, and 15% FA. The 7-day unconfined compressive strength (UCS) of MSWSS-stabilized soil was 1.7 MPa, which was 36% higher than stabilization with ordinary portland cement (OPC) and met the construction requirements of highways. After 7 days of curing, the UCS of MSWSS-stabilized soil was significantly higher than that in the OPC group. X-ray powder diffraction (XRD), thermogravimetric analysis (TGA), and scanning electron microscopy (SEM) analysis indicated that the prominent hydration products were ettringite (AFt) and C-S-H gel. The results showed that an amount of AFt and C-S-H were formed in the initial stage of curing, resulting in a rapid improvement in early UCS. As the curing proceeded, the content of AFt and C-S-H increased constantly and grew intertwined with each other, which lead to the denser microstructure of stabilized soil and better mechanical strength.

Keywords: soil stabilization; industrial solid waste; D-optimal mixture approach; hydration mechanism

1. Introduction

Rapid urbanization has led to an exponential increase in the amount of solid waste that is generated. According to the report of the World Bank Group, global cities generate 2.01 billion tons of solid waste each year, including industrial waste, agricultural waste, and domestic waste. It is expected to increase to 3.40 billion tons per year by 2050 [1]. According to the statistics, only about 13.5% of solid waste is recycled globally [1]. Landfills and incineration are the two main solid waste disposal methods. However, these conventional treatments waste land resources, release harmful gases, and cause a negative impact on the economy, environment, and human health [2–4].

On the other hand, the development of urban transportation has brought challenges to the road construction industry [5,6]. The quality of pavement depends on the mechanical properties of the underlying soil, including strength, compressibility, permeability, etc. [7–9]. One of the effective ways to improve road quality is to use soil stabilizers. Traditional soil stabilizers are cement, lime, and coal fly ash, which are characterized by lower costs and being less time consuming [8,10,11]. However, although the soil solidified using cement exhibits high early strength, its later strength development is limited. In contrast, soil solidified with lime and coal fly ash shows lower early compressive strength [12,13]. Additionally, the production process of these materials generates a large amount of carbon dioxide, accelerating global warming. The carbon dioxide emissions from cement production account for approximately 8% of global carbon dioxide emissions [14–17]. Therefore, environmentally friendly, efficient, and sustainable soil stabilizers should be explored as alternatives for pavement construction.

Recently, non-traditional stabilizers, especially cementitious materials based on multiple industrial solid wastes, have received more attention due to their low carbon emissions, low cost, low energy consumption, and better sustainability [18–22].

Steel slag (SS) and blast furnace slag (BFS) are the industrial solid waste generated in the process of steelmaking, and their major components are CaO, SiO_2, and Al_2O_3 [23–25]. In 2021, the SS output in China exceeded 120 million tons, but its comprehensive utilization was only about 20% [26]. Carbide slag (CS) is the predominant by-product resulting from the hydrolysis of calcium carbide during acetylene production, which is mainly composed of CaOH [23]. It is predicted that the annual production of CS exceeds 50 million tons in China, with a utilization rate is less than 30% [27]. Desulfurized gypsum (DG) is an industrial waste generated from smelters, large-scale enterprise boilers, and the desulfurization of flue gas in thermal power plants [28]. The output of DG in 2019 was approximately 130 million tons [29]. The main component of DG is $CaSO_4{\cdot}2H_2O$, which also contains impurities such as $CaCO_3$ and $CaSO_3$. Fly ash (FA) is a major solid waste emitted from thermal power plants [30,31]. The global annual production is estimated at 360 million tons [32], while China's annual output is about 800–900 million tons [33]. FA contains many toxic elements including Pb, As, Se, and Cr, and is considered to be a hazardous waste to the environment and human health [34,35].

From a chemical composition point of view, these industrial solid wastes (including SS, BFS, CS, DG, and FA) exhibit higher pozzolanic activity and alkalinity and can be used as raw materials for the preparation of geopolymers [21,36]. This is especially the case with CS, which, as an alkaline industrial solid waste, has the same main components and can be considered as a substitute for lime [37]. A geopolymer can interact with soil particles to form a cementitious gel through a pozzolanic reaction. The addition of CS, as an alkaline resource, provides OH^- and Ca^{2+}. Under alkaline conditions, the dissolution of aluminum and silica mineral components can be activated and react with $Ca(OH)_2$ to generate C-S-H, which can reduce pores and increase the compressive strength and stiffness of soil [38]. Several studies have been conducted on the utilization of industrial solid waste as soil stabilizers [19,29,39–43]. However, few studies investigated the action mechanisms of multiple industrial solid wastes as soil solidifiers, e.g., five different industrial solid wastes. The advantages of various solid wastes can be complemented to improve the utilization rate effectively and reduce costs.

In the present study, a novel low-carbon and environmentally friendly soil stabilizer is proposed in order to improve the comprehensive utilization efficiency of industrial solid waste and reduce environmental pollution. Steel slag (SS), carbide slag (CS), blast furnace slag (BFS), fly ash (FA), and desulfurized gypsum (DG) were used as raw materials to formulate a multiple industrial solid waste-based soil stabilizer (MSWSS). The D-optimal mixture design approach was utilized to determine the optimal formulation of the various raw materials for the MSWSS, which can reduce the number of experiments and costs [44]. The mechanical characteristics of the stabilized soil were evaluated using the unconfined compressive strength (UCS) test and compared with conventional soil stabilizers. The micromorphology of solidified soil and hydration mechanisms were investigated by using X-ray diffraction (XRD), scanning electron microscopy (SEM), and thermogravimetric analysis (TGA).

2. Materials and Methods

2.1. Materials

The test soil used in this study was collected from Cangzhou, Hebei Province, China, and had such characteristics as a small plasticity index, poor stability when exposed to water, low strength, and the inability to be directly used as base for the construction of pavement. The characteristics of the test soil are provided in Table 1.

Table 1. Physical and mechanical properties of the test soil.

Natural Water Content /%	Optimum Water Content/%	Maximum Dry Density/g·cm^{-3}	Plastic Limit /%	Liquid Limit /%	Plasticity Index
7.22	14.40	1.85	25.10	40.28	15.18

The raw materials used to produce the MSWSS included CS, BFS, SS, FA, and DG, which were collected from the following companies: Hebei Jinniu Chemical Co., Ltd., Xingtai, China; Cangzhou Zhongtie Equipment Manufacturing Material Co., Ltd., Cangzhou, China; and Shenhua Group Guohua Power Plant, Yongzhou, China. The chemical compositions of the test soil and the raw materials are listed in Table 2, while the XRD patterns of the raw materials are shown in Figure 1. It can be seen from the results that SiO_2 (52.93%) comprised the largest proportion of the test soil, and the main phases were quartz, plagioclase, muscovite, and calcite. SS, CS, and FA had a high content of CaO (39.92%, 93.83%, 41.82%), while BFS had a high content of SiO_2 (39.83%). Additionally, SS contained 27.26% Fe_2O_3.The major components of DG were CaO (47.43%) and SO_3 (43.32%). As shown in Figure 1, the main phases of FA were quartz, mullite, CaO, and corundum. BFS is an amorphous phase structure. The dominant phases of SS were tetracalcium aluminoferrite C_4AF ($4CaO{\cdot}Al_2O_3{\cdot}Fe_2O_3$), dicalcium silicate C_2S ($2CaO{\cdot}SiO_2$), and tricalcium silicate C_3S ($3CaO{\cdot}SiO_2$).

Table 2. Chemical composition of the test soil and the raw materials (wt. %).

	SiO_2	Al_2O_3	CaO	Fe_2O_3	MgO	K_2O	Na_2O	Cl	TiO_2	SO_3	LOI
test soil	52.93	18.55	10.86	7.94	3.45	3.66	0.88	0.08	0.94	0.1	1.21
SS	15.04	2.36	39.92	27.26	7.84	0.03	0.44	0.23	0.82	0.23	2.92
CS	2.46	1.71	93.83	0.30	0	0	0.53	0.67	0.05	0.36	26.84
BFS	39.83	30.33	14.52	7.92	1.01	1.42	1.06	0	1.49	1.12	2.15
DG	3.19	1.21	47.43	1.19	1.42	0.29	0.39	0.84	0.11	43.32	20.27
FA	28.06	15.73	41.82	1.60	7.91	0.36	0.35	0.04	1.17	2.47	3.49

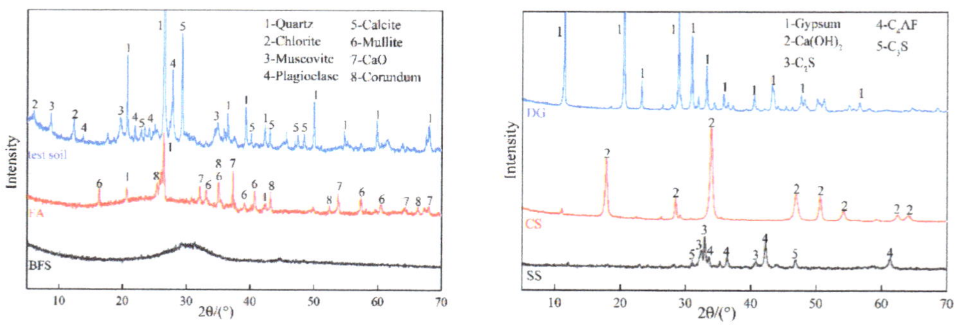

Figure 1. XRD patterns of the test soil and the raw materials.

2.2. Sample Preparation

The test soil was dried at a temperature of 55 °C and ground to a specific surface area reached 450–500 cm^2/g. The MSWSS was produced by mixing CS, BFS, SS, FA, and DG in a certain proportion obtained using the D-optimal mixture design approach. The amount of the addition of MSWSS was fixed at 10% of the total mass of the test soil. Then, a certain amount of water was added, the MSWSS and test soil were mixed with the water, and the mixture moisture content was 17.86%. Afterwards, the samples were prepared in accordance with the "Test Methods of Materials Stabilized with Inorganic Binders for Highway Engineering" (JTGE51-2009). The specimens were placed within a cylindrical mold with a diameter and height of 50mm and shaped by applying a stable pressure using a TYA-2000 digital display pressure testing machine at a loading rate of 1 mm/min for 2 min. Afterwards, the samples were weighed, enclosed within sealed bags, and moved to a standard environment with a relative humidity of over 95% and a temperature maintained at 20 ± 2 °C for the specified curing time. After reaching the set time, the specimens were tested for compressive strength and microstructure characterization.

2.3. D-Optimal Mixture Design

The D-optimal mixture design is commonly used in experimental design to optimize multiple variables simultaneously. In this study, the optimal formulation of raw materials was determined using D-optimal design. The steel slag (SS), carbide slag (CS), blast furnace slag (BFS), desulfurized gypsum (DG), and fly ash (FA) were labelled as A, B, C, D, and E. The lower and upper limits of each component were proposed based on preliminary experiments: A (5–20 wt. %), B (20–50 wt. %), C (5–15 wt. %), D (5–20 wt. %), and E (10–25 wt. %). A 35-run D-optimal mixture design with multiple constraints on the component proportions was implemented by using Design-Expert software version 13.0 (Table 3). The combination of the components and the experimental results of 7d UCS are listed in Table 4.

Table 3. The variables and their constraints for D-optimal mixture design.

Variables	Name	Content/wt. %	
		Low	High
A	Steel slag	5	20
B	Carbide slag	20	50
C	Blast furnace slag	5	15
D	Desulfurized gypsum	5	20
E	Fly ash	10	25

Table 4. Experimental design of the five components and the results for 7-day unconfined compressive strength (7d UCS).

Number	The Ratio of Each Component/wt. %					7d UCS/MPa
	A	B	C	D	E	
1	12.6	50.0	10.3	9.2	17.9	1.28
2	13.0	47.7	9.3	20.0	10.0	1.23
3	12.6	50.0	10.3	9.2	17.9	1.23
4	20.0	32.4	10.1	12.5	25.0	1.35
5	13.8	31.4	15.0	18.3	21.5	1.15
6	20.0	50.0	15.0	5.0	10.0	1.17
7	8.2	37.7	11.6	17.5	25.0	1.28
8	8.2	37.7	11.6	17.5	25.0	1.27
9	12.6	50.0	10.3	9.2	17.9	1.26
10	13.8	41.7	8.6	11.0	25.0	1.27
11	17.5	41.5	15.0	5.0	21.0	1.22
12	5.0	50.0	5.0	17.3	22.8	1.19
13	20.0	32.8	11.6	20.0	15.6	1.10
14	5.0	45.5	14.8	20.0	14.8	1.59
15	17.3	23.0	15.0	19.7	25.0	1.01
16	15.3	42.7	15.0	15.2	11.7	1.58
17	20.0	41.0	10.5	11.4	17.1	1.15
18	20.0	40.5	8.6	20.0	10.9	1.12
19	15.3	42.7	15.0	15.2	11.7	1.53
20	20.0	47.8	5.0	13.0	14.2	1.05

Table 4. *Cont.*

Number	The Ratio of Each Component/wt. %					7d UCS/MPa
	A	B	C	D	E	
21	15.6	35.7	5.0	20.0	23.6	0.95
22	17.5	41.5	15.0	5.0	21.0	1.21
23	5.0	50.0	15.0	5.0	25.0	1.25
24	5.0	44.4	15.0	12.9	22.7	1.63
25	20.0	45.0	5.0	5.0	25.0	1.30
26	20.0	25.0	10.0	20.0	25.0	1.06
27	12.8	40.0	9.4	20.0	17.8	1.28
28	11.7	50.0	5.0	8.3	25.0	1.03
29	10.0	40.0	15.0	10.0	25.0	1.33
30	20.0	50.0	8.3	5.0	16.7	1.12
31	12.8	40.0	9.4	20.0	17.8	1.26
32	5.0	50.0	15.0	15.0	15.0	1.64
33	18.3	41.7	5.0	10.0	25.0	1.19
34	12.5	50.0	15.0	12.5	10.0	1.54
35	13.2	46.8	8.7	13.2	18.2	1.29

2.4. Compressive Strength Test

The unconfined compressive strengths (UCS) were tested according to Chinese standard JTG E51-2009 [45] using an electro-hydraulic universal testing machine (CTS-E200) with a loading rate of 1mm/min. The measurements were performed on at least three representative samples to obtain an average value.

2.5. X-ray Powder Diffraction (XRD)

The phase compositions of the samples were detected using the X-ray powder diffraction (XRD) method using a Bruker Advance D8 X-ray diffractometer with Cu Kα radiation (λ = 0.1542 nm) operating at 20 kV and 200 mA. The scanning angle ranged from 5° to 80°, with a scanning speed of 5°/min and a step size of 0.02°.

2.6. Thermogravimetric Analysis (TGA)

Thermal analyses were performed on a thermogravimetric differential scanning calorimeter (TG-DSC, NETZSCH STA-409, Selb, Germany) with a nitrogen protective gas. A 30 mg sample was used for each measurement, with a heating rate of 10 °C/min in the temperature range of 25 to 1000 °C.

2.7. Scanning Electron Microscopy (SEM)

Microstructural analysis of the specimens was performed using scanning electron microscopy (SEM) alongside an energy dispersive spectra (EDS) analyzer. EDS was utilized to determine the elemental composition of hydration products. SEM-EDS was conducted using a Zeiss SUPRA 55 field emission-scanning electron microscope (FE-SEM) outfitted with a LinkNA1000 energy spectrometer.

3. Results and Discussion

3.1. Optimization of Mixture Composition for MSWSS

3.1.1. Fitting Model and Response Surface Analysis

The interaction between the five components was mathematically modeled against the response to 7d UCS. A comparison and evaluation of different fitting models was performed using the Design-Expert 13 program. The assessment results of the accuracy of the four fitting models are shown in Table 5. The model selection criteria were a higher F-value, the adjusted and predicted R^2 value, and a low p-value (<0.05). According to the result, the special cubic model was selected. The p-value of the special cubic model was below 0.0001, suggesting that the model adequately describes the response. With a coefficient of determination of $R^2 = 0.9971$, it can be inferred that the experimental data closely align with the predicted values, affirming the model's high significance.

Table 5. Accuracy evaluation of different fitting models.

Model	Sequential *p*-Value	Lack of Fit *p*-Value	Adjusted R^2	Predicted R^2	Evaluate
Linear	0.0002	<0.0001	0.4607	0.2808	
Quadratic	0.0002	0.0008	0.8162	0.5376	
Special Cubic	<0.0001	0.8173	0.9893	0.8363	Suggested
Cubic	0.8173		0.9862		Aliased

Analyses of variance (ANOVAs) were conducted on the response (7d UCS) using the Design-Expert software. The results of the ANOVAs for 7d UCS are shown in Table 6. The *p*-values and *F*-values represent the significance of variable factors. In general, smaller *p* -values (*p*-value > 0.05 insignificant, *p*-value < 0.05 significant, *p*-value < 0.01 extremely significant), along with larger *F*-values, denote the more significant impact of the variable factors on the response [46]. It can be seen from Table 6 that except for the interaction terms CD, CE, DE, ACD, and BCE, the *p*-values of other interaction terms were all < 0.01, which means that they had a notable influence on the 7d UCS. In contrast, interactive items CD, CE, DE, ACD, and BCE had no significant effects on 7d UCS because their *p*-values were greater than 0.05. Multiple quadratic regression equations were used to fit the functional relationship between the influencing variable factors and the responses. Equation (1) reflects the mathematical relations between 7d UCS and factors A, B, C, D, and E. The coefficients of the equation could be used to describe the effect of these variables on the response (7d UCS). A positive coefficient in the equation signifies a positive effect, while a negative coefficient indicates a negative effect. According to the results, the interactive terms AB, AD, AE, BC, BE, ABC, ABD, ABE, ADE, BCD, and BDE had an extremely significant effect (*p*-value < 0.01) on the 7d UCS.

$$\begin{aligned} Y_{7d\ UCS} = {} & -16.15A + 2.32B - 4.69C + 2.33D - 6.08E + 38.77AB + 49.56AC + 33.14AD + 69.80AE + 20.40BC + \\ & 8.38BD + 18.29BE - 4.67CD + 8.21CE + 8.67DE - 94.45ABC - 37.40ABD - 113.92ABE + 17.71ACD - \\ & 44.35ACE - 90.85ADE + 38.52BCD - 4.88BCE + 21.40BDE + 3.81CDE \end{aligned} \tag{1}$$

Table 6. ANOVA table of experimental results.

Source	Sum of Squares	Free Degree	Mean Square	*F*-Value	*p*-Value	
Model	1.07	24	0.0445	128.53	<0.0001	significant
Linear Mixture	0.5633	4	0.1408	406.92	<0.0001	
AB	0.0436	1	0.0436	125.97	<0.0001	
AC	0.0083	1	0.0083	24.05	0.0008	
AD	0.0174	1	0.0174	50.15	<0.0001	
AE	0.0543	1	0.0543	156.97	<0.0001	
BC	0.0068	1	0.0068	19.72	0.0016	
BD	0.0028	1	0.0028	7.95	0.0200	
BE	0.0112	1	0.0112	32.42	0.0003	
CD	0.0001	1	0.0001	0.2386	0.6369	
CE	0.0002	1	0.0002	0.4495	0.5194	
DE	0.0012	1	0.0012	3.45	0.0960	
ABC	0.0171	1	0.0171	49.50	<0.0001	
ABD	0.0099	1	0.0099	28.74	0.0005	
ABE	0.0658	1	0.0658	190.23	<0.0001	
ACD	0.0004	1	0.0004	1.30	0.2841	
ACE	0.0014	1	0.0014	4.11	0.0733	
ADE	0.0284	1	0.0284	82.04	<0.0001	

Table 6. *Cont.*

Source	Sum of Squares	Free Degree	Mean Square	*F*-Value	*p*-Value
BCD	0.0038	1	0.0038	10.86	0.0093
BCE	0.0000	1	0.0000	0.1022	0.7565
BDE	0.0037	1	0.0037	10.83	0.0094

Table 7 presents the findings of the model reliability test analysis. R^2 represents the correlation coefficient between the predicted and actual values. A higher R^2 value indicates a stronger correlation. If the values of the adjusted R^2 and predicted R^2 are high and the difference between them is close (adjusted R^2-predicted $R^2 < 0.2$), then the regression model can further prove this conclusion. As shown in Table 7, the values of R^2, adjusted R^2, predicted R^2, and adequate precision were 0.9971, 0.9893, 0.8363, and 43.2860, respectively. This indicated that the model equation can represent the relationship between the components and the 7d UCS well. The value of C.V. % was 1.48, indicating that the experiment shows high reliability and accuracy. The results showed that this model was able to predict the 7d UCS relatively accurately.

Table 7. Model reliability test analysis.

Model	Std. Dev.	Mean /MPa	C.V./%	R^2	Adjusted R^2	Predicted R^2	Adequate Precision
$Y_{7d\ UCS}$	0.0186	1.26	1.48	0.9971	0.9893	0.8363	43.2860

The response surface interaction analysis was used to evaluate the effect of each raw material on the 7d UCS of the solidified soil. Figure 2 shows the 3D surface graphs and contour illustrating the impact of the independent variables on the 7d UCS. Generally, when the response surface shows a curved form, it indicates that there is a degree of interaction among the raw material components [46]. As seen in Figure 2, an increase in the dosage of desulfurized gypsum and blast furnace slag resulted in an increase in the 7d UCS, whereas an increase in the dosage of steel slag, carbide slag, and fly ash had the opposite effect.

3.1.2. Optimization of Mixture Composition

The maximum unconfined compressive strength value for 7 days (7d UCS) is used as the target optimization value. Based on regression model analysis combined with the 3D response surface plots and contour plots, the optimal ratio of the soil solidifier can be obtained. According to the predicted result, the optimal raw material proportion for preparing the MSWSS was: 5% steel slag, 50% carbide slag, 15% blast furnace slag, 15% desulfurized gypsum and 15% fly ash, respectively. The predicted value of the 7d UCS was 1.65 MPa. To confirm the precision of the predicted value, experiments on the compressive strength for 7 days were carried out. The outcome is displayed in Table 8. The experimental value closely aligned with the predicted value, which demonstrated the reliability of the model. The value of the absolute relative deviations was 1.81% (<5%), suggesting that the prediction model exhibits high accuracy.

Table 8. Comparison of the observed values and the predicted values after optimization.

Proportion of Components/wt. %					7d UCS/MPa		Absolute Relative Deviations/%
A	B	C	D	E	Predicted Value	Observed Value	
5	50	15	15	15	1.65	1.62	1.81

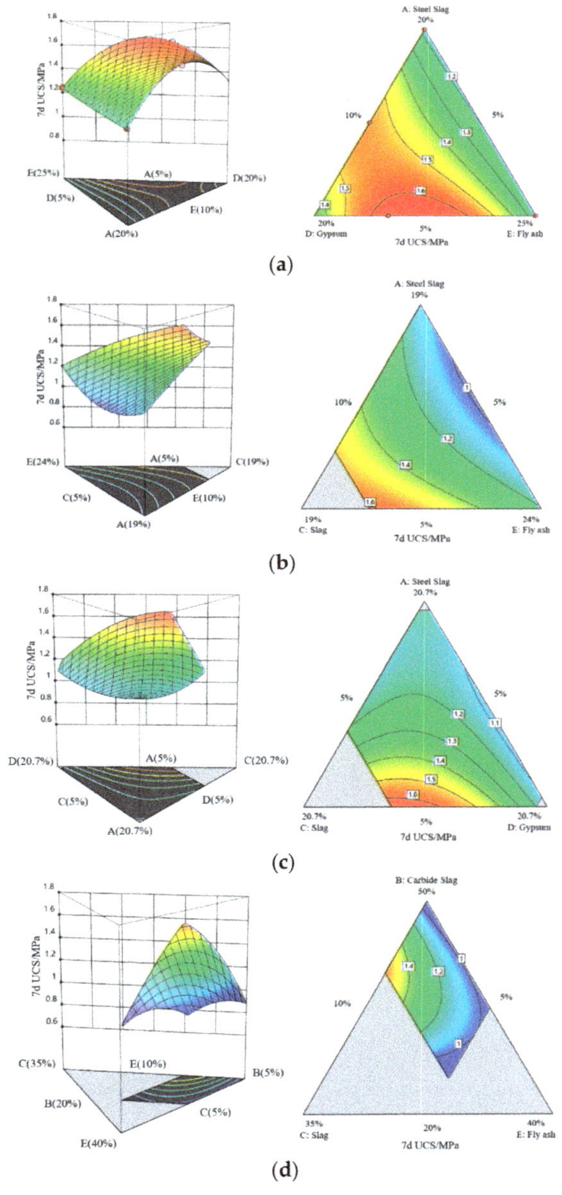

Figure 2. 3D surface graphs and contour plots for the effects of the independent variables on 7d UCS (A: Steel slag, B: Carbide slag, C: Blast furnace slag, D: Desulfurized gypsum, E: Fly ash): (**a**) the effect of A, D, E on 7d UCS; (**b**) the effect of A, C, E on 7d UCS; (**c**) the effect of A, C, D on 7d UCS; (**d**) the effect of B, C, E on 7d UCS.

3.2. Compressive Strength Test

MSWSS was prepared from multiple solid waste according to the optimal ratio obtained using the D-optimal mixture design method (SS:CS:BFS:DG:FA = 5:50:15:15:15). To compare the solidification effects of multiple solid waste-based stabilizer and ordinary portland cement (OPC) on the test soil, the unconfined compressive strength (UCS) of these two stabilized soils was tested and analyzed. Standard specimens were prepared using the test soil plus 10% MSWSS, with a moisture content of 17.65%. In addition, 10% P.O32.5 OPC was used instead of MSWSS as a comparative experiment.

The unconfined compressive strength (UCS) of the stabilized soil at various curing ages is presented in Figure 3. The UCS of the stabilized soil showed a progressive increase with the curing time. The soil solidified using MSWSS has a UCS ranging from 0.35 MPa to 3.27 MPa. At the same time, soil solidified using OPC has a UCS ranging from 0.47 MPa to 2.89 MPa. The 7d UCS results of the MSWSS group and of the OPC group were 1.7 MPa and 1.25 MPa, respectively, which met the standard requirements of Chinese technical standards for the Technical Guidelines for Construction of Highway Roadbases. The 7d UCS of MSWSS-stabilized soil is 36% higher than that of the OPC group. It can be seen from Figure 3 that the UCS of OPC stabilized soil was higher than that of MSWSS-stabilized soil when the curing age is in the first 3 days. Afterwards, with the prolongation of curing time, the UCS of soil stabilized using MSWSS was significantly higher than that of the OPC group. After curing for 180 days, the compressive strength exhibited a notable increase to 3.27 MPa and 2.89 MPa, respectively.

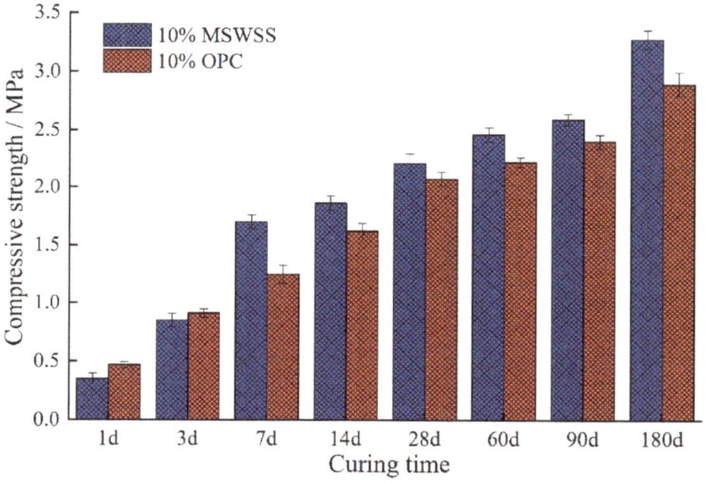

Figure 3. Compressive strength of stabilized soil at different curing times.

3.3. *XRD Analysis*

The XRD patterns of the solidified test soil at curing times of 7, 28, and 90 days are presented in Figure 4, and the primary phase compositions have been identified and labeled. Based on the results, the main phases were quartz, calcite, and plagioclase, and the primary hydration products were ettringite (AFt) and calcium silicate hydrate (C-S-H). This is consistent with the results of the SEM analysis (Figure 5). In addition, it also contained small amounts of chlorite and microcline. After curing for 7 days, ettringite and C-S-H can be observed from XRD pattern. As the curing time progressed, the intensity of AFt increased gradually, and the intensity of C-S-H also showed a weak increase, indicating that the hydration reaction was constantly developing. Meanwhile, a small amount of C_2S was detected, and the peak intensity decreased from 7 to 28 days of curing time, consistent with previous research [4,47]. Moreover, no distinct peaks of C_2S were observed after 90 days of curing. This showed that that C_2S is involved in the hydration process. Generally, it was assumed that the hydration reaction of C_2S generated C-S-H.

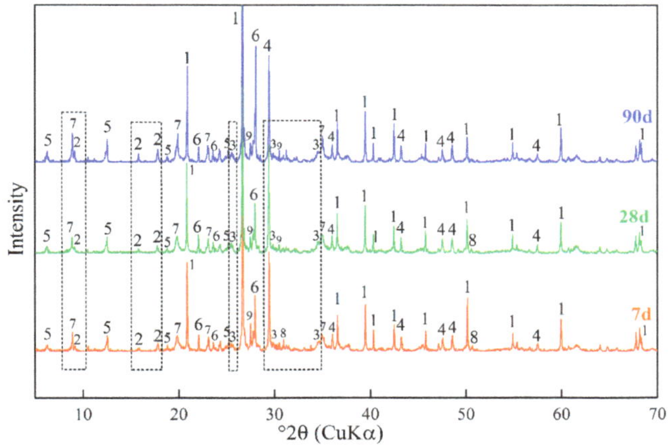

Figure 4. XRD patterns of stabilized soil with different curing times. 1-Quartz; 2-Ettringite (AFt); 3-C-S-H; 4-Calcite; 5-Chlorite; 6-Plagioclase; 7-MUCSovite; 8-C_2S; 9-Microcline.

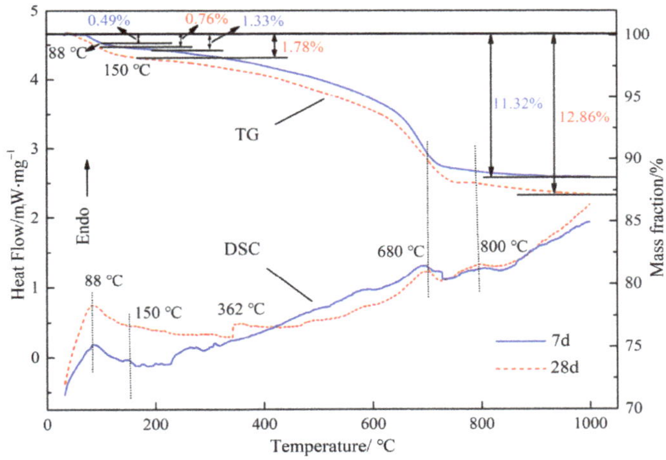

Figure 5. TG-DSC diagram of solidified soil at different curing times.

3.4. TG–DSC Analysis

The TG-DSC curves of the samples after curing for 7 and 28 days are displayed in Figure 5. The endothermic peaks at 50–400 °C could be primarily attributed to the thermal decomposition of C-S-H gel and AFt [24]. The endothermic peaks at 88 °C and 150 °C of the DSC curves were due to the release of free water [48]. The mass losses at 88 °C and 150 °C of the stabilized soil at 7 days and 28 days were 0.49% and 1.33%, and 0.76% and 1.78%, respectively. The endothermic peak at 362 °C of the DSC curves at a curing age of 28 days can be attributed to the dehydration of the crystal water of C-S-H and ettringite. The endothermic peaks at 680 °C and 800 °C were due to the decarburization and decomposition of calcite [25]. The results of the thermal analysis showed that the total mass loss of the samples after curing for 7 days and 28 days was 11.32% and 12.86%, respectively. This indicated that the hydration reaction continues as the curing age increases. The XRD analysis results also showed that the aggregate mass of the hydration products increases with the increase in curing time.

3.5. SEM-EDS Analysis

Figure 6 illustrates the microstructure of the solidified soils after 7 and 28 days of hydration. Figure 6a,b represent the samples at 7 curing days and 28 curing days. After curing for 7days, the sample presented a loose arrangement and showed the typical needle-rod crystal structures of AFt (Figure 6a). Simultaneously, C-S-H gel was detected, which covered the needle-rod crystals of the AFt. The AFt crystals and C-S-H grew intertwined with each other, and this kind of structure could effectively improve the early compressive strength of the stabilized soil (Figure 3).

Figure 6. SEM morphology of the solidified soil for (**a**) 7 days and (**b**) 28 days curing times.

As the curing time progressed to 28 days (Figure 6b), the content of the needle-rod structures representing ettringite was increased, which corresponds with the findings of the XRD analysis. Meanwhile, the loosely structured solidified soil became denser, which might be attributed to the gradually increasing output of the C-S-H gel filling the pore space. At this stage, the AFt is surrounded by C-S-H gel, resulting in a significant enhancement in the compressive strength.

3.6. Discussion

Acçording to characterization of the solidified soil, the primary hydration products were C-S-H gel ($Ca_5Si_6O_{16}(OH){\cdot}4H_2O$) and AFt ($3CaO{\cdot}Al_2O_3{\cdot}3CaSO_4{\cdot}32H_2O$). Carbide slag (CS) plays a crucial role in the hydration process because it provides Ca^{2+} ions and an alkaline environment. At the initial phase of hydration, the C_2S ($2CaO{\cdot}SiO_2$) and C_3S ($3CaO{\cdot}SiO_2$) in SS were hydrolyzed to form C-S-H gel and calcium hydroxide ($Ca(OH)_2$) (Equations (2) and (3)), which improved the early compressive strength (Figure 3). Meanwhile, a notable quantity of $Ca(OH)_2$ was produced, raising the alkalinity of the binder.

$$2C_3S + 6H_2O \rightarrow 3C - S - H + 3Ca(OH)_2 \tag{2}$$

$$2C_2S + 4H_2O \rightarrow 3C - S - H + Ca(OH)_2 \tag{3}$$

In alkaline environments, OH^- reacts with the amphoteric oxide Al_2O_3 and acidic oxide SiO_2 which can accelerate the hydrolysis of the glass phase, destroying the Ca-O, Si-O, and Al-O bonds and form $[Al(OH)_6]^{3-}$ and $[Si(OH)_3]^-$ [49]. Subsequently, with the participation of the Ca^{2+} and $SO_4{}^{2+}$ produced by the hydrolysis of DG, it reacts with $[Al(OH)_6]^{3-}$ and $[Si(OH)_3]^-$ to generate AFt. At the same time, $[Si(OH)_3]^-$ could react with Ca^{2+} to generate C-S-H gel. With the progression of the hydration reaction, the content of C-S-H and AFt increased significantly (Figure 4). The chemical reaction can be expressed in Equations (4)–(7). Consequently, large amounts of C-S-H and AFt grew interlacedly, filling the gaps between soil particles and improving the compactness and strength of the solidified soil. The hydration mechanism of the MSWSS is shown in Figure 7.

$$Al_2O_3 + 6OH^- + 3H_2O \rightarrow 2[Al(OH)_6]^{3-} \tag{4}$$

$$SiO_2 + OH^- + H_2O \rightarrow [Si(OH)_3]^- \quad (5)$$

$$2[Al(OH)_6]^{3-} + 3CaSO_4 \cdot H_2O + 23H_2O \rightarrow CaO \cdot Al_2O_3 \cdot 3CaSO_4 \cdot 32H_2O \quad (6)$$

$$[Si(OH)_3]^- + Ca^{2+} \rightarrow C-S-H \quad (7)$$

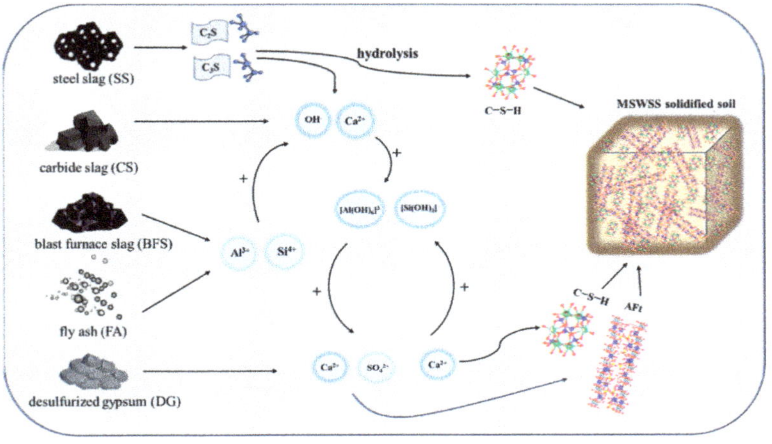

Figure 7. Hydration mechanism of the multiple industrial solid waste-based soil stabilizer (MSWSS).

4. Conclusions

This study proposed the preparation of a multiple industrial solid waste-based soil stabilizer (MSWSS) using steel slag (SS), carbide slag (CS), blast furnace slag (BFS), fly ash (FA), and desulfurized gypsum (DG). The D-optimal mixture approach was utilized in the experimental design to obtain the optimal formulation of raw materials. The unconfined compressive strength (UCS) of the solidified test soil was evaluated, and the hydration mechanism was analyzed using the XRD, TG-DSC, and SEM methods. In summary, the following conclusions can be drawn:

(1) Interaction between the five components was modeled mathematically against the responses of 7d UCS. According to the result of the ANOVA and the model reliability analysis, the prediction model showed a lower p-value (<0.0001) and lower absolute relative deviations (1.81%), and a higher F-value (406.92), R2 value (0.9971), and adjusted R2 value (0.9893). This indicated that the model had high reliability and was able to predict 7d UCS relatively accurately.

(2) The D-optimal mixture approach indicated that the optimal ratio of MSWSS should be 5% steel slag (SS), 50% carbide slag (CS), 15% blast furnace slag (BFS), 15% desulfurized gypsum (DG), and 15% fly ash (FA). The UCS of the stabilized soil increased with the curing time. The 7d UCS of soil stabilized using the multiple industrial solid waste-based soil stabilizer was 1.7 MPa and the 180d UCS increased to 3.27 MPa, which all met the requirements for the "Technical Guidelines for Construction of Highway Roadbases". The UCS of the MSWSS stabilized test soil group was significantly higher than that of the OPC group.

(3) XRD, TG-DSC, and SEM results revealed that the primary hydration products were AFt and C-S-H gel. As the curing progressed, the content of the AFt and C-S-H increased and the microstructure of stabilized soil exhibited a denser structure, leading to better mechanical strength.

The research proposed a novel low-carbon and environmentally friendly soil stabilizer, MSWSS, based on multiple industrial solid waste products, which could achieve better mechanical properties than traditional OPC stabilizers and meet the requirements of highway construction. This technology has the potential to enhance the comprehensive utilization of industrial solid waste and contribute to environmental protection.

Author Contributions: Conceptualization, X.W. (Xiaoli Wang), X.W (Xiancong Wang), and P.F.; methodology, investigation, data curation, X.W. (Xiaoli Wang), X.W (Xiancong Wang), B.L., J.S. and M.X.; writing—original draft preparation, X.W. (Xiaoli Wang) and X.W (Xiancong Wang); writing—review and editing, X.W. (Xiaoli Wang) and P.F.; supervision, project administration, P.F.; funding acquisition, X.W. (Xiaoli Wang) and P.F. All authors have read and agreed to the published version of the manuscript.

Funding: This research was funded by Key Research and Development Program of Hebei Province (grant numbers 22373809D), National Key Research and Development Program of China (grant numbers 2018YFC1900604), and Open Foundation of State Key Laboratory of Mineral Processing (grant numbers BGRIMM-KJSKL-2022-22).

Data Availability Statement: The data presented in this study are available on request from the corresponding author.

Conflicts of Interest: Authors Jinjin Shi and Miao Xu were employed by the Cangzhou Municipal Engineering Company Limited. The remaining authors declare that the research was conducted in the absence of any commercial or financial relationships that could be construed as a potential conflict of interest.

References

1. World Bank Group. *Trends in Solid Waste Management*; The World Bank: Bretton Woods, NH, USA, 2020; Available online: https://datatopics.worldbank.org/what-a-waste/trends_in_solid_waste_management.html (accessed on 9 February 2024).
2. Guo, Y.; Glad, T.; Zhong, Z.Z.; He, R.N.; Tian, J.P.; Chen, L.J. Environmental life-cycle assessment of municipal solid waste incineration stocks in Chinese industrial parks. *Resour. Conserv. Recycl.* **2018**, *139*, 387–395. [CrossRef]
3. Liu, A.; Ren, F.; Lin, W.L.Y.; Wang, J.Y. A review of municipal solid waste environmental standards with a focus on incinerator residues. *Int. J. Sustain. Built Environ.* **2015**, *4*, 165–188. [CrossRef]
4. Wang, X.; Fu, P.; Deng, W.; Shi, J.; Xu, M. Study on Mechanism of MSWI Fly Ash Solidified by Multiple Solid Waste-Based Cementitious Material Using the Rietveld Method. *Processes* **2023**, *11*, 2311. [CrossRef]
5. Kumar, R.A.; Singh, G.; Kumar, T.A. Comparative study of soil stabilization with glass powder, plastic and e-waste: A review. *Mater. Today Proc.* **2020**, *32*, 771–776. [CrossRef]
6. Lu, J.Z.; Jiang, S.; Chen, J.J.; Lee, C.H.; Cai, Z.W.; Ruan, H.D. Fabrication of superhydrophobic soil stabilizers derived from solid wastes applied for road construction: A review. *Transp. Geotech.* **2023**, *40*, 100974. [CrossRef]
7. Soltani-Jigheh, H.; Tahaei, Y.S.N. Effect of liquid polymer on properties of fine-grained soils. *Geotech. Eng.* **2019**, *50*, 9–21.
8. Hassan, N.; Hassan, W.H.W.; Rashid, A.S.A.; Latifi, N.; Yunus, N.Z.M.; Horpibulsuk, S.; Moayedi, H. Microstructural characteristics of organic soils treated with biomass silica stabilizer. *Environ. Earth Sci.* **2019**, *78*, 367. [CrossRef]
9. Zhang, M.; Guo, H.; El-Korchi, T.; Zhang, G.; Tao, M. Experimental feasibility study of geopolymer as the next-generation soil stabilizer. *Constr. Build. Mater.* **2013**, *47*, 1468–1478. [CrossRef]
10. Latifi, N.; Vahedifard, F.; Ghazanfari, E.; Rashid, A.S.A. Sustainable Usage of Calcium Carbide Residue for Stabilization of Clays. *J. Mater. Civ. Eng.* **2018**, *30*, 2313. [CrossRef]
11. Latifi, N.; Vahedifard, F.; Siddiqua, S.; Horpibulsuk, S. Solidification–Stabilization of Heavy Metal–Contaminated Clays Using Gypsum: Multiscale Assessment. *Int. J. Géoméch.* **2018**, *18*, 1283. [CrossRef]
12. Kavak, A.; Akyarlı, A. A field application for lime stabilization. *Environ. Geol.* **2007**, *15*, 987–997. [CrossRef]
13. Ninov, J.; Donchev, I. Lime stabilization of clay from the 'Mirkovo' deposit. *J. Therm. Anal. Calorim.* **2008**, *19*, 487–490. [CrossRef]
14. Tayeh, B.A.; Hasaniyah, M.W.; Zeyad, A.M.; Awad, M.M.; Alaskar, A.; Mohamed, A.M.; Alyousef, R. Durability and mechanical properties of seashell partially-replaced cement. *J. Build. Eng.* **2020**, *31*, 101328. [CrossRef]
15. Marvila, M.T.; Alexandre, J.; Azevedo, A.R.G.; Zanelato, E.B.; Xavier, G.C.; Monteiro, S.N. Study on the replacement of the hydrated lime by kaolinitic clay in mortars. *Adv. Appl. Ceram.* **2019**, *118*, 373–380. [CrossRef]
16. Latifi, N.; Vahedifard, F.; Ghazanfari, E.; Horpibulsuk, S.; Marto, A.; Williams, J. Sustainable Improvement of Clays Using Low-Carbon Nontraditional Additive. *Int. J. Géoméch.* **2018**, *18*, 1086. [CrossRef]
17. Wang, L.; Chen, S.S.; Tsang, D.C.W.; Poon, C.; Shih, K. Recycling contaminated wood into eco-friendly particleboard using green cement and carbon dioxide curing. *J. Clean. Prod.* **2016**, *137*, 861–870. [CrossRef]
18. Gu, J.; Liu, X.; Zhang, Z. Road base materials prepared by multi-industrial solid wastes in China: A review. *Constr. Build. Mater.* **2023**, *373*, 130860. [CrossRef]
19. Rafiean, A.H.; Kani, E.N.; Haddad, A. Mechanical and Durability Properties of Poorly Graded Sandy Soil Stabilized with Activated Slag. *J. Mater. Civ. Eng.* **2020**, *32*, 2990.

20. Ghaffar, H.; Al-Kheetan, M.; Ewens, P.; Wang, T.; Zhuang, J. Investigation of the interfacial bonding between flax/wool twine and various cementitious matrices in mortar composites. *Constr. Build. Mater.* **2020**, *239*, 117833. [CrossRef]
21. Thirumalai, R.; Babu, S.S.; Naveennayak, V.; Nirmal, R.; Lokesh, G. A Review on Stabilization of Expansive Soil Using Industrial Solid Wastes. *Engineering* **2017**, *9*, 1008–1017. [CrossRef]
22. Zhang, J.; Li, S.; Li, Z. Investigation the synergistic effects in quaternary binder containing red mud, blast furnace slag, steel slag and flue gas desulfurization gypsum based on artificial neural networks. *J. Clean. Prod.* **2020**, *273*, 122972. [CrossRef]
23. Traven, K.; Češnovar, M.; Ducman, V. Particle size manipulation as an influential parameter in the development of mechanical properties in electric arc furnace slag-based AAM. *Ceram. Int.* **2019**, *45*, 22632–22641. [CrossRef]
24. Li, W.; Yi, Y. Use of carbide slag from acetylene industry for activation of ground granulated blast-furnace slag. *Constr. Build. Mater.* **2020**, *238*, 117713. [CrossRef]
25. Gao, D.; Zhang, Z.; Meng, Y.; Tang, J.; Yang, L. Effect of Flue Gas Desulfurization Gypsum on the Properties of Calcium Sulfoaluminate Cement Blended with Ground Granulated Blast Furnace Slag. *Materials* **2021**, *14*, 382. [CrossRef] [PubMed]
26. Wang, J.; Fu, H.; Yan, X.; Wang, L.; Wang, P. Research status of comprehensive utilization of steel slag. *China Nonferr. Metall.* **2021**, *50*, 77–82.
27. Wang, S.J.; Bu, H.T.; Chen, H.J.; Hu, T.; Chen, W.Z.; Wu, J.H.; Hu, H.J.; Lin, M.Z.; Li, Y.T.; Jiang, G.B. Floatable magnetic aerogel based on alkaline residue used for the convenient removal of heavy metals from wastewater. *Chem. Eng. J.* **2020**, *399*, 125760. [CrossRef]
28. Nan, J.K.; Chen, X.M.; Wang, X.Y.; Lashari, M.S.; Wang, Y.M.; Guo, Z.C.; Du, Z.J. Effects of applying flue gas desulfurization gypsum and humic acid on soil physicochemical properties and rapeseed yield of a saline-sodic cropland in the eastern coastal area of China. *J. Soils Sediments* **2016**, *16*, 38–50. [CrossRef]
29. Wang, X.; Kim, S.; Wu, Y.; Liu, Y.; Liu, T.; Wang, Y. Study on the optimization and performance of GFC soil stabilizer based on response surface methodology in soft soil stabilization. *Soils Found.* **2023**, *63*, 101278. [CrossRef]
30. Amran, M.; Debbarma, S.; Ozbakkaloglu, T. Fly ash-based eco-friendly geopolymer concrete: A critical review of the long-term durability properties. *Constr. Build. Mater.* **2021**, *270*, 121857. [CrossRef]
31. Makarichi, L.; Jutidamrongphan, W.; Techato, K.A. The evolution of waste-toenergy incineration: A review. *Renew. Sustain. Energy Rev.* **2018**, *91*, 812–821. [CrossRef]
32. Mao, Y.; Wu, H.; Wang, W.; Jia, M.; Che, X. Pretreatment of municipal solid waste incineration fly ash and preparation of solid waste source sulphoaluminate cementitious material. *J. Hazard. Mater.* **2020**, *385*, 121580. [CrossRef]
33. Chen, R.; Zhang, R.; Han, H. Climate neutral in agricultural production system: A regional case from China. *Environ. Sci. Pollut. Res.* **2021**, *28*, 33682–33697. [CrossRef]
34. Marieta, C.; Guerrero, A.; Leon, I. Municipal solid waste incineration fly ash to produce eco-friendly binders for sustainable building construction. *Waste Manag.* **2021**, *120*, 114–124. [CrossRef]
35. Wang, N.N.; Sun, X.Y.; Zhao, Q.; Yang, Y.; Wang, P. Leachability and adverse effects of coal fly ash: A review. *J. Hazard. Mater.* **2020**, *396*, 122725. [CrossRef]
36. Ren, C.; Wang, W.; Li, G. Preparation of high-performance cementitious materials from industrial solid waste. *Constr. Build. Mater.* **2017**, *152*, 39–47. [CrossRef]
37. Yang, J.; Zeng, J.; He, X.; Zhang, Y.; Su, Y.; Tan, H. Sustainable clinker-free solid waste binder produced from wet-ground granulated blast-furnace slag, phosphogypsum and carbide slag. *Constr. Build. Mater.* **2022**, *330*, 127218. [CrossRef]
38. Kampala, A.; Horpibulsuk, S. Engineering Properties of Silty Clay Stabilized with Calcium Carbide Residue. *J. Mater. Civ. Eng.* **2013**, *25*, 632–644. [CrossRef]
39. Yi, Y.; Gu, L.; Liu, S.; Puppala, A.J. Carbide slag–activated ground granulated blastfurnace slag for soft clay stabilization. *Can. Geotech. J.* **2015**, *52*, 656–663. [CrossRef]
40. He, J.; Shi, X.K.; Li, Z.X.; Zhang, L.; Feng, X.Y.; Zhou, L.R. Strength properties of dredged soil at high water content treated with soda residue, carbide slag, and ground granulated blast furnace slag. *Constr. Build. Mater.* **2020**, *242*, 118126. [CrossRef]
41. Ren, P.; Ling, T.C. Roles of chlorine and sulphate in MSWIFA in GGBFS binder: Hydration, mechanical properties and stabilization considerations. *Environ. Pollut.* **2021**, *284*, 117175. [CrossRef]
42. Wen, N.; Zhao, Y.; Yu, Z.; Liu, M. A sludge and modified rice husk ash-based geopolymer: Synthesis and characterization analysis. *J. Clean. Prod.* **2019**, *226*, 805–814. [CrossRef]
43. Phummiphan, I.; Horpibulsuk, S.; Phoongernkham, T.; Arulrajah, A.; Shen, S.L. Marginal Lateritic Soil Stabilized with Calcium Carbide Residue and Fly Ash Geopolymers as a Sustainable Pavement Base Material. *J. Mater. Civ. Eng.* **2017**, *29*, 1708. [CrossRef]
44. Venkatesan, M.; Zaib, Q.; Shah, I.H.; Park, H.S. Optimum utilization of waste foundry sand and fly ash for geopolymerconcrete synthesis using D-optimal mixture design of experiments. *Resour. Conserv. Recycl.* **2019**, *148*, 114–123. [CrossRef]
45. Lei, B.; Fu, P.; Wang, X.; Ni, W.; Zhang, S.; Wang, X.; Shi, J.; Xu, M. The Hydration Mechanisms of Co-Stabilization Saline Soils by Using Multiple Solid Wastes. *Processes* **2023**, *11*, 2679. [CrossRef]
46. Asadzadeh, S.; Khoshbayan, S. Multi-objective optimization of influential factors on production process of foamed concrete using Box-Behnken approach. *Constr. Build. Mater.* **2018**, *170*, 101–110. [CrossRef]
47. Xu, C.; Ni, W.; Li, K.; Zhang, S.; Xu, D. Activation mechanisms of three types of industrial by-product gypsums on steel slag–granulated blast furnace slag-based binders. *Constr. Build. Mater.* **2021**, *288*, 123111. [CrossRef]

48. Durdziński, P.T. *Hydration of Multi-Component Cements Containing Cement Clinker, Slag, Calcareous Fly Ash and Limestone*; EPFL: Lausanne, Switzerland, 2016; p. 189.
49. Zhang, N.; Li, H.; Liu, X. Hydration mechanism and leaching behavior of bauxitecalcination-method red mud-coal gangue based cementitious materials. *J. Hazard. Mater.* **2016**, *314*, 172–180. [CrossRef]

processes

MDPI

Article

Advancing Plastic Waste Management for a Circular Economy: Comparative Insights from Plovdiv (Bulgaria) and Kostanay (Kazakhstan)

Ekaterina Todorova [1], Gulnara Yunussova [2,*], Xeniya Formazyuk [3], Aleksandrina Kostadinova-Slaveva [1,*], Gulzhan Kazkenova [2] and Gulzat Jetpysbai [2]

1 Faculty of Ecology and Landscape Architecture, University of Forestry, 1797 Sofia, Bulgaria; etodorova@ltu.bg
2 Akhmet Baitursynuly Kostanay Regional University, Kostanay 110000, Kazakhstan; gulzhan_kazkenova@mail.ru (G.K.); gulzatzhetpisbai@mail.ru (G.J.)
3 Tobol-Torgay Basin Inspection for Regulation, Protection and Use of Water Resources, Astana 020000, Kazakhstan; ksenyakrivopusk2018@gmail.com
* Correspondence: gulnara_yun@mail.ru (G.Y.); aslaveva@ltu.bg (A.K.-S.)

Abstract: Plastic waste management is a critical challenge worldwide, influenced by the socio–economic conditions and policy frameworks of individual countries. This study evaluates the plastic waste management practices in two cities, Kostanay, Kazakhstan, and Plovdiv, Bulgaria, through a comparative analysis of household waste data, recycling systems, and legislative frameworks. The cities are similar in their main features, but the practice of plastic waste management is developed to varying degrees. In the period from 2017 to 2023, the population of Plovdiv decreased by 6.77% and the amount of municipal waste decreased by 3.55%. In Kostanay during 2021–2023, the population of the city increased by about 5%, and the amount of waste generated by more than 25%. Using morphological analysis and municipal records, this research identifies gaps in current practices and explores actionable strategies for enhancing plastic waste recycling. The qualitative analysis of the practice of urban plastic waste management shows that, compared to Plovdiv, the Kazakh city has underused the managerial and technological capabilities in the process of developing the plastic waste management industry and the transition to a circular economy. This study highlights the potential for implementing circular economy principles in both cities, with targeted recommendations to address existing challenges and maximize waste treatment and recycling. In Plovdiv, thermochemical processes for recycling plastic waste into new products are promising, while in Kostanay, mechanical methods for optimizing waste collection and transportation are promising.

Keywords: plastic waste; circular economy; green solution; sustainable management

1. Introduction

Plastic, as a light, convenient, flexible, and durable material, occupies an increasingly important place in the production of widespread waste, and its quantity in the market continues to grow. The global production of plastic was estimated at 400.3 million tons in 2022. Also called the "material with 1000 uses", plastic finds application throughout human life with its versatile properties. Since a large amount of plastic products become waste within just a few months or years, they are also one of the priority sectors for implementing a circular economy. The sustainable management of plastic waste is a critical component of the broader waste management strategies aimed at achieving a

 https://doi.org/10.3390/pr13030888

circular economy. This requires not only an understanding of the sources and collection methods for plastic waste, but also the integration of legislative, technological, and socio–economic factors that influence its effective recycling and recovery. Despite global efforts, a significant portion of plastic waste continues to be discarded in mixed municipal waste, complicating the subsequent recycling and recovery processes [1,2]. Previous studies of urban solid waste management systems focused on the analysis and assessment of sustainable management of both the system as a whole, using individual cities as examples, and individual influencing external and internal factors. Modern factors that ensure the growth of plastic waste generation include such socio–economic factors as an increase in the standard of living [3], GDP [4], population density [4,5], urbanization [6], and urban population growth [3]. At the same time, a decrease in the volume of plastic waste occurs if investments in waste management infrastructure grow [7] and waste collection routes and schedules are optimized [8].

Factors influencing the formation rate of municipal solid waste have been classified into seven types [9,10]. Most often, studies show that the composition and quantity of municipal waste are most strongly influenced by the number and level of income [10].

Analysis of waste management systems in the literature is usually carried out within a specific country or city [11–14]. A significant part of the studies is devoted to the analysis of practices in one region: the Global South [15], developing countries [16], EU countries [17–19], etc. Authors of country-wide and regional studies critically rethink the experience of municipal solid waste management and specialize in their needs. As a result, driving factors are persistently identified; a search is underway for those relationships that improve the practices in municipalities, preferred recycling methods that increase the efficiency of waste management systems, and methodologies that allow a full assessment of the work of waste management systems using an optimal set of indicators.

Papers on Kazakhstan in the English-language sector of scientific research in waste management are few. They examine various aspects of plastic waste management in Kazakhstan. The sustainability of household waste management in Kazakhstan, including plastic waste, is limited by the fact that the infrastructure for waste processing and disposal is not developed, insufficient investment, and weak coordination between government agencies and businesses [20]. It is also noted that, although there is a Green Economy Transition Plan, a plan for the transition to a circular economy is needed [21]. More complex strategies for the sustainable operation of waste management systems, Zero Waste, Integrated Waste Systems, and Industrial Symbiosis, are useful, but will not be applied in Kazakhstan until later.

Since 2007, the European Union and Central Asia (CA) have been developing a partnership in five areas, including environmental protection [22]. At the same time, there are no cross-sectoral studies in these regions in the area of waste management, which reduces the potential for transferring European experiences. Such studies will help countries at the early stages of waste management system development to highlight gaps in this activity and contribute to the formation of a more effective waste management system, and on this basis, to achieve progress in a shorter period of time.

This study aims to evaluate and compare the plastic waste management practices in two cities—Kostanay in Kazakhstan and Plovdiv in Bulgaria—by analyzing the household waste data, recycling policies, and the readiness for transition to a circular economy. This study considers the integration of circular economy principles, paying specific attention to the differences in separate collection systems. The research objectives are to: (1) identify gaps in current waste management systems, (2) evaluate the economic and legislative frameworks influencing plastic waste recycling, and (3) propose strategies for enhancing the sustainable management of plastic waste in both cities.

2. Materials and Methods

For the purposes of this study, a comparative analysis was conducted between the cities of Kostanay in Kazakhstan and Plovdiv in Bulgaria to evaluate their plastic waste management practices and readiness for a circular economy.

Study Area

In order to turn waste into a resource, the goals of waste management must be aligned with the goals of the transition to a circular economy. Sustainable solid waste management (SWM) contributes to achieving 11 United Nations Sustainable Development Goals (SDGs) [23,24]. The Republic of Kazakhstan and the Republic of Bulgaria have implemented their waste management targets for achieving recycling levels that are relatively similar and aimed at the transition to a circular economy. For Kazakhstan, the goal is for the share of recycled waste to reach 50% by 2050 [25]. Although the country has not yet set specific targets for plastic waste collection and recycling, its recent policy initiatives such as the 2019 ban on the disposal of dry fractions of waste (plastic waste, waste paper, cardboard and paper waste, glass) into solid waste landfills indicate a growing commitment to improving waste management. Kazakhstan was chosen for comparison because it represents an example of a country in the early stages of implementing circular economy principles. This comparison allows for an analysis of how different levels of policy maturity influence the effectiveness of plastic waste management, Figure 1.

Figure 1. Map of the location of the cities under study. Map source—https://upload.wikimedia.org/wikipedia/commons/4/47/Bulgaria_Kazakhstan_Locator.png (accessed on 17 September 2024).

In contrast, Bulgaria fulfills the ambitious pan-European goals for waste recycling and reuse, with the aim that by 2035, 65% of household waste in the country must be prepared for reuse or recycling, and only up to 10% of household waste will be landfilled [26]. Special attention in the regulatory framework of Bulgaria is paid to waste from packaging, which is one of the common mass-distributed household wastes. In packaging waste regulations, recycling targets have been set at 55% of plastic packaging in the market by 2030 [27].

In the Republic of Kazakhstan, no specific goals have been set for the collection and recycling of individual waste streams and their fractions. One of the shortcomings of the waste management policy in the country is the lack of clear and accurate information regarding the quantities of individual types of waste in the mixed household waste. Clarity about the type and quantity of recyclable and recoverable waste is key to sustainable waste management and the selection of appropriate waste treatment methods within the context of the circular economy.

Bulgaria is a member of the EU and Kazakhstan belongs to the Central Asian region (CA). In 2023, Bulgaria's GDP was $37,410 per person, while Kazakhstan's GDP was $38,515 per person [28]. There is no informal waste collection sector in either country.

Plovdiv is located in the south of Bulgaria. The city is the administrative center of the region of the same name, the second most populous city, the second industrial center of the country, and the center of an agricultural region; tourism is developed here, including global tourism. The area of the city is 102 km^2.

Kostanay is located in the north of Kazakhstan. The city is the administrative center of the Kostanay region and one of the three largest agricultural regions in Kazakhstan. Industry and tourism are developed in the city. The area of the city is 240 km^2.

The cities of Plovdiv and Kostanay were chosen due to their similar population sizes, economic profiles, and university presence, which ensures a comparable socio–economic context. Additionally, both cities represent diverse regulatory frameworks for waste management, offering valuable insights into the influence of policy maturity on circular economy implementation.

This analysis focused specifically on household waste, with particular attention on plastic waste as a key component. The research methodology includes an analysis of the most important quantitative parameters for a waste management system's performance, as well as a comparison of the qualitative parameters for plastic waste management in the cities under consideration. The quantitative parameters in this study were the total amount of household waste generated and the population number. The qualitative parameters were the components of the organizational and institutional structure, economic measures, and legislative initiatives (Table 1).

Table 1. Aspects of waste management systems in the studied cities.

Aspects of Waste Management Systems	Plovdiv City	Kostanay City
Legislation	2008—EU Waste Directive [29] 2018—European Strategy for Plastics in a Circular Economy (recycle 50% of plastic packaging waste by 2025 and 55% by 2030) [30,31] 2019—Directive on reducing the environmental impact of certain plastic products [32]	2007—Environmental Code of the Republic of Kazakhstan (RK), on waste—chapters 42–44 2010—Comprehensive Waste Management Plan in the RK, now no longer in effect 2012—Concept of the transition of the RK to a green economy. The main indicators of the Comprehensive Waste Management Plan of the RK were included in the Action Plan for the implementation of the Concept of the transition of the RK to a green economy. 2021—Environmental Code new edition on waste, chapters 23–31
	The EPR principle is applied to plastic packaging	The EPR principle has been applied in a limited scope (2016, cars and their components; 2017, packaging at a "zero rate", manufacturers and importers do not pay a recycling fee; 2020, agricultural machinery)
	2020—Ban on eight single-use plastic products in the EU Requirements for PET bottles—from 1 January 2025 at least 25% by weight must be made of recycled plastic, from 1 January 2030 at least 30% EU introduces mandatory transition to 100% recyclable packaging by 2030	2019—Landfilling of plastic waste banned

Table 1. *Cont.*

Aspects of Waste Management Systems	Plovdiv City	Kostanay City
Plastic waste management	2015 EU Circular Economy Action Plan Targets by 2035: 65% recycling or recycling, only 10% of plastic waste will be landfilled	Green Economy Concept for the Republic of Kazakhstan The goal of the Republic of Kazakhstan by 2040 is to recycle 40% of solid waste, in 2050—to recycle 50% of solid waste 100% coverage of solid waste collection in large cities 2024—3.5% of plastic recycled in Kazakhstan
	Plastic packaging targets: by 2030, 55% of plastic packaging, 90% of single-use plastic will be recycled 2018 Bulgaria—59.2% recycling of plastic packaging, 3rd place in the EU	There are no targets in the Republic of Kazakhstan for the plastic waste cycle
	Plovdiv Municipality Waste Management Program for 2021–2028	Development plan for the territory of Kostanay region for 2021–2026 2024 Development of the Program for handling solid municipal waste in the territory of Kostanay Plastic bags—packaging in stores to choose from, at the request of buyers Recycling of plastic in Kostanay region—into 9 types of new products
	Financial incentives for municipalities to meet established targets Deposit system for types of plastic packaging	No deposit-collateral packaging system—planned, not implemented
	Separate collection of plastic (plastic packaging, household plastic)	Separate collection of plastic waste in mesh containers, at the same time mixed solid waste is collected, which also contains plastic waste Sorting—separation of commercial types of plastic (PET bottles, HDPE, LDPE) Collection of solid waste in Kostanay region: 2007—0.7%, 2023—18.2%
	The infrastructure includes containers for separate collection, garbage trucks, sorting (transfer) stations in the form of reception points, solid waste landfills	The infrastructure includes containers for separate collection, garbage trucks, solid waste landfills Sorting (transfer) stations—no
Organizational structure	Municipality—executive authority, waste flow management Collection of packaging in containers Organization for the recycling of packaging—collection, sorting, transportation Separately, there are centers for the purchase of plastic bottles from the population	Government agencies authorized in the field of waste management at the regional/provincial level: Department of Ecology (legislative functions), Department of Natural Resources and Nature Management (executive authority) Local level: city akimats (executive authority), waste flow management Collection and sorting of plastic waste is performed by waste collection and/or waste removal organizations Recycling is performed by recycling organizations Landfills—for the disposal of mixed solid municipal waste and plastic waste in their composition

3. Results and Discussion

3.1. Identification of the Amount of Household Waste Generated

Data for the city of Plovdiv show that there has been a trend of decreasing the generated mixed municipal waste in recent years. According to the Waste Management Program for the Municipality of Plovdiv 2021–2028, mixed municipal waste collected by municipal systems decreased by 3.55% for the period 2017–2020. The trend is expected to continue, with the quantities of mixed waste collected decreasing while preventing waste generation and expanding the separate collection system in the municipality, with the aim of permanently remaining below 125 thousand tons per year. This trend is also associated with a gradual decrease in the population of the city of Plovdiv (Figure 2). According to Eurostat data, in most countries households generate an average of up to 90% of municipal waste, with the rest generated by commercial sources and the administration. For the period from 2017 to 2023, the city's population decreased by 6.77% and this is also reflected in the amount of waste generated [33].

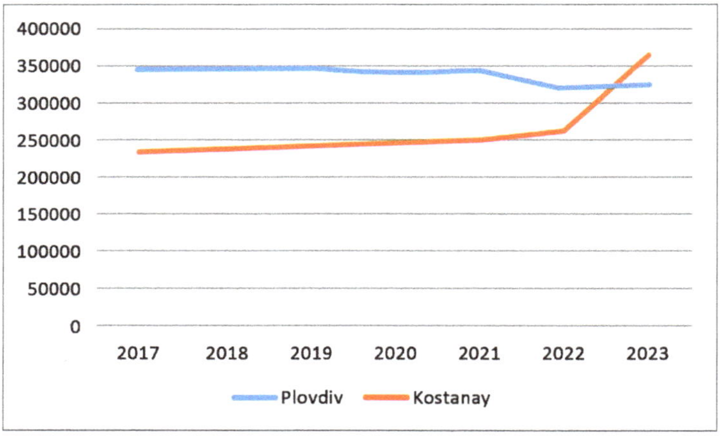

Figure 2. Plovdiv and Kostanay populations 2017–2023 [34,35].

The amount of household waste generated directly depends on the population of the locality.

In Plovdiv, a well-developed infrastructure supports the collection of household plastic waste through designated bins and recycling companies. In this regard, a more accurate reflection of the amount of mixed household waste generated is given by the accumulation rate per inhabitant. The reported accumulation rate for mixed household waste in the municipality of Plovdiv in 2020 was 369.58 kg/inhabitant/year or 126,413.38 tons of waste generated by the 342,048 inhabitants of the municipality per year.

In Kostanay (Figure 2), there is also an increasing trend in the city's population, which is associated with the larger amounts of household waste generated.

For the period 2021–2023, the city's population increased by about 5% and the generated waste by over 25%, which can be attributed to the increasing standard of living in the city. The rate of accumulation of mixed household waste calculated for the city of Kostanay is significantly higher than that determined for the city of Plovdiv, and amounts to 736 kg/h/y [36,37].

A review of the generated mixed household waste for the city of Kostanay (Figure 3) shows an increasing trend, both for mixed household waste and for solid household waste. In contrast, in Plovdiv (Figure 4) we see a downward trend.

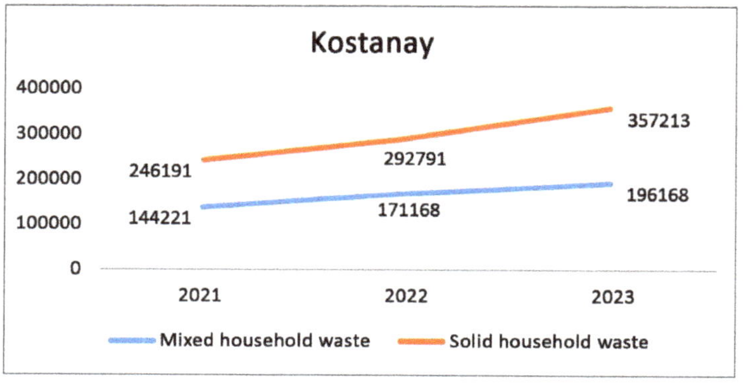

Figure 3. Trends in the quantities of mixed household waste and solid household waste in the city of Kostanay, t [38].

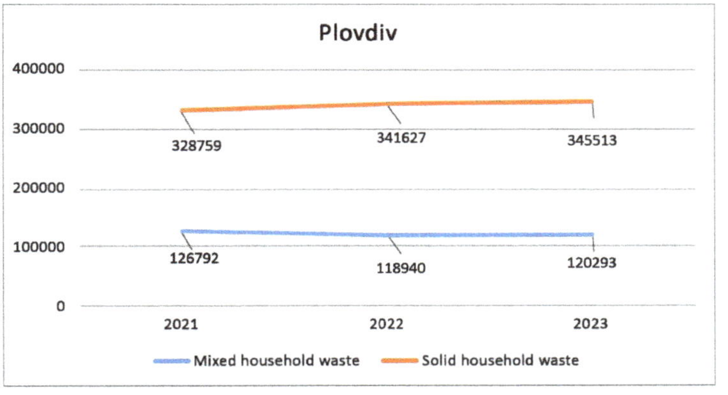

Figure 4. Trends in the quantities of mixed household waste and solid household waste in the city of Plovdiv, t [34].

Municipal waste is the waste generated in populated areas, including as a result of human activity as well as industrial waste, that is similar in its composition and nature of formation. Solid municipal waste includes only the municipal waste in solid form [37].

In 2024 in Kostanay, public hearings were held regarding the adopted city Solid Waste Management Program. The current coverage of 80% to 100% of the population with solid waste collection and removal is calculated for the period 2024–2031. Separately collected waste will be additionally sorted with the allocation of the most economically valuable fractions. The existing system of separate collection will be expanded by collecting food waste and textiles. Research will be conducted on the morphological composition of the solid waste to select a technology for deep processing, the use of an automated geoinformation system "Waste Management". Assessing the current state of the solid waste system in Kostanay, experts say that the waste recycling market in the city is characterized by low competitiveness [39].

3.2. Amount of Plastic in Household Waste

Of the mixed municipal waste generated in the city of Plovdiv, according to a morphological analysis of the waste (Figure 5), 16.81% is plastic, or about 21,250 tons per year.

In the two cities under consideration, different systems for the separate collection of plastic waste operate. An organized system for the separate collection of packaging waste has been introduced in Bulgaria. The system for the separate collection of packaging waste, organized by recovery organizations and based on joint activity contracts with the mayors of the municipalities, includes at least green and yellow (and blue) containers for the collection of packaging waste from households. This is in addition to those for household waste, ensuring a minimum total container volume for settlements of over 100,000 inhabitants: for every 750 inhabitants—containers with a minimum total volume of 3300 L [27]. The system is based on the principle of "Extended Producer Responsibility"—an environmental principle applied as a set of measures to reduce the overall environmental impact of a product after the use of which widespread waste is formed. In addition to the commitments related to packaging waste, which is considered widespread waste, the mayors of the municipalities are obliged to organize the separate collection of household waste in the territory of the municipality, at least for the following waste materials: paper and cardboard, metals, plastics, and glass, and to provide sites for the free transfer of separately collected household waste in all settlements with a population greater than 10,000 inhabitants [26]. A system for the separate collection of packaging waste has been established in the territory of the municipality of Plovdiv, which is operated by packaging recovery organizations, and six sites have been designated for the free delivery of separately collected waste from households in different parts of the city. Various types of waste are collected at the designated sites, including plastic packaging (150,102) and household plastics (200,139) [33]. These requirements for waste management in the municipality of Plovdiv are legally established; in addition to them, there are over 30 waste collection sites and six plastic waste recycling companies in the territory of the municipality, established as private initiatives of legal entities, to which plastic waste can also be transferred [6]. Based on all the established separate collection systems in the municipalities in Bulgaria, including the private initiatives, municipalities receive financial incentives for meeting the set targets which aim to increase the recycled and recovered municipal waste.

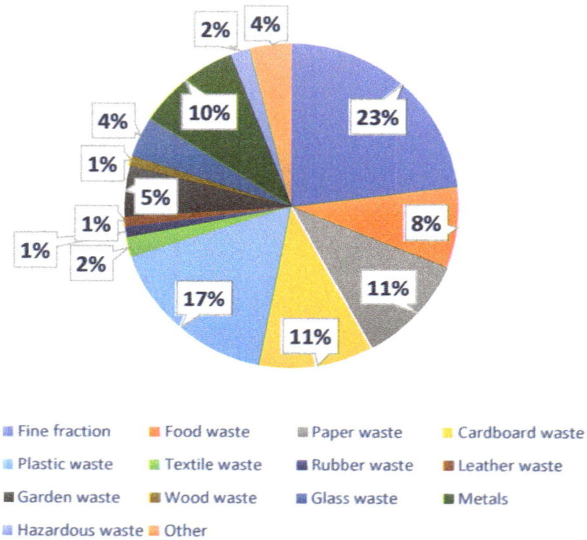

Figure 5. Morphological analysis of municipal waste in the city of Plovdiv [33].

Morphological analysis of waste in the city of Plovdiv (Bulgaria) was carried out in accordance with the current methodology used in Bulgaria [33,40]. In the territory of the municipality of Plovdiv, the generated plastic waste, calculated based on the morphological analysis, is about 21,250 tons per year. The coverage of the population of Plovdiv by the household waste collection system is approximately 99%. Plastic occupies a significant part of the generated municipal waste, ranking second after the fine fraction in the composition of mixed municipal waste.

The collection of household waste in Kostanay is carried out in two ways: in containers located on container sites, and in a container-less way by travelling around the territory and collecting waste in bags placed in certain places, according to a schedule. Collection in bags (roadside collection) is carried out mainly in the private sector. Then, the waste is transported by garbage trucks to the site of the sorting organization. The coverage of the population of Kostanay by the household waste collection system is 80%. For the collection of plastic waste from the population, mesh containers are installed [41]. Despite the established separate plastic collection system, large quantities of plastic are also found in containers intended for mixed solid waste. The data are similar to the results from

Plovdiv, where large quantities of plastic waste are found in the mixed household waste. This shows that people in both Plovdiv and Kostanay still do not have sufficiently well-established habits for the separate collection of plastic waste; as it is contaminated with food waste, part of the plastic waste ends up in landfills.

Plastic accounts for 13% of the total waste volume (Figure 6).

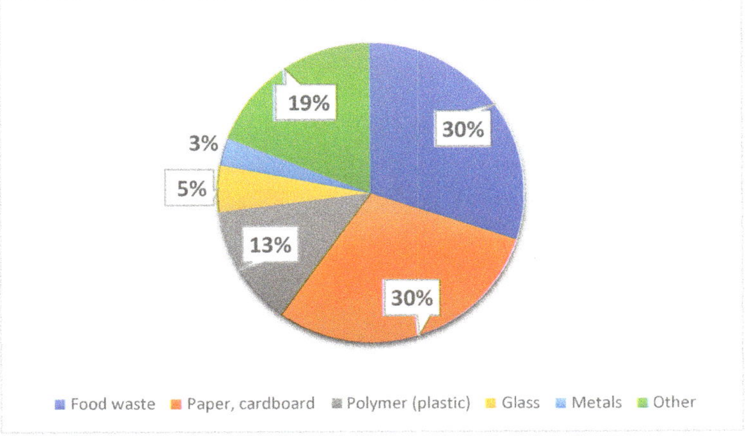

Figure 6. Morphological analysis of municipal waste in the city of Kostanay [42].

According to the sorting company, it separates about 3–4 tons of recyclable waste daily, of which about 70% is plastic; about 2.1–2.8 tons of plastic is separated from mixed household waste daily. With 246 working days a year, this amounts to about 520 tons of plastic waste per year that could be transferred for further processing or recovery. However, a large portion of the plastic waste from mixed waste is heavily contaminated and unsuitable for subsequent recycling [41]. This leads to the loss of valuable materials suitable for use in the country's industry.

In the city of Kostanay, there are six enterprises that carry out the separate collection and sorting of plastic waste, and three that recycle the plastic waste. As a result of recycling the plastic waste, nine types of finished products are produced from plastic waste: hatches for inspection wells, geogrid, geotextile, geocomposite, geoframes, polyethylene gas and water pipes, cable pipes, and polymer sand pavers [41].

Although in both cities—Plovdiv and Kostanay—there are currently separate collections and some of the plastics have found their application in new products, there is still significant potential for the implementation of a circular economy for plastics, which remains unrealized. For the city of Plovdiv, this amounts to over 21 thousand tons per year. For the city of Kostanay, assessing the potential for a circular economy is more difficult, since there is no practice of regularly conducting a morphological analysis of the composition of household waste in this country. For the purposes of this article, we calculated the amount of plastic waste in the mixed flow of household waste in the city of Kostanay, based on a percentage content of 13%. Therefore, if the 196,168 tons of annually generated mixed household waste has a plastic content of 13%, 25,502 tons of plastic waste will go to landfills. Thus, in the mixed municipal waste of each of the two cities in Bulgaria and Kazakhstan, there is a hidden potential for the introduction of a circular economy for approximately 20 thousand tons of plastic waste. Since the average price of plastics is about 100 euros per ton, the waste from these two cities contains a potential 2 million euros worth of lost value buried in landfills.

3.3. Green Solutions for a Circular Economy of Household Plastic Waste

In recent years, sustainable waste management has been largely associated with achieving compliance with the waste hierarchy. The waste hierarchy is implemented as a priority order in legislation and in waste prevention and management policy. It is the final corner of the European Union (EU) policy and legislation in the field of waste and is set out in the EU Waste Framework Directive [29]. Its objectives are to minimize the adverse impacts of production and waste management and to improve resource efficiency. In the field of plastic waste, the EU has set targets for achieving the waste hierarchy. The targets set are also applied in Bulgaria through the country's regulatory documents (Figure 7).

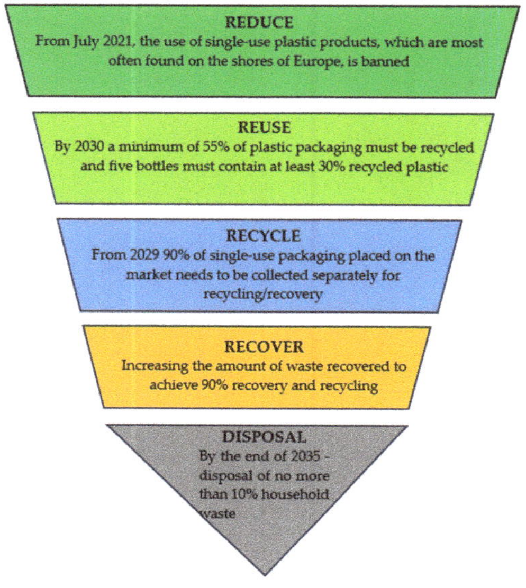

Figure 7. Targets in Bulgaria for achieving compliance with the waste hierarchy.

In accordance with the waste management hierarchy, for already generated plastic waste it is advisable to seek methods for its recycling or recovery—raw material or energy recovery—and to direct that part of the plastic waste that is not suitable for recovery to a landfill. Part of the plastic waste is often contaminated and not suitable for recycling. In order to meet the requirements of the waste management hierarchy, various options for waste recovery are sought.

Options for recovery of MSW, including plastics, through which primary and secondary products can be obtained are given in Figure 8 [43].

The capacity of the installation can hold from 10,000 t/year to 100,000 t/year of processed waste. The main facility is a thermocatalytic reactor in which triboactivated waste is converted into "synthetic oil" in the presence of a catalyst, an inert or low-oxygen gas environment, a temperature of 400–500 °C, gaseous and liquid hydrocarbons, intensive mixing with gaseous, liquid, and solid products for 30–40 min. Physicochemical processes, cracking, "thermolysis", and Fischer–Tropsch synthesis take place in the reactor, which converts the waste into a mixture of liquid hydrocarbons or "synthetic oil". There are two outlet streams from the reactor:

- Primary—consisting of a mixture of gaseous hydrocarbons (C9–C20) which, after passing through a catalyst, are converted into "synthetic oil" with impurities of aromatic and cyclic hydrocarbons.

- Secondary—consisting of solid particles of graphite (carbon), spent catalyst, and inert (inorganic) waste. These are separated and used in commercial products.

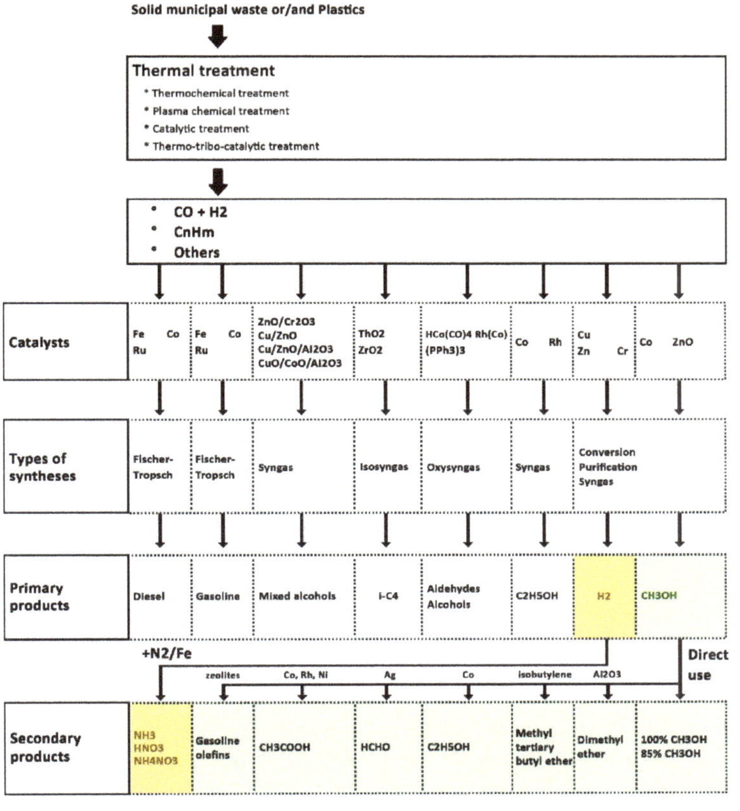

Figure 8. Possible solutions for the treatment of solid municipal waste or/and plastics to generate various chemical products.

The demonstrated technological possibilities for treating solid municipal waste, including plastics, through thermochemical and thermocatalytic treatments are applicable in practice and convincingly show that waste can also be an alternative raw material resource for obtaining organic and inorganic compounds with wide applications.

In the cities of Plovdiv and Kostanay, as shown in the previous sections, the plastic waste management systems are at different levels of development, although they are evolving in one direction—a circular economy. In Kostanay, further development is associated with increased efficiency at the stage of collecting and transporting plastic waste to processors. In Plovdiv, the stage of processing plastic waste into new products requires development. Therefore, the proposed scheme of thermal technological changes (Figure 6) for the production of primary and secondary products is promising for Plovdiv. In Kostanay, technological changes in the city's plastic waste management system will be more associated with mechanical processing technologies.

In general, the results of the comparison of the urban management systems in the cities of Plovdiv (Bulgaria) and Kostanay (Kazakhstan) can be presented in the form of Table 1, which demonstrates the similarities and differences in the studied aspects of the waste management systems.

3.4. Discussion

The results of this study allow us to consider the urban plastic waste management systems in Kostanay and Plovdiv from the perspective of a circular economy, revealing both similarities and differences that determine their development trajectory towards sustainability. Despite significant territorial distance, both cities demonstrate similar characteristics, namely lower GDP and higher levels of waste generation [17]. This allows us to classify them into the same category of city, which is confirmed by the comparable quantitative indicators for the performance of the waste management systems. Both cities have the necessary technical infrastructure; however, the potential of this requires further development. In particular, for Kostanay, a key step towards increasing the degree of circularity will be the creation of collection points/transfer stations, which will optimize the collection and sorting of waste. The presence of plastic recycling plants in both cities, as well as attempts to economically assess the potential of plastic waste, indicate the recognition of plastic as a valuable resource, and not just as garbage. Separating the plastic waste stream into a separate category and creating a dedicated infrastructure will help to maximize the value of these materials through reuse, recycling, and other strategies, which is a key principle in the circular economy.

Studies in Africa [44], South Asia [45], and the EU [Swedish and Bulgarian] have shown intra-regional differences in the maturity of the waste management regulations. A comparison of waste management in Kostanay and Plovdiv (Table 1) demonstrates this trend at the inter-regional level, in particular for the EU and Central Asian regions.

The comparison with Plovdiv, which has significant experience in the circular economy, serves as a valuable source of knowledge and best practice for Kostanay. Implementing proven tools and approaches, such as developing targets for plastic waste, identifying hidden waste volumes, and implementing effective information strategies, will help accelerate the transition to a circular model. It is important to highlight that the circular economy and the Sustainable Development Goals (SDGs) are closely interrelated and complementary. The circular economy is a powerful tool for achieving many of the SDGs, while the SDGs create favorable conditions for its development. The circular economy is based on the principles of responsible consumption and production, which include waste reduction, efficient use of resources, and reuse and recycling of materials. Systematic consideration of these principles when improving waste management systems, as shown in [24], will allow the economic potential of plastic waste to be translated into reality for both cities.

Similar studies in Eastern Europe [46,47] highlight the importance of regulating packaging waste to achieve higher recycling rates, which is also true for Plovdiv. Studies in Central Asia, including Bishkek and Almaty [48], reveal gaps in data availability and public awareness, which is also true for Kostanay.

These studies highlight the need to develop tailor-made strategies for urban waste management systems that take into account local socio–economic and legislative conditions. Kazakhstan's lag in this area is explained by the later development of waste management compared to EU countries, which determines legislative, technological, and organizational features. Thus, both cities have already moved away from the linear model of waste management and are taking active steps towards circularity.

4. Conclusions

As a result of this study, the following conclusions can be drawn:

1. The compared cities of Kostanay in Kazakhstan and Plovdiv in Bulgaria have a similar population, a well-developed economy and tourism, and are university cities. Nevertheless, the rate of MSW accumulation for Kostanay is 736 kg/person/year, and

for Plovdiv it is 369.58 kg/person. One of the factors in this difference is probably the need to purchase drinking water in Kostanay, which is available in plastic packaging.
2 The methodology used is applicable for comparing settlements close in population, especially when no morphological analysis has been done for one of them.
3. An information strategy is needed to explain to the population in both the city of Plovdiv and the city of Kostanay the good practices for the separate collection of plastic waste so that it is not contaminated with other waste, including food waste, so it is not recycled but landfilled.
4. The analysis shows that in the mixed household waste of only two cities in Bulgaria and Kazakhstan, there is a hidden potential for the implementation of a circular economy for over 50 thousand tons of plastic waste; with the average price of plastic at about 100 euros per ton, the waste in these two cities has a potential value of over 5.4 million euros.
5. In both cities, in the context of green solutions for a circular economy, the collected plastic waste is recycled into products, but this does not exhaust all possibilities.
6. A comparative analysis of the practices in the two cities showed differences in the level of circularity and in the methods of recycling plastic waste. If the system of solid waste management, including plastic waste, is developing according to the 3R scheme, then in Plovdiv the current level corresponds to the 5R scheme. Improving the collection of municipal solid waste, especially plastics, can also lead to the implementation of thermochemical and thermocatalytic technologies for the production of organic and inorganic compounds for subsequent use as basic raw materials or reagents. This is another green solution for a circular economy in Plovdiv. In Kostanay, the acceleration of progress in plastic waste management is linked to the development of infrastructure and mechanical recycling methods.
7. Future studies on the development of plastic waste management systems towards a circular economy in the cities under consideration should include other areas of technological change for plastic waste recycling, and characterize the development efficiency with indicators acceptable at different levels of system development for their comparability and evaluation.

Author Contributions: Conceptualization, E.T. and X.F.; Methodology, E.T. and G.Y.; Software, E.T., G.Y. and G.J.; Validation, E.T. and G.Y.; Formal analysis, E.T. and G.J.; Investigation, X.F.; Resources, X.F. and G.K.; Data curation, X.F. and G.K.; Writing—original draft, E.T., X.F., G.K. and G.J.; Writing—review & editing, G.Y. and A.K.-S.; Visualization, A.K.-S.; Supervision, G.Y. and A.K.-S.; Project administration, A.K.-S.; Funding acquisition, A.K.-S. All authors have read and agreed to the published version of the manuscript.

Funding: This research received no external funding.

Data Availability Statement: The data presented in this study are available on request from the corresponding authors.

Acknowledgments: The scientific studies mentioned in the article are under the "International Scientist" program of A. Baitursynuly Kostanay Regional University, Kazakhstan and the participation of Ekaterina Todorova, Eng. from University of Forestry, Bulgaria.

Conflicts of Interest: The authors declare no conflicts of interest.

References

1. Plastic Waste and Recycling in the EU: Data and Facts. Available online: https://www.europarl.europa.eu/pdfs/news/expert/2018/12/story/20181212STO21610/20181212STO21610_bg.pdf (accessed on 10 November 2024).
2. Pilapitiya, P.G.C.; Amila, R. The world of plastic waste: A review. *Clean. Mater.* **2024**, *11*, 100220. [CrossRef]
3. Hoornweg, D.; Bhada-Tata, P.; Kennedy, C. Environment: Waste production must peak this century. *Nature* **2013**, *502*, 615–617. [CrossRef] [PubMed]
4. Zambrano-Monserrate, M.A.; Ruano, M.A.; Ormeño-Candelario, V. Determinants of municipal solid waste: A global analysis by countries' income level. *Environ. Sci. Pollut. Res.* **2021**, *28*, 62421–62430. [CrossRef] [PubMed]
5. Estay-Ossandón, C.; Mena-Nieto, A.; Harsch, N.; Bahamonde-Garcia, M. Analysis of the Key Variables That Affect the Generation of Municipal Solid Waste in the Balearics Islands (2000–2014). In *Project Management and Engineering Research*; Lecture Notes in Management and Industrial Engineering; Ayuso Muñoz, J.L., Yagüe Blanco, J.L., Capuz-Rizo, S.F., Eds.; Springer: Cham, Switzerland, 2021. [CrossRef]
6. National Waste Information System. Available online: https://nwms.eea.government.bg/app/base/home (accessed on 1 December 2024).
7. Lebreton, L.; Andrady, A. Future scenarios of global plastic waste generation and disposal. *Palgrave Commun.* **2019**, *5*, 6. [CrossRef]
8. Alshaikh, R.; Abdelfatah, A. Optimization Techniques in Municipal Solid Waste Management: A Systematic Review. *Sustainability* **2024**, *16*, 6585. [CrossRef]
9. Bruvoll, A. Factors influence solid waste generation and management. *J. Solid Waste Technol. Manag.* **2001**, *27*, 156–162.
10. Masebinu, S.O.; Akinlabi, E.T.; Muzenda, E.; Aboyade, A.O.; Mbohwa, C.; Manyuchi, M.; Naidoo, P. A Review on Factors Affecting Municipal Solid Waste Generation. 2017. Available online: https://ujcontent.uj.ac.za/esploro/outputs/conferencePaper/A-review-on-factors-affecting-municipal/9910222007691#file-0 (accessed on 19 October 2017).
11. Ahsan, A.; Alamgir, M.; El-Sergany, M.M.; Shams, S.; Rowshon, M.K.; Daud, N.N. Assessment of municipal solid waste management system in a developing country. *Chin. J. Eng.* **2014**, *2014*, 561935. [CrossRef]
12. Wilson, D.C.; Rodic, L.; Scheinberg, A.; Velis, C.A.; Alabaster, G. Comparative analysis of solid waste management in 20 cities. *Waste Manag. Res.* **2012**, *30*, 237–254. [CrossRef]
13. Fatty, F.N.; Komma, L. In-Depth Analysis of Municipal Solid Waste Management in Kanifing Municipality, The Gambia. *Int. J. Sci. Res. Publ.* **2019**, *9*, 95114. [CrossRef]
14. Edalatpour, M.A.; Mirzapour Al-e-hashem, S.M.J.; Karimi, B.; Bahli, B. Investigation on a novel sustainable model for waste management in megacities: A case study in Tehran municipality. *Sustain. Cities Soc.* **2018**, *36*, 286–301. [CrossRef]
15. Awino, F.B.; Apitz, S.E. Solid waste management in the context of the waste hierarchy and circular economy frameworks: An international critical review. *Integr. Environ. Assess. Manag.* **2024**, *20*, 9–35. [CrossRef] [PubMed]
16. Wilson, D.C.; Velis, C.A.; Rodic, L. Integrated sustainable waste management in developing countries. *Proc. Inst. Civil. Eng. Waste Resour. Manag.* **2013**, *166*, 52–68. [CrossRef]
17. Blagoeva, N.; Georgieva, V.; Dimova, D. Relationship between GDP and Municipal Waste: Regional Disparities and Implication for Waste Management Policies. *Sustainability* **2023**, *15*, 15193. [CrossRef]
18. Pires, A.; Martinho, G.; Chang, N.-B. Solid waste management in European countries: A review of systems analysis techniques. *J. Environ. Manag.* **2011**, *92*, 1033–1050. [CrossRef]
19. Stoeva, K.; Alriksson, S. Influence of recycling programmes on waste separation behaviour. *Waste Manag.* **2017**, *68*, 732–741. [CrossRef] [PubMed]
20. Inglezakis, V.J.; Moustakas, K.; Khamitovac, G.; Tokmurzin, D.; Rakhmatulina, R.; Serik, B.; Abikak, Y.; Poulopoulos, S.G. Municipal solid waste management in Kazakhstan: Astana and Almaty case studies. *Chem. Eng. Trans.* **2017**, *56*, 565–570. Available online: https://www.cetjournal.it/index.php/cet/article/view/CET1756095 (accessed on 20 March 2017).
21. Zhidebekkyzy, A.; Temerbulatova, Z.; Bilan, Y. The Improvement of The Waste Management System In Kazakhstan: Impact Evaluation. *Pol. J. Manag. Stud.* **2022**, *25*, 423–439. [CrossRef]
22. The Europian Union and Central Asia: The New Partnership in Action. 2009. Available online: https://www.eeas.europa.eu/sites/default/files/the_european_union_and_central_asia_the_new_partnership_in_action.pdf (accessed on 1 April 2008).
23. United Nations Sustainable Development Goals (SDGs). Available online: https://sdgs.un.org/goals (accessed on 29 December 2024).
24. Wilson, D.C.; Rodic, L.; Modak, P.; Soos, R.; Carpintero, A.; Velis, K.; Iyer, M.; Simonett, O. Global Waste Management Outlook. Report. UN Environment Programme, the International Solid Waste Association. 2015. Available online: https://eprints.whiterose.ac.uk/110512/1/UNEP%20ISWA%20GWMO%20-%20Chapter%203%20-%20Waste%20Management%20-%20Global%20Status.pdf (accessed on 1 February 2015).
25. On the Concept of the Transition of the Republic of Kazakhstan to a "Green Economy". Decree of the President of the Republic of Kazakhstan Dated 30 May 2013 No. 577. Available online: https://adilet.zan.kz/rus/docs/U1300000577 (accessed on 28 December 2024). (In Kazakh and Russian)

26. Waste Management Act in Force from 13.07.2012. Available online: https://www.moew.government.bg/static/media/ups/tiny/%D0%A3%D0%9E%D0%9E%D0%9F/%D0%97%D0%B0%D0%BA%D0%BE%D0%BD%D0%BE%D0%B4%D0%B0%D1%82%D0%B5%D0%BB%D1%81%D1%82%D0%B2%D0%BE/WASTE%20MANAGEMENT%20ACT_13.pdf (accessed on 13 July 2012).
27. Ordinance on Packaging and Packaging Waste, Adopted on November 6, 2012. Available online: https://www.moew.government.bg/static/media/ups/tiny/%D0%A3%D0%9E%D0%9E%D0%9F/%D0%97%D0%B0%D0%BA%D0%BE%D0%BD%D0%BE%D0%B4%D0%B0%D1%82%D0%B5%D0%BB%D1%81%D1%82%D0%B2%D0%BE/%D0%9D%D0%90%D0%A0%D0%95%D0%94%D0%91%D0%98/%D0%9F%D0%9C%D0%A1%202022/NAREDBA_za_opakovkite_i_otpadacite_ot_opakovki.pdf (accessed on 16 December 2022). (In Bulgarian)
28. GDP Based on Purchasing-Power-Parity (PPP) per Capita. World Bank. Available online: https://data.worldbank.org/indicator/NY.GDP.PCAP.PP.CD (accessed on 1 July 2024).
29. *2008/98/EC*; Directive 2008/98/EC of the European Parliament and of the Council of 19 November 2008 on waste and repealing certain Directives. European Union: Brussels, Belgium, 2008; L312/3.
30. Decree No 354 of 26 October 2021 on the Adoption of Ordinance on Reducing the Impact of Certain Plastic Products on the Environment. Available online: https://www.fao.org/faolex/results/details/en/c/LEX-FAOC207012/ (accessed on 26 October 2021).
31. *52018DC0028–EN*; Communication from the Commission to the European Parliament, the Council, the European Economic and Social Committee and the Committee of the Regions, A European Strategy for Plastics in a Circular Economy. European Union: Brussels, Belgium, 2018.
32. *EU 2019/904*: Directive (EU) 2019/904 of the the European Parliament and of the Council of 5 June 2019 on the Reduction of the Impact of Certain Plastic Products on the Rnvironment. European Union: Brussels, Belgium, 2019; L155/1.
33. Waste Management Program of the Municipality of Plovdiv 2021–2028. Available online: https://www.plovdiv.bg/wp-content/uploads/2024/03/%D0%9F%D1%80%D0%BE%D0%B3%D1%80%D0%B0%D0%BC%D0%B0_%D1%83%D0%BF%D1%80_%D0%BE%D1%82%D0%BF%D0%B0%D0%B4%D1%8A%D1%86%D0%B8_%D1%80%D0%B5%D1%88%D0%B5%D0%BD%D0%B8%D0%B5_2023.pdf (accessed on 5 March 2021). (In Bulgarian).
34. National Statistical Institute, Bulgaria. Available online: https://nsi.bg/en/content/2974/population (accessed on 12 November 2024).
35. Population of the Republic of Kazakhstan by Gender and Type of Locality. Available online: https://stat.gov.kz/en/industries/social-statistics/demography/publications/ (accessed on 1 July 2024).
36. United Nations Statistics Division. Demographic Statistics Database. Available online: https://data.un.org/Data.aspx?d=POP&f=tableCode 240 (accessed on 10 November 2024).
37. *Environmental Protection in the Republic of Kazakhstan/Statistical Compilation*; Agency for Strategic Planning and Reforms of the Republic of Kazakhstan Bureau of National Statistics: Astana, Kazakhstan, 2024; 142p. (In English)
38. Green Economy Indicators. I-6 Generation of Solid Waste, Municipal Waste and the Level of Their Processing. Available online: https://stat.gov.kz/en/green-economy-indicators/25825/ (accessed on 10 November 2024).
39. Zimmer, P. Waste Management Program Put Up for Discussion in Kostanay. Available online: https://top-news.kz/v-kostanae-utverdili-programmu-upravlenija-othodami/ (accessed on 10 July 2024).
40. *Methodology for Determining the Morphological Composition of Household Waste*; Ministry of Environment and Water: Sofia, Bulgaria, 2019.
41. Krivopusk, X. Assessment of Plastic Waste Management in Kostanay Region [In Russian], «Baitursynov Readings-2024» on the Topic: «The Concept of Development of Higher Education and Science of Kazakhstan—The Basis for the Growth of Human Capital and Innovation»: Materials of the International Scientific and Practical Conference on April 19, 2024.—Kostanay: Kostanay Regional University Named after Akhmet Baitursynuly, 2024; ; pp. 435–439. Available online: https://ksu.edu.kz/en/science-and-innovation/materials-of-scientific-conferences/ (accessed on 25 April 2024).
42. Strategy for the Development of the Activities of EPR Operator LLP for 2020–2024. Available online: https://recycle.kz/storage/app/media/main/Strategiya-Operatora.pdf (accessed on 13 July 2020). (In Russian)
43. Kyoseva, V.; Todorova, E.; Dombalov, I. *The Most Frequently Asked Questions Related to the Conversion of Household Waste into Raw Materials and Energy Resources*; High End Publishing Ltd.: Sofia, Bulgaria, 2011; ISBN 978-954-92844-1-6.
44. Wilson, D.C.; Velis, C.; Cheeseman, C. Role of informal sector recycling in waste management in developing countries. *Habitat Int.* **2012**, *30*, 797–808. [CrossRef]
45. Kaza, S.; Yao, L.; Bhada-Tata, P.; Van Woerden, F. *What a Waste 2.0: A Global Snapshot of Solid. Waste Management to 2050*; World Bank: Washington, DC, USA, 2018. [CrossRef]
46. Zhang, Z.; Chen, Z.; Zhang, J.; Liu, Y.; Chen, L.; Yang, M.; Osman, A.I.; Farghali, M.; Liu, E.; Hassan, D.; et al. Municipal solid waste management challenges in developing regions: A comprehensive review and future perspectives for Asia and Africa. *Sci. Total Environ* **2024**, *930*, 172794. [CrossRef]

47. Di Foggia, G.; Beccarello, M. An Overview of Packaging Waste Models in Some European Countries. *Recycling* **2022**, *7*, 38. [CrossRef]
48. Abylkhani, B.; Guney, M.; Aiymbetov, B.; Yagofarova, A.; Sarbassov, Y.; Zorpas, A.A.; Venetis, C.; Inglezakis, V. Detailed municipal solid waste composition analysis for Nur-Sultan City, Kazakhstan with implications for sustainable waste management in Central Asia. *Environ. Sci. Pollut. Res.* **2021**, *28*, 24406–24418. [CrossRef] [PubMed]

processes

MDPI

Review

Tiny Particles, Big Problems: The Threat of Microplastics to Marine Life and Human Health

Goutam Saha and Suvash C. Saha *

School of Mechanical and Mechatronic Engineering, University of Technology Sydney, Sydney, NSW 2007, Australia; goutam.saha@uts.edu.au
* Correspondence: suvash.saha@uts.edu.au

Abstract: Microplastics, primarily derived from plastic waste, are pervasive environmental pollutants found across aquatic and terrestrial ecosystems. This review investigates microplastics' presence, distribution, and impacts in marine ecosystems, with a particular focus on fish species. Research indicates that microplastics are present in various anatomical parts of fish, including the gastrointestinal tracts and gills, with significant implications for marine biodiversity and human health through seafood consumption. The review also highlights the sources of microplastics, such as synthetic textiles, packaging, and personal care products, and explores the pathways through which these particles enter marine environments. Advanced detection techniques have identified microplastics in human tissues, underscoring the urgency of addressing this environmental threat. Comprehensive strategies are essential to mitigate microplastic pollution and protect both marine life and human health.

Keywords: microplastics; fish species; marine environment; sources; human health

1. Introduction

Microplastics infiltrate water, air, and soil worldwide. Scientists have provided various insights into their harmful impacts, yet a clear understanding of the extent of their damage to human and animal health remains elusive [1,2]. Nonetheless, there is a consensus that microplastics significantly harm marine ecosystems and are recognised as hazardous pollutants [3]. The primary contributor to microplastic pollution in oceans is the fragmentation of plastic debris originating from waste that enters the ocean from various sources [4].

1.1. Sources of Microplastics

The world is full of plastics, which generate microplastics and nanoplates that are available everywhere, such as water, air, and soil. They have become significant environmental pollutants, posing risks to wildlife and human health. In the following, we will discuss the sources of microplastics and nanoplastics.

The release of tiny synthetic particles or fibres from clothing, textiles, or other synthetic materials into the environment is often very small and can easily become airborne or wash into water systems [5,6]. Similarly, plastic packaging in modern consumer goods contributes to microplastic pollution during both opening and disposal processes [7]. Personal care and cosmetic products, especially prevalent in densely populated urban areas, are also implicated as sources of environmental contamination by microplastics [8,9]. Also, the wear and tear of tyres release microplastics into the environment, a factor often overlooked but significant in exacerbating contamination [10]. Commercial fishing activities inadvertently generate microplastics through processes like netting and gear abrasion [11,12]. Even seemingly harmless items like toy building bricks can be unsuspected reservoirs of microplastics and nanoplastics, posing risks to children who are more susceptible to exposure [13]. Furthermore, industrial activities such as the deterioration of building materials [14] and the transformation of plastic furniture by fire contribute to microplastic and nanoplastic pollution [15]. Advancements in technology, such as Raman imaging, have facilitated the detection and characterisation of these tiny particles across various matrices, such as kitchen blenders [16], smartphones [17], printed toner powders [18], chopping boards [19], cut Polyvinyl Chloride (PVC) pipes [20], non-stick cookware [21], kitchen sponges [22], burned disposable gloves [23], PPE masks [24], rubber bands [25], and haemodialysis waters [26]. Recently, microplastics have also been found in bottled drinking water [27,28] as well as cookware [29]. Details are given in Figure 1.

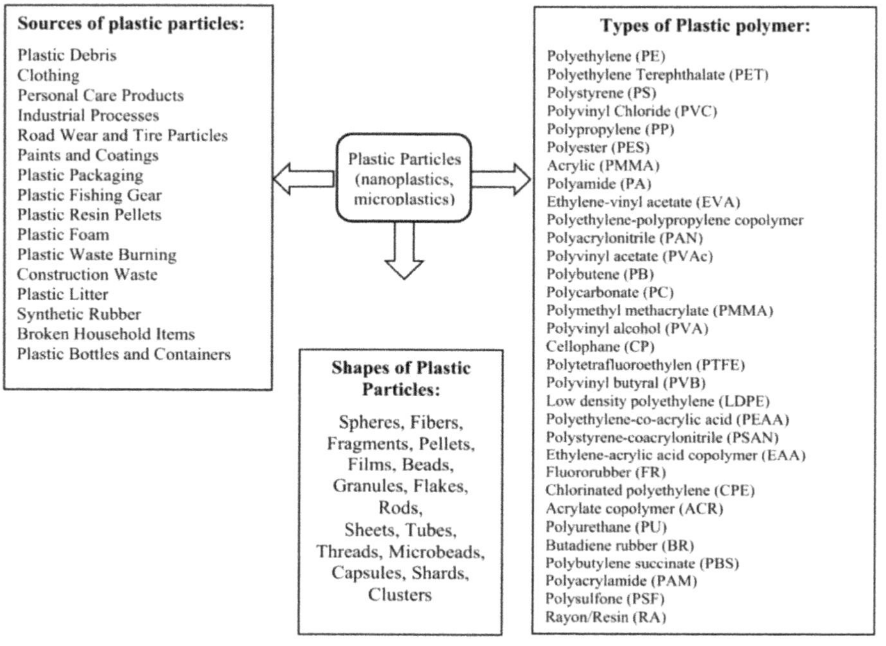

Figure 1. Sources of plastic particles, its types and shapes.

1.2. Microplastics in Fish Species

Over the years, researchers have shown that microplastics are found in different fish species, and the details of this literature are discussed below.

In 2016, it was seen that Mussels from China's coastline contained microplastics, with higher levels in wild mussels and those from areas with intensive human activities, primarily fibres and fragments, with a significant proportion smaller than 250 μm, suggesting that mussels could be used as a bio-indicator of microplastic pollution [30]. Also, microplastics were found in 77% of Japanese anchovy in Tokyo Bay, with an average of 2.3 pieces per fish, mostly polyethylene and polypropylene fragments, with some microbeads and a size range of 150 μm to 1000 μm, indicating that microplastics have penetrated the marine ecosystem [31].

In 2017, it was observed that adult grass shrimp exposed to microplastics of various sizes and shapes experienced acute toxicity, with mortality rates ranging from 0% to 55%, and that the shape and size of the particles significantly influenced ingestion and residence time in the gut and gills, with fibres causing significantly higher mortality [32]. Also, only 1 out of 400 North Sea fish (0.25%) had ingested microplastics, with two particles found in a single Sprat, and the particles were identified as polymethylmethacrylate [33]. Moreover, microplastics were found in the livers of the European anchovy, European pilchard, and Atlantic herring, with 80% of anchovy livers containing large microplastics (124–438 μm) [34].

In 2018, it was found that the windward beaches along Trindade Island's eastern coast exhibited a notable concentration of debris, primarily consisting of small plastic fragments. Furthermore, this debris already engaged with the island's wildlife, including seabirds and endangered land-dwelling crabs, adversely affecting these populations [35]. Later, microplastics and other debris were found in all seawater and mussel samples from U.K. coastal waters and supermarkets, with higher levels in pre-cooked supermarket mussels, presenting a route for human exposure [36]. Also, microplastics were found in various tissues of fish and a crustacean from five sites of the Musa Estuary and the Persian Gulf, with a total of 828 microplastics detected, mainly fibrous fragments of various colour and sizes, and some suspected to be paint fragments, raising concerns about the potential transfer of synthetic materials into humans [37]. Moreover, fish from the Persian Gulf contained microplastics and metals, with concentrations increasing with fish size [38]. In addition, microplastics were found in 26 individual fish from 26 species along the Saudi Arabian coast of the Red Sea. Most microplastics were films (61.5%) and fishing thread (38.5%), made of polypropylene and polyethylene [39]. Also, only 2.1% of 292 planktivorous fish (6 individuals) in the Southeastern Pacific coast had microplastics in their digestive tract. It was found that the microplastics were degraded fragments and threads, 1.1–4.9 mm long, and of various colours [40].

In 2019, it was observed that marine plastic litter posed a significant risk to the fin whale in the Mediterranean, with the highest risk of plastic ingestion in the Central Ligurian Sea, and all three sources of plastic litter contributing to impacting cetaceans in the Pelagos Sanctuary [41]. Also, 9 out of 11 marine fish species found in Seri Kembangan, Malaysia, contained plastic debris in their viscera and gills. Up to 76.8% of isolated particles were plastic polymers, with sizes ranging from 200 to 34,900 μm [42]. Moreover, it was seen that different fish species from Haizhou Bay, China, had microplastics in their tissues, where the microplastics were mostly fibres, black or grey, and made of cellophane. It was also observed that the skin and gills had more microplastics than the gut, and scaleless fish had higher microplastic abundances in their skin [43].

In 2020, it was observed that sea anemones, abundant along the Amazon coast, ingest meso- and microplastics, with 75.6% of the examined individuals containing plastic particles, primarily fibres (84%), followed by fragments and films, with a mean of 1.6 items per individual, and a weak positive correlation between anemone weight and plastic particles [44]. Also, microplastics were found in 12 fish species from the Beibu Gulf, with 0.027–1 item per individual, mostly transparent fibres, polyester, and nylon, with demersal fish having a higher microplastic abundance [45]. Additionally, microplastic contamination was found in the seawater and fish from Tuticorin, the southeastern coast of India, with epipelagic fish having higher levels; most microplastics were small blue fibres, with polyethylene being the most common type [46]. Moreover, microplastics were found in 49% of 150 fish from the Northeast Atlantic Ocean, and the estimated human intake through fish consumption ranged from 518 to 3078 microplastic items per year per capita [47]. Also, microplastic pollution was found in the Han River and its tributaries, with varying concentrations and types. Polyethylene, silicone, and polystyrene were most common in the river, while polytetrafluoroethylene, polyethylene, and polyester dominated in the tributaries. Microplastics were found in fish intestines and gills, but not flesh, with fragments being the most common form [48]. In addition, microplastics were found in brown shrimp and tiger shrimp from the Northern Bay of Bengal, Bangladesh. A total of 33 and 39 microplastic items were found, respectively, with an average of 3.40 and 3.87 items per gram of gastrointestinal tract. Filament and fibre shapes were most common, with black being the dominant colour [49]. Also, microplastics were present in the guts of rabbitfish (*Siganus fuscescens*) in coastal Philippines, with semi-synthetic microfibers (rayon) being dominant in sediment samples from Silliman Beach but absent in the fish guts [50].

In 2021, it was observed that Longnose stingrays in the Western Atlantic Ocean had ingested microplastics, with almost a third of the examined specimens containing microplastics in their stomach contents, primarily fibres (82%), blue in colour (47%), and made of polyethylene terephthalate (PET) (35%) [51]. Also, microplastics were found in the water, sediment, and marine organisms (shellfish and finfish) in the Sal Estuary, Goa, India, with high concentrations in finfish, as well as small-sized microplastics dominant in biota [52]. Moreover, 85.4% of commercial fish from the Bohai Sea had ingested microplastics, with an average of 2.14 items per individual, mostly fibrous and PET [53].

In 2022, it was observed that microplastic contamination was widespread in dolphinfish from the Eastern Pacific Ocean, averaging 9.3 pieces per individual, with the majority being polyester (46.8%) and polyethylene terephthalate (38.1%) [54]. Also, microplastics were found in 14 marine dried fish products from seven Asian countries, mostly fibres, with polyethylene, polyethylene terephthalate, and polystyrene being the main polymers, and the highest count was in *Etrumeus micropus* from Japan [55]. Moreover, microplastics were found in dried Bombay duck and ribbon fish from the Bay of Bengal, with higher levels in samples from Kuakata; fibres were the most common type, followed by fragments and other types, with polyethylene, polystyrene, and polyamide polymers identified [56]. Furthermore, microplastics were found in the water, sediment, and fish samples from Mumbai's coastal waters, where fibres were the most common shape, and eleven types of plastic polymers were identified [57]. In addition, microplastics were found in 47.8% of 180 fish specimens from the Northern Adriatic Sea, with a total of 233 fragments identified. The mean concentration ranged from 1.75 to 4.11 items/individual across six species, and polyethylene and polypropylene microplastics were found, ranging in size from 0.054 to 0.765 mm [58]. Also, microplastics were found in the gills and guts of 26 fish species in Haizhou Bay and the adjacent waters, with blue fibre being the most common form [59].

In 2023, it was found that 30% of fish from 24 beaches in the Machado River, Western Brazilian Amazon, had microplastics in their digestive tracts, with 617 microplastics found; contamination was higher in fish from beaches closer to urban settlements, particularly carnivorous fish [60]. Also, microplastics were found in three fish species from the Bay of Bengal, with dried fish having significantly higher amounts than fresh fish; fibres were the most common type found, followed by fragments and other types, with most being small and red, and low-density polyethylene was the most common polymer [61]. Moreover, 35 freshwater fish species in India were analysed, and the highest abundance of microplastic contamination was found in *Channa punctatus*. Fiber-type microplastics were the most dominant, while polyethylene-type polymer microplastics were found mainly in edible tissue [62]. In addition, microplastics were found in fish species from the Pasig and Marikina Rivers in the Philippines, and polypropylene and polyethylene fragments were the most common microplastics identified [63]. Also, a separate research study observed that the Sundarbans mangrove forests in Bangladesh were highly contaminated with microplastics. It was noted that nine fish species had microplastics in their gastrointestinal tract and muscles, where most particles were smaller than 1 mm and black in colour, with polyamide being the most abundant polymer type [64].

1.3. Microplastics Detected in Different Regions and Different Fish Species

The infiltration of microplastics into marine habitats is a pervasive issue affecting a multitude of species. Research has identified their presence in the digestive systems of numerous marine organisms, spanning from the majestic whales to the humble yellow crabs and green turtles [35,41]. Among the affected species, the Longnose stingray, Cangicum anemone, shrimp, mussels, dolphin, various molluscs, and even Japanese anchovies have been found to ingest these harmful particles [30–32,36,44,51,52,54,65]. Notably, microplastics have also been detected in fish populations from diverse regions, such as the Beibu Gulf in the South China Sea, the Machado River in the Western Brazilian Amazon, and the Bohai Sea in China [45,53,60]. Furthermore, the global scope of this issue is evident in the presence of microplastics in dried marine fish sourced from different countries [55,56]. Table 1 provides a comprehensive overview of the fish species affected by microplastic contamination, emphasising the widespread nature of this environmental concern. These findings collectively highlight microplastic pollution's global distribution and their impact on marine ecosystems.

Table 1. List of regions and fish specifies where microplastics are detected.

Ref.	Region	Fish Species
[52]	Goa, India	Gray mullet, Catfish, Whipfin silver-biddy, Pearlspot
[37]	Persian Gulf	Mackerel, Longfin lizardfish, Barramundi, Tongue sole
[66]	Tyrrhenian Sea, Italy	Gray mullet, Annular seabream, Red mullet
[67]	Western Pacific Ocean	Highfin seabream, Flying gurnard
[61]	Bay of Bengal	Bombay duck, Ribbon fish, Hairfin anchovy
[46]	Tuticorin, India	Bombay duck, Goldspot herring, White sardine, Indian mackerel, Skipjack tuna, Sailfish
[47]	Northeast Atlantic Ocean	European seabass, Atlantic horse mackerel, Atlantic chub mackerel
[48]	Han River, South Korea	Carp, Crucian carp, Bluegill, Bass, Catfish, Snakehead
[57]	Mumbai coast, India	White sardine, Shrimp, Belanger croaker, Bombay duck, Malabar sole fish
[49]	Northern Bay of Bengal, Bangladesh	Brown shrimp, Tiger shrimp
[68]	South America	Brown hoplo
[62]	India	Spotted snakehead, Rohu, Bata labeo, Spotted mahseer, Amphibious barb
[63]	Pasig and Marikina Rivers, Philippines	Nile tilapia, Manila Sea catfish, Armored catfish
[58]	Adriatic Sea, Italy	European pilchard or sardine, European anchovy, European hake, Spotted flounder, Striped red mullet, Rock goby
[33]	North Sea, Netherlands	Atlantic herring, Sprat, Common dab, Whiting
[38]	Persian Gulf	Shrimp scad, Orange-spotted Grouper, Pickhandle barracuda, Bartail flathead
[64]	Mongla port, Bangladesh	Ilish, Bhetki, Poa, Tengra, Payra, Loitta, Chemo, Bele
[43,59]	Haizhou Bay, China	Kamala River sprat, Red-finned mudskipper, Half-smooth tongue sole, Blackbarred sandperch, Chinese silver pomfret
[50]	Central Philippines	Rabbitfish

1.4. *Microplastics Detected in Different Body Parts of Fishes*

Table 2 illustrates the various research endeavours focusing on marine fish species, detailing the anatomical regions examined for microplastic presence, the geographical origins of the specimens, and the data analysis methodologies employed. The investigations predominantly prioritise the gastrointestinal tracts and gills of fish, with statistical analysis emerging as most commonly utilised approach by researchers.

Table 2. List of observed parts of fish, region of collection of fish, data analysis techniques, and name of the software used.

Ref.	Observed Parts	Region of Collection	Microscope	Data Analysis Approach and Software
[36]	Mussel tissues	United Kingdom	Olympus SZX10 (Tokyo, Japan)	Statistical
[54]	Gills, stomachs, intestinal tracts, and muscles	Eastern Pacific Ocean	Leica M205A, OPUS 7.8 (Wetzlar, Germany)	Statistical
[52]	Gastrointestinal tracts	Goa, west coast of India	AIM-3800, Olympus SX10	Statistical, PAST
[45]	Gastrointestinal tracts, and gills	Beibu Gulf, South China Sea	Olympus SZX10, Nicolet iN 10	Experimental
[53]	Gastrointestinal tracts	Bohai Sea, China	Olympus, SZX10, Nicolet™ iN10	Statistical, SPSS v. 20
[55]	Gastrointestinal tracts, muscles and gills	Taiwan, Thailand, Japan, China, South Korea, Vietnam, Sri Lanka	Olympus SZX16, JobinYvon LabRAM HR800	Statistical
[56]	-	Cox's Bazar and Kuakata, Bay of Bengal, Bangladesh	Daffodil MCX100 (Gurgaon, India), Nicolet iS5 FT-IR (Green Bay, WI, USA)	Statistical, SPSS v. 22
[66]	Muscles and gills	Tyrrhenian sea, Italy	Nicolet™ iN10, Omnic™ Picta™	Statistical
[61]	Muscles and gills	Chattogram and Kuakata, Bay of Bengal, Bangladesh	Daffodil MCX100, Nicolet iS5 FT-IR	Statistical
[46]	Gastrointestinal tracts	Tuticorin, Southeast coast of India	Thermo Nicolet model iS5 (Waltham, MA, USA)	Statistical
[47]	Gastrointestinal tracts, muscles and gills	Northeast Atlantic Ocean	LEICA S9i	Statistical, SPSS v. 24
[48]	Gastrointestinal tract, gills, and fillets	Han River, South Korea	FTIR Microscope, NicoletTM iN10TM MX	Experimental
[57]	Gastrointestinal tracts	Mumbai coast, India	SZX16 Model	Statistical, SPSS v. 20
[49]	Gastrointestinal tracts	Northern Bay of Bengal, Bangladesh	XSZ-107BN, IR Affinity-1, Model-8900	Statistical, R software
[68]	Gastrointestinal tracts, and Stomachs	Pajeú river, Northeast of Brazil	dissecting microscope (45×)	Statistical, R v. 3.2.1
[62]	Gastrointestinal tracts, muscles and gills	Lucknow, Uttar Pradesh, India	Leica, EZ4, Witec Alpha 300RA	Statistical, GraphPad PRISM v. 8.4.0
[63]	-	Pasig River, Marikina River, Philippines	Olympus Microscope BX41, Origin-Prov2021	Experimental
[58]	-	Adriatic Sea	Nikon SMZ745T, LabSpec 6 (Tokyo, Japan)	Experimental
[42]	Viscera and gills	Malaysia	Motic SMZ-140 (Hong Kong), Horiba LabRam HR (Tokyo, Japan)	Statistical, SPSS v. 24
[33]	Digestive tract	Coasts of the Netherlands, Belgium, France and Great Britain	Scimitar 1000 FT-IR	Experimental
[38]	Muscles	Northeast of Persian Gulf, Iran	Inductively coupled plasma mass spectrometry	Statistical

Table 2. *Cont.*

Ref.	Observed Parts	Region of Collection	Microscope	Data Analysis Approach and Software
[34]	Livers	Mediterranean Sea, Europe	Olympus Provis AX-70, LabRam 300	Experimental
[64]	Gastrointestinal tracts, and muscles	Pasur river, Mongla, Bangladesh	Motic B410E, Stemi 508	Statistical
[43]	Gut, skin and gills	Haizhou Bay, China	Nikon SMZ 1500N, Thermo Nicolet iN10 MX	Statistical, SPSS v. 23
[59]	Gills and guts	Haizhou Bay, China	Olympus SZX2-FOF	Statistical, SPSS v. 25
[39]	-	Saudi Arabian coast of the Red Sea	Stemi 2000 Zeiss (Oberkochen, Germany)	Statistical, RStudio v. 1.1.419
[40]	Digestive tract	Coasts of Panama, Colombia, Ecuador, Peru, and Chile	Agilent Handheld 4300 FTIR (Santa Clara, CA, USA)	Experimental
[69]	-	West coast of India	Olympus DSX 110, LUMOS II	Statistical, SPSS v. 22

Different types of plastic polymers have been found in different fish species, and their percentage presence in fish species is shown in Table 3. PE, PET, PS, PP, and PES are observed to be the most detected polymers in fish species. Additionally, CP is not a plastic polymer but a natural polymer, which is also highly found in fish species.

Table 3. Percentage of different types of polymers found in fish species.

	References									
Types of Polymers (Microplastics)	[54]	[45]	[53]	[55]	[61]	[46]	[63]	[43]	[39]	[69]
PE	0.7	6	0.5	36	38	54	30.95	13	42	33
PET	38.1	0	16.9	26	0	0	2.38	4.5	0	4
PS	5	0	0.4	18	22	7	2.38	0	4	14.5
PVC	0	0	0	12	16	0	0	0	8	11.5
PP	7.9	6	2.5	8	0	7	57.14	15	42	21.5
PES	46.8	44	0	0	0	14	0	0	0	0
PMMA	0	6	0	0	0	3	0	0	0	0
PA	0	38	0.4	0	13	15	0	8	0	0
EVA	0	0	0	0	9	0	0	0	0	0
Polyethylene-polypropylene copolymer	1.4	0	0	0	0	0	7.14	0	0	0
PAN	0	0	0.9	0	0	0	0	0	4	0
PVAc	0	0	0.5	0	0	0	0	0	0	0
PB	0	0	0.2	0	0	0	0	0	0	0
PC	0	0	0.2	0	0	0	0	0	0	6.5
PMMA	0	0	0	0	0	0	0	0	0	4
PVA	0	0	0	0	0	0	0	0	0	5
Unidentified	0.1	0	0	0	2	0	0.01	19.5	0	0
Non-plastic particles	0	0	0	0	0	0	0	6.5	0	0
CP	0	0	77.5	0	0	0	0	33.5	0	0

1.5. Microplastics in Human Body

Humans are exposed to microplastics through various sources, including air, water, food, and soil. In particular, fish is a significant component of the human diet, and the presence of microplastics in fish means that humans are indirectly exposed to microplastic pollution. In 2024, researchers have discovered that these microplastics are potentially associated with cardiovascular diseases [70] and have identified microplastics in different types of human arteries [71]. Researchers have also uncovered tiny plastic particles, including microplastics, within bodily fluids [72]. Laser direct infrared spectroscopy has enabled scientists to detect and measure microplastics in tissues like the endometrium [73], gallstones [74], placenta [75], and even in the amniotic fluid of preterm births [76]. Furthermore, microplastics have been found in human urine and kidney tissue [77,78], lower limb joints [79], vitreous humor [80], and lung tissue [81]. Advanced detection methods have revealed microplastics in the testes and semen [82,83], and even blood [84] itself. Even samples of human stool [85], and from patients undergoing heart surgery [86], have shown traces of microplastics. Details are presented in Table 4.

Table 4. Presence of microplastics in human organs, fluids, and waste products.

Ref.	Sample Type	Detection Method	Type of Microplastics	Average Concentrations
[71]	Arteries	Pyrolysis–Gas Chromatography–Mass Spectrometry	PET (73.70%), PA-66 (15.54%), PVC (9.69%), PE (1.07%)	118.66 ± 53.87 μg/g tissue
[72]	Bodily fluids	Raman Microspectroscopy	PP (13.04%), PS (43.48%), PTFE(4.35%), PVB (8.70%), PA-6 (8.70%), LDPE (8.70%), PEAA (4.35%), PSAN (4.35%), PVA (4.35%)	-
[73]	Endometrium	Laser Direct Infrared Spectroscopy	EAA (34.58%), FR (14.87%), CPE (11.47%), PE (9.95%), ACR (7.76%), PET (6.63%), PP (6.68%), PS (0.85%), PVC (0.98%), EVA (0.41%), PU (2.09%), BR (2.61%)	0 to 117 particles/100 mg
[74]	Gallstones	Pyrolysis–Gas Chromatography–Mass Spectrometry and Laser Direct Infrared Spectroscopy	PS, PE, PP, PET, EVA	-
[75]	Placenta	Laser Direct Infrared Spectroscopy	PVC (43.27%), PP (14.55%), PBS (10.90%), PET (7.27%), PC (6.91%), PS (5.82%), PA (5.45%), polyester fibre (2.91%), PE (1.45%), PAM (0.73%), PSF (0.73%)	2.70 ± 2.65 particles/g
[78]	Urine	Micro-Fourier Transform Infrared Spectroscopy	Healthy donors: PE (27%), PS (16%), PP (12%), Endometriosis participants: PTFE (59%), PE (16%)	-
[79]	Lower limb joints	Micro-Fourier Transform Infrared Spectroscopy	PET (27.1%), PE (21.9%), RA (12.0%), PES (11.1%), PP (9.3%), PA (8.5%), PVC (4.7%), PS (4.4%), PC (2.0%)	5.24 ± 2.07 particles/g
[80]	Vitreous humor	Pyrolysis–Gas Chromatography–Mass Spectrometry and Laser Direct Infrared Spectroscopy	PA (74.8%), PVC (7.3%)	-
[81]	Lung tissue	Micro-Fourier Transform Infrared Spectroscopy	PP (23%), PET (18%), RA (15%), PE (10%), PTFE (10%), PS (8%), PAN (2%), PES (2%), PMMA (3%), PUR (3%)	1.42 ± 1.50 MP/g of tissue
[82]	Testes and semen	Pyrolysis–gas Chromatography–Mass Spectrometry	Semen: PVC (25%), PE (25%), PA (17%), PS (13%), PP (13%), PET (7%) Testis: PS (67.7%), PVC (12.9%), PE (12.9%), PP (6.5%)	Semen: 0.23 ± 0.45 particles/mL, Testis: 11.60 ± 15.52 particles/g
[84]	Blood	Pyrolysis–gas Chromatography–Mass Spectrometry	PE, PS, PET, PMMA	1.6 μg/mL
[85]	Stool	Fourier-Transform Infrared Microspectroscopy	PP (62.8%), PET (17.0%), PS (11.2%), PE (4.8%), PVC (0.54%), PU (0.40%), PA (0.54%), PC (0.67%)	-

2. Conclusions and Future Outlook

Microplastic pollution in marine environments represents a significant and growing environmental challenge with extensive implications for marine biodiversity, ecosystem functionality, and human health. This review has underscored the ubiquity of microplastics in marine ecosystems, tracing their origins to various sources such as synthetic textiles, packaging, and personal care products. Once introduced into the environment, these microplastics infiltrate marine food webs, impacting a wide array of marine organisms, particularly fish.

The ingestion of microplastics by fish has been well documented, with particles being found in their gastrointestinal tracts, gills, and other tissues. These ingested microplastics can cause physical harm, such as blockages and injuries, as well as physiological stress, including reduced feeding, impaired growth, and reproductive issues. Furthermore, microplastics can act as vectors for harmful chemicals and pathogens, exacerbating their detrimental effects on marine life. Human exposure to microplastics primarily occurs through the consumption of seafood. The detection of microplastics in human tissues and their potential link to health issues such as inflammation, cellular damage, and endocrine disruption raises significant concerns. Despite the advancements in detection techniques and growing evidence of the adverse effects, the complete extent of microplastic pollution's impact on human health remains underexplored. Comprehensive and long-term studies are essential to mitigate microplastic pollution's impacts. Future research should focus on the chronic effects of microplastic exposure on marine organisms and the subsequent implications for human health. Such studies should consider various species, developmental stages, and environmental conditions to provide a holistic understanding.

Furthermore, identifying and quantifying the primary sources and pathways through which microplastics enter marine environments is crucial for developing targeted mitigation strategies. Enhanced detection methods and data collection and analysis standardisation are necessary to accurately assess microplastic concentrations across different matrices and studies. Innovation in material science to develop biodegradable alternatives to conventional plastics can offer a sustainable solution to reducing plastic pollution. However, understanding these alternatives' degradation process and environmental impact is vital for ensuring their effectiveness. The development of effective policies and regulations is critical to addressing the issue of microplastic pollution. Interdisciplinary research that integrates environmental science, public health, and policy studies can inform the creation of regulations to reduce plastic production and improve waste management practices. Evaluating the effectiveness of existing policies will also be beneficial in shaping future interventions.

Public awareness and education play a pivotal role in combating microplastic pollution. Educating communities about the sources, impacts, and mitigation strategies can foster behavioural changes that reduce plastic waste. Innovative and effective educational campaigns can engage the public in environmental stewardship. Therefore, addressing the complex issue of microplastic pollution requires a multifaceted approach involving comprehensive research, innovative solutions, effective policies, and public engagement. By advancing our understanding and implementing targeted actions, we can protect marine ecosystems and safeguard human health from the pervasive threat of microplastics.

Author Contributions: Conceptualization, G.S.; methodology, G.S.; formal analysis, G.S. and S.C.S.; investigation, G.S. and S.C.S.; resources, G.S.; data curation, G.S.; writing—original draft preparation, G.S.; writing—review and editing, S.C.S.; visualization, G.S.; project administration, S.C.S. All authors have read and agreed to the published version of the manuscript.

Funding: This research received no external funding.

Data Availability Statement: All data are freely available in the manuscript.

Acknowledgments: The authors used AI-assisted technology (ChatGPT 3.5) for language editing and grammar checking.

Conflicts of Interest: The authors declare no conflicts of interest.

References

1. Saha, S.C.; Saha, G. Effect of microplastics deposition on human lung airways: A review with computational benefits and challenges. *Heliyon* **2024**, *10*, e24355. [CrossRef]
2. Huang, X.; Saha, S.C.; Saha, G.; Francis, I.; Luo, Z. Transport and deposition of microplastics and nanoplastics in the human respiratory tract. *Environ. Adv.* **2024**, *16*, 100525. [CrossRef]
3. Miller, M.E.; Hamann, M.; Kroon, F.J. Bioaccumulation and biomagnification of microplastics in marine organisms: A review and meta-analysis of current data. *PLoS ONE* **2020**, *15*, e0240792. [CrossRef] [PubMed]
4. Guo, X.; Wang, J. The chemical behaviors of microplastics in marine environment: A review. *Mar. Pollut. Bull.* **2019**, *142*, 1–14. [CrossRef] [PubMed]
5. De Falco, F.; Di Pace, E.; Cocca, M.; Avella, M. The contribution of washing processes of synthetic clothes to microplastic pollution. *Sci. Rep.* **2019**, *9*, 6633. [CrossRef] [PubMed]
6. Weis, J.S. Most Microplastics Come from Clothes. *Bioscience* **2021**, *71*, 321. [CrossRef]
7. Sobhani, Z.; Lei, Y.; Tang, Y.; Wu, L.; Zhang, X.; Naidu, R.; Megharaj, M.; Fang, C. Microplastics generated when opening plastic packaging. *Sci. Rep.* **2020**, *10*, 4841. [CrossRef]
8. Bashir, S.M.; Kimiko, S.; Mak, C.W.; Fang, J.K.H.; Gonçalves, D. Personal Care and Cosmetic Products as a Potential Source of Environmental Contamination by Microplastics in a Densely Populated Asian City. *Front. Mar. Sci.* **2021**, *8*, 683482. [CrossRef]
9. Dąbrowska, A.; Mielańczuk, M.; Syczewski, M. The Raman spectroscopy and SEM/EDS investigation of the primary sources of microplastics from cosmetics available in Poland. *Chemosphere* **2022**, *308*, 136407. [CrossRef]
10. Jan Kole, P.; Löhr, A.J.; Van Belleghem, F.G.A.J.; Ragas, A.M.J. Wear and tear of tyres: A stealthy source of microplastics in the environment. *Int. J. Environ. Res. Public Health* **2017**, *14*, 1265. [CrossRef]
11. Wright, L.S.; Napper, I.E.; Thompson, R.C. Potential microplastic release from beached fishing gear in Great Britain's region of highest fishing litter density. *Mar. Pollut. Bull.* **2021**, *173 Pt B*, 113115. [CrossRef]
12. Syversen, T.; Lilleng, G. *Microplastics Derived from Commercial Fishing Activities*; IntechOpen: London, UK, 2023. [CrossRef]
13. Luo, Y.; Naidu, R.; Fang, C. Toy building bricks as a potential source of microplastics and nanoplastics. *J. Hazard. Mater.* **2024**, *471*, 134424. [CrossRef] [PubMed]
14. Yuk, H.; Jo, H.H.; Nam, J.; Kim, Y.U.; Kim, S. Microplastic: A particulate matter (PM) generated by deterioration of building materials. *J. Hazard. Mater.* **2022**, *437*, 129290. [CrossRef]
15. Luo, Y.; Naidu, R.; Fang, C. Accelerated transformation of plastic furniture into microplastics and nanoplastics by fire. *Environ. Pollut.* **2023**, *317*, 120737. [CrossRef] [PubMed]
16. Luo, Y.; Awoyemi, O.S.; Naidu, R.; Fang, C. Detection of microplastics and nanoplastics released from a kitchen blender using Raman imaging. *J. Hazard. Mater.* **2023**, *453*, 131403. [CrossRef] [PubMed]
17. Luo, Y.; Naidu, R.; Fang, C. Raman imaging to capture microplastics and nanoplastics carried by smartphones. *Sci. Total Environ.* **2023**, *864*, 160959. [CrossRef] [PubMed]
18. Luo, Y.; Zhang, Z.; Naidu, R.; Zhang, X.; Fang, C. Raman imaging of microplastics and nanoplastics released from the printed toner powders burned by a mimicked bushfire. *Sci. Total Environ.* **2022**, *849*, 157686. [CrossRef] [PubMed]
19. Luo, Y.; Chuah, C.; Amin Md, A.; Khoshyan, A.; Gibson, C.T.; Tang, Y.; Naidu, R.; Fang, C. Assessment of microplastics and nanoplastics released from a chopping board using Raman imaging in combination with three algorithms. *J. Hazard. Mater.* **2022**, *431*, 128636. [CrossRef] [PubMed]
20. Luo, Y.; Al Amin, M.; Gibson, C.T.; Chuah, C.; Tang, Y.; Naidu, R.; Fang, C. Raman imaging of microplastics and nanoplastics generated by cutting PVC pipe. *Environ. Pollut.* **2022**, *298*, 118857. [CrossRef]
21. Luo, Y.; Gibson, C.T.; Chuah, C.; Tang, Y.; Naidu, R.; Fang, C. Raman imaging for the identification of Teflon microplastics and nanoplastics released from non-stick cookware. *Sci. Total Environ.* **2022**, *851*, 158293. [CrossRef]
22. Luo, Y.; Qi, F.; Gibson, C.T.; Lei, Y.; Fang, C. Investigating kitchen sponge-derived microplastics and nanoplastics with Raman imaging and multivariate analysis. *Sci. Total Environ.* **2022**, *824*, 153963. [CrossRef] [PubMed]
23. Luo, Y.; Gibson, C.T.; Chuah, C.; Tang, Y.; Ruan, Y.; Naidu, R.; Fang, C. Fire releases micro- and nanoplastics: Raman imaging on burned disposable gloves. *Environ. Pollut.* **2022**, *312*, 120073. [CrossRef] [PubMed]
24. Luo, Y.; Naidu, R.; Zhang, X.; Fang, C. Microplastics and nanoplastics released from a PPE mask under a simulated bushfire condition. *J. Hazard. Mater.* **2022**, *439*, 129621. [CrossRef] [PubMed]
25. Fang, C.; Awoyemi, O.S.; Luo, Y.; Naidu, R. Investigating Microplastics and Nanoplastics Released from a Rubber Band Used for Orthodontic Treatment with Improved Raman Imaging Algorithms. *Environ. Health* **2023**, *1*, 63–71. [CrossRef]
26. Passos, R.S.; Davenport, A.; Busquets, R.; Selden, C.; Silva, L.B.; Baptista, J.S.; Barceló, D.; Campos, L.C. Microplastics and nanoplastics in haemodialysis waters: Emerging threats to be in our radar. *Environ. Toxicol. Pharmacol.* **2023**, *102*, 104253. [CrossRef] [PubMed]
27. Hossain, M.B.; Yu, J.; Banik, P.; Noman, M.A.; Nur, A.A.U.; Haque, M.R.; Rahman, M.M.; Albeshr, M.F.; Arai, T. First evidence of microplastics and their characterization in bottled drinking water from a developing country. *Front. Environ. Sci.* **2023**, *11*, 1232931. [CrossRef]

28. Chen, Y.; Xu, H.; Luo, Y.; Ding, Y.; Huang, J.; Wu, H.; Han, J.; Du, L.; Kang, A.; Jia, M.; et al. Plastic bottles for chilled carbonated beverages as a source of microplastics and nanoplastics. *Water Res.* **2023**, *242*, 120243. [CrossRef] [PubMed]
29. Cole, M.; Gomiero, A.; Jaén-Gil, A.; Haave, M.; Lusher, A. Microplastic and PTFE contamination of food from cookware. *Sci. Total Environ.* **2024**, *929*, 172577. [CrossRef]
30. Li, J.; Qu, X.; Su, L.; Zhang, W.; Yang, D.; Kolandhasamy, P.; Li, D.; Shi, H. Microplastics in mussels along the coastal waters of China. *Environ. Pollut. (1987)* **2016**, *214*, 177–184. [CrossRef]
31. Tanaka, K.; Takada, H. Microplastic fragments and microbeads in digestive tracts of planktivorous fish from urban coastal waters. *Sci. Rep.* **2016**, *6*, 34351. [CrossRef]
32. Gray, A.D.; Weinstein, J.E. Size- and shape-dependent effects of microplastic particles on adult daggerblade grass shrimp (*Palaemonetes pugio*): Uptake and retention of microplastics in grass shrimp. *Environ. Toxicol. Chem.* **2017**, *36*, 3074–3080. [CrossRef] [PubMed]
33. Hermsen, E.; Pompe, R.; Besseling, E.; Koelmans, A.A. Detection of low numbers of microplastics in North Sea fish using strict quality assurance criteria. *Mar. Pollut. Bull.* **2017**, *122*, 253–258. [CrossRef] [PubMed]
34. Collard, F.; Gilbert, B.; Compère, P.; Eppe, G.; Das, K.; Jauniaux, T.; Parmentier, E. Microplastics in livers of European anchovies (*Engraulis encrasicolus*, L.). *Environ. Pollut.* **2017**, *229*, 1000–1005. [CrossRef] [PubMed]
35. Andrades, R.; Santos, R.G.; Joyeux, J.C.; Chelazzi, D.; Cincinelli, A.; Giarrizzo, T. Marine debris in Trindade Island, a remote island of the South Atlantic. *Mar. Pollut. Bull.* **2018**, *137*, 180–184. [CrossRef] [PubMed]
36. Li, J.; Green, C.; Reynolds, A.; Shi, H.; Rotchell, J.M. Microplastics in mussels sampled from coastal waters and supermarkets in the United Kingdom. *Environ. Pollut.* **2018**, *241*, 35–44. [CrossRef] [PubMed]
37. Abbasi, S.; Soltani, N.; Keshavarzi, B.; Moore, F.; Turner, A.; Hassanaghaei, M. Microplastics in different tissues of fish and prawn from the Musa Estuary, Persian Gulf. *Chemosphere* **2018**, *205*, 80–87. [CrossRef] [PubMed]
38. Akhbarizadeh, R.; Moore, F.; Keshavarzi, B. Investigating a probable relationship between microplastics and potentially toxic elements in fish muscles from northeast of Persian Gulf. *Environ. Pollut.* **2018**, *232*, 154–163. [CrossRef] [PubMed]
39. Baalkhuyur, F.M.; Bin Dohaish, E.-J.A.; Elhalwagy, M.E.A.; Alikunhi, N.M.; AlSuwailem, A.M.; Røstad, A.; Coker, D.J.; Berumen, M.L.; Duarte, C.M. Microplastic in the gastrointestinal tract of fishes along the Saudi Arabian Red Sea coast. *Mar. Pollut. Bull.* **2018**, *131 Pt A*, 407–415. [CrossRef]
40. Ory, N.; Chagnon, C.; Felix, F.; Fernández, C.; Ferreira, J.L.; Gallardo, C.; Ordóñez, O.G.; Henostroza, A.; Laaz, E.; Mizraji, R.; et al. Low prevalence of microplastic contamination in planktivorous fish species from the southeast Pacific Ocean. *Mar. Pollut. Bull.* **2018**, *127*, 211–216. [CrossRef]
41. Guerrini, F.; Mari, L.; Casagrandi, R. Modeling plastics exposure for the marine biota: Risk maps for fin whales in the Pelagos Sanctuary (North-Western Mediterranean). *Front. Mar. Sci.* **2019**, *6*, 1–10. [CrossRef]
42. Karbalaei, S.; Golieskardi, A.; Hamzah, H.B.; Abdulwahid, S.; Hanachi, P.; Walker, T.R.; Karami, A. Abundance and characteristics of microplastics in commercial marine fish from Malaysia. *Mar. Pollut. Bull.* **2019**, *148*, 5–15. [CrossRef] [PubMed]
43. Feng, Z.; Zhang, T.; Li, Y.; He, X.; Wang, R.; Xu, J.; Gao, G. The accumulation of microplastics in fish from an important fish farm and mariculture area, Haizhou Bay, China. *Sci. Total Environ.* **2019**, *696*, 133948. [CrossRef] [PubMed]
44. Morais, L.M.S.; Sarti, F.; Chelazzi, D.; Cincinelli, A.; Giarrizzo, T.; Martinelli Filho, J.E. The sea anemone *Bunodosoma cangicum* as a potential biomonitor for microplastics contamination on the Brazilian Amazon coast. *Environ. Pollut.* **2020**, *265*, 114817. [CrossRef] [PubMed]
45. Koongolla, J.B.; Lin, L.; Pan, Y.-F.; Yang, C.-P.; Sun, D.-R.; Liu, S.; Xu, X.-R.; Maharana, D.; Huang, J.-S.; Heng-Xiang Li, H.-X. Occurrence of microplastics in gastrointestinal tracts and gills of fish from Beibu Gulf, South China Sea. *Environ. Pollut. (1987)* **2020**, *258*, 113734. [CrossRef] [PubMed]
46. Sathish, M.N.; Jeyasanta, I.; Patterson, J. Occurrence of microplastics in epipelagic and mesopelagic fishes from Tuticorin, Southeast coast of India. *Sci. Total Environ.* **2020**, *720*, 137614. [CrossRef] [PubMed]
47. Barboza, L.G.A.; Lopes, C.; Oliveira, P.; Bessa, F.; Otero, V.; Henriques, B.; Raimundo, J.; Caetano, M.; Vale, C.; Guilhermino, L. Microplastics in wild fish from North East Atlantic Ocean and its potential for causing neurotoxic effects, lipid oxidative damage, and human health risks associated with ingestion exposure. *Sci. Total Environ.* **2020**, *717*, 134625. [CrossRef] [PubMed]
48. Park, T.-J.; Lee, S.-H.; Lee, M.-S.; Lee, J.-K.; Lee, S.-H.; Zoh, K.-D. Occurrence of microplastics in the Han River and riverine fish in South Korea. *Sci. Total Environ.* **2020**, *708*, 134535. [CrossRef] [PubMed]
49. Hossain, M.S.; Rahman, M.S.; Uddin, M.N.; Sharifuzzaman, S.M.; Chowdhury, S.R.; Sarker, S.; Chowdhury, M.S.N. Microplastic contamination in Penaeid shrimp from the Northern Bay of Bengal. *Chemosphere* **2020**, *238*, 124688. [CrossRef]
50. Bucol, L.A.; Romano, E.F.; Cabcaban, S.M.; Siplon, L.M.D.; Madrid, G.C.; Bucol, A.A.; Polidoro, B. Microplastics in marine sediments and rabbitfish (*Siganus fuscescens*) from selected coastal areas of Negros Oriental, Philippines. *Mar. Pollut. Bull.* **2020**, *150*, 110685. [CrossRef]
51. Pegado, T.; Erabo, L.; Schmid, K.; Sarti, F.; Gava, T.T.; Nunes, J.; Chelazzi, D.; Cincinelli, A.; Giarrizzo, T. Ingestion of microplastics by *Hypanus guttatus* stingrays in the Western Atlantic Ocean (Brazilian Amazon Coast). *Mar. Pollut. Bull.* **2021**, *162*, 111799. [CrossRef]
52. Saha, M.; Naik, A.; Desai, A.; Nanajkar, M.; Rathore, C.; Kumar, M.; Gupta, P. Microplastics in seafood as an emerging threat to marine environment: A case study in Goa, west coast of India. *Chemosphere* **2021**, *270*, 129359. [CrossRef] [PubMed]

53. Wang, Q.; Zhu, X.; Hou, C.; Wu, Y.; Teng, J.; Zhang, C.; Tan, H.; Shan, E.; Zhang, W.; Zhao, J. Microplastic uptake in commercial fishes from the Bohai Sea, China. *Chemosphere* **2021**, *263*, 127962. [CrossRef] [PubMed]
54. Li, W.; Pan, Z.; Xu, J.; Liu, Q.; Zou, Q.; Lin, H.; Wu, L.; Huang, H. Microplastics in a pelagic dolphinfish (*Coryphaena hippurus*) from the Eastern Pacific Ocean and the implications for fish health. *Sci. Total Environ.* **2022**, *809*, 151126. [CrossRef] [PubMed]
55. Piyawardhana, N.; Weerathunga, V.; Chen, H.-S.; Guo, L.; Huang, P.-J.; Ranatunga, R.R.M.K.P.; Hung, C.-C. Occurrence of microplastics in commercial marine dried fish in Asian countries. *J. Hazard. Mater.* **2022**, *423 Pt B*, 127093. [CrossRef]
56. Hasan, J.; Islam, S.M.M.; Alam, M.S.; Johnson, D.; Belton, B.; Hossain, M.A.R.; Shahjahan, M. Presence of microplastics in two common dried marine fish species from Bangladesh. *Mar. Pollut. Bull.* **2022**, *176*, 113430. [CrossRef] [PubMed]
57. Gurjar, U.R.; Xavier, K.A.M.; Shukla, S.P.; Jaiswar, A.K.; Deshmukhe, G.; Nayak, B.B. Microplastic pollution in coastal ecosystem off Mumbai coast, India. *Chemosphere* **2022**, *288 Pt 1*, 132484. [CrossRef]
58. Mistri, M.; Sfriso, A.A.; Casoni, E.; Nicoli, M.; Vaccaro, C.; Munari, C. Microplastic accumulation in commercial fish from the Adriatic Sea. *Mar. Pollut. Bull.* **2022**, *174*, 113279. [CrossRef] [PubMed]
59. Gao, S.; Li, Z.; Wang, N.; Lu, Y.; Zhang, S. Microplastics in different tissues of caught fish in the artificial reef area and adjacent waters of Haizhou Bay. *Mar. Pollut. Bull.* **2022**, *174*, 113112. [CrossRef] [PubMed]
60. da Costa, I.D.; Costa, L.L.; da Silva Oliveira, A.; de Carvalho, C.E.V.; Zalmon, I.R. Microplastics in fishes in amazon riverine beaches: Influence of feeding mode and distance to urban settlements. *Sci. Total Environ.* **2023**, *863*, 160934. [CrossRef]
61. Hasan, J.; Dristy, E.Y.; Mondal, P.; Hoque, M.S.; Sumon, K.A.; Hossain, M.A.R.; Shahjahan, M. Dried fish more prone to microplastics contamination over fresh fish—Higher potential of trophic transfer to human body. *Ecotoxicol. Environ. Saf.* **2023**, *250*, 114510. [CrossRef]
62. Pandey, N.; Verma, R.; Patnaik, S.; Anbumani, S. Abundance, characteristics, and risk assessment of microplastics in indigenous freshwater fishes of India. *Environ. Res.* **2023**, *218*, 115011. [CrossRef] [PubMed]
63. Espiritu, E.Q.; Rodolfo, R.S.; Evangelista, S.M.J.; Feliciano, J.J.G.; Sumaway, A.M.N.; Pauco, J.L.R.; Alvarez, K.V.N.; Enriquez, E.P. Microplastics contamination in the fishes of selected sites in Pasig River and Marikina River in the Philippines. *Mar. Pollut. Bull.* **2023**, *187*, 114573. [CrossRef] [PubMed]
64. Nawar, N.; Rahman, M.M.; Chowdhury, F.N.; Marzia, S.; Ali, M.M.; Akbor, M.A.; Siddique, M.A.B.; Khatun, M.A.; Shahjalal, M.; Huque, R.; et al. Characterization of microplastic pollution in the Pasur river of the Sundarbans ecosystem (Bangladesh) with emphasis on water, sediments, and fish. *Sci. Total Environ.* **2023**, *868*, 161704. [CrossRef] [PubMed]
65. Li, J.; Yang, D.; Li, L.; Jabeen, K.; Shi, H. Microplastics in commercial bivalves from China. *Environ. Pollut. (1987)* **2015**, *207*, 190–195. [CrossRef] [PubMed]
66. Squillante, J.; Scivicco, M.; Ariano, A.; Nolasco, A.; Esposito, F.; Cacciola, N.A.; Severino, L.; Cirillo, T. Occurrence of phthalate esters and preliminary data on microplastics in fish from the Tyrrhenian sea (Italy) and impact on human health. *Environ. Pollut. (1987)* **2023**, *316 Pt 1*, 120664. [CrossRef] [PubMed]
67. FishBase. Psenopsis Anomala Summary Page. 2019. Available online: https://www.fishbase.se/summary/497 (accessed on 1 April 2024).
68. Silva-Cavalcanti, J.S.; Silva, J.D.B.; de França, E.J.; de Araújo, M.C.B.; Gusmão, F. Microplastics ingestion by a common tropical freshwater fishing resource. *Environ. Pollut. (1987)* **2017**, *221*, 218–226. [CrossRef] [PubMed]
69. Rukmangada, R.; Naidu, B.C.; Nayak, B.B.; Balange, A.; Chouksey, M.K.; Xavier, K.A.M. Microplastic contamination in salted and sun dried fish and implications for food security—A study on the effect of location, style and constituents of dried fish on microplastics load. *Mar. Pollut. Bull.* **2023**, *191*, 114909. [CrossRef] [PubMed]
70. Afzal, Z.; Basit, A.; Habib, U.; Aman, M.; Azim, M.; Cao, H. Emerging risks of Microplastics and Nanoplastics (MNPs): Is a silent threat for cardiovascular disease? *Int. J. Cardiol. Cardiovasc. Risk Prev.* **2024**, *21*, 200280. [CrossRef]
71. Liu, S.; Wang, C.; Yang, Y.; Du, Z.; Li, L.; Zhang, M.; Ni, S.; Yue, Z.; Yang, K.; Wang, Y.; et al. Microplastics in three types of human arteries detected by pyrolysis-gas chromatography/mass spectrometry (Py-GC/MS). *J. Hazard. Mater.* **2024**, *469*, 133855. [CrossRef]
72. Guan, Q.; Jiang, J.; Huang, Y.; Wang, Q.; Liu, Z.; Ma, X.; Yang, X.; Li, Y.; Wang, S.; Cui, W.; et al. The landscape of micron-scale particles including microplastics in human enclosed body fluids. *J. Hazard. Mater.* **2023**, *442*, 130138. [CrossRef]
73. Sun, J.; Sui, M.; Wang, T.; Teng, X.; Sun, J.; Chen, M. Detection and quantification of various microplastics in human endometrium based on laser direct infrared spectroscopy. *Sci. Total Environ.* **2024**, *906*, 167760. [CrossRef] [PubMed]
74. Zhang, D.; Wu, C.; Liu, Y.; Li, W.; Li, S.; Peng, L.; Kang, L.; Ullah, S.; Gong, Z.; Li, Z.; et al. Microplastics are detected in human gallstones and have the ability to form large cholesterol-microplastic heteroaggregates. *J. Hazard. Mater.* **2024**, *467*, 133631. [CrossRef] [PubMed]
75. Zhu, L.; Zhu, J.; Zuo, R.; Xu, Q.; Qian, Y.; Lihui, A.N. Identification of microplastics in human placenta using laser direct infrared spectroscopy. *Sci. Total Environ.* **2023**, *856*, 159060. [CrossRef] [PubMed]
76. Halfar, J.; Čabanová, K.; Vávra, K.; Delongová, P.; Motyka, O.; Špaček, R.; Kukutschová, J.; Šimetka, O.; Heviánková, S. Microplastics and additives in patients with preterm birth: The first evidence of their presence in both human amniotic fluid and placenta. *Chemosphere* **2023**, *343*, 140301. [CrossRef] [PubMed]
77. Massardo, S.; Verzola, D.; Alberti, S.; Caboni, C.; Santostefano, M.; Eugenio Verrina, E.; Angeletti, A.; Lugani, F.; Ghiggeri, G.M.; Bruschi, M.; et al. MicroRaman spectroscopy detects the presence of microplastics in human urine and kidney tissue. *Environ. Int.* **2024**, *184*, 108444. [CrossRef] [PubMed]

78. Rotchell, J.M.; Austin, C.; Chapman, E.; Atherall, C.A.; Liddle, C.R.; Dunstan, T.S.; Ben Blackburn, B.; Mead, A.; Filart, K.; Beeby, E.; et al. Microplastics in human urine: Characterisation using μFTIR and sampling challenges using healthy donors and endometriosis participants. *Ecotoxicol. Environ. Saf.* **2024**, *274*, 116208. [CrossRef] [PubMed]
79. Li, Z.; Zheng, Y.; Maimaiti, Z.; Fu, J.; Yang, F.; Li, Z.-Y.; Shi, Y.; Hao, L.-B.; Chen, J.-Y.; Xu, C. Identification and analysis of microplastics in human lower limb joints. *J. Hazard. Mater.* **2024**, *461*, 132640. [CrossRef] [PubMed]
80. Zhong, Y.; Yang, Y.; Zhang, L.; Ma, D.; Wen, K.; Cai, J.; Cai, Z.; Wang, C.; Chai, X.; Zhong, J.; et al. Revealing new insights: Two-center evidence of microplastics in human vitreous humor and their implications for ocular health. *Sci. Total Environ.* **2024**, *921*, 171109. [CrossRef] [PubMed]
81. Jenner, L.C.; Rotchell, J.M.; Bennett, R.T.; Cowen, M.; Tentzeris, V.; Sadofsky, L.R. Detection of microplastics in human lung tissue using μFTIR spectroscopy. *Sci. Total Environ.* **2022**, *831*, 154907. [CrossRef]
82. Zhao, Q.; Zhu, L.; Weng, J.; Jin, Z.; Cao, Y.; Jiang, H.; Zhang, Z. Detection and characterization of microplastics in the human testis and semen. *Sci. Total Environ.* **2023**, *877*, 162713. [CrossRef]
83. Montano, L.; Giorgini, E.; Notarstefano, V.; Notari, T.; Ricciardi, M.; Piscopo, M.; Motta, O. Raman Microspectroscopy evidence of microplastics in human semen. *Sci. Total Environ.* **2023**, *901*, 165922. [CrossRef] [PubMed]
84. Leslie, H.A.; van Velzen, M.J.M.; Brandsma, S.H.; Vethaak, A.D.; Garcia-Vallejo, J.J.; Lamoree, M.H. Discovery and quantification of plastic particle pollution in human blood. *Environ. Int.* **2022**, *163*, 107199. [CrossRef] [PubMed]
85. Schwabl, P.; Koppel, S.; Konigshofer, P.; Bucsics, T.; Trauner, M.; Reiberger, T.; Liebmann, B. Detection of various microplastics in human stool: A prospective case series. *Ann. Intern. Med.* **2019**, *171*, 453–457. [CrossRef] [PubMed]
86. Yang, Y.; Xie, E.; Du, Z.; Peng, Z.; Han, Z.; Li, L.; Zhao, R.; Qin, Y.; Xue, M.; Li, F.; et al. Detection of Various Microplastics in Patients Undergoing Cardiac Surgery. *Environ. Sci. Technol.* **2023**, *57*, 10911. [CrossRef] [PubMed]

processes

MDPI

Article

Raman Technology for Process Control: Waste Shell Demineralization for Producing Transparent Polymer Foils Reinforced with Natural Antioxidants and Calcium Acetate By-Products

Simona Cîntă Pînzaru [1,2,*], Iuliana-Cornelia Poplăcean [1,*], Karlo Maškarić [1,2,*], Dănuț-Alexandru Dumitru [1], Lucian Barbu-Tudoran [3,4], Tudor-Liviu Tămaș [5], Fran Nekvapil [1,2,3] and Bogdan Neculai [6]

1 Biomolecular Physics Department, Babeş–Bolyai University, Kogălniceanu 1, 400084 Cluj-Napoca, Romania; danut.dumitru@stud.ubbcluj.ro (D.-A.D.); fran.nekvapil@itim-cj.ro or fran.nekvapil@ubbcluj.ro (F.N.)
2 Institute for Research, Development and Innovation in Applied Natural Sciences, Babeș-Bolyai University, Fântânele 30, 400327 Cluj-Napoca, Romania
3 National Institute for Research and Development of Isotopic and Molecular Technologies, Donath 67-103, 400293 Cluj-Napoca, Romania; lucian.barbu@itim-cj.ro or lucianbarbu@yahoo.com
4 Electron Microscopy Center, Babeș-Bolyai University, Clinicilor 5-7, 400006 Cluj-Napoca, Romania
5 Department of Geology, Babeș-Bolyai University, M. Kogălniceanu 1, 400084 Cluj-Napoca, Romania; tudor.tamas@ubbcluj.ro
6 Metrohm Analytics Romania SRL, E. Racoviţă 5, 041753 Bucharest, Romania; bogdan.neculai@metrohm.ro
* Correspondence: simona.pinzaru@ubbcluj.ro (S.C.P.); iuliana.poplacean@stud.ubbcluj.ro (I.-C.P.); karlo.maskaric@ubbcluj.ro (K.M.)

Abstract: Waste biogenic materials derived from seafood exploitation represent valuable resources of new compounds within the blue bioeconomy concept. Here, we describe the effectiveness of Raman technology implementation as an in-line tool for the demineralization process control of crustaceans or gastropods. Transparent chitin polymeric foils and calcium acetate by-products were obtained from three waste crustacean shells (*C. sapidus*, *S. mantis*, and *M. squinado*) using a slow, green chemical approach employing acetic acid. Progressive mineral dissolution and increasing of the Raman characteristic signal of chitin is shown in a time-dependent manner using NIR-Raman spectroscopy, while resonance Raman shows intact carotenoids in reacted shells after 2 weeks. Chitin foil products are species-specific, and the demineralization bath of the waste shell mixture can be effectively tracked by Raman tools for solvent control and decision making for the recovery of calcium acetate by-products. Comparatively obtained calcium acetate from *Rapana venosa* snail shells, the subject of Raman analyses, allowed assessing by-product identity, hydration status, purity, and suitability as recrystallized material for further use as a pharmaceutical compound derived from different crustaceans or gastropod species. Cross validation of the results was done using FT-IR, XRD, and SEM-EDX techniques. A hand-held flexible TacticID Raman system with 1064 nm excitation demonstrated its effectiveness as a rapid, in-line decision making tool during process control and revealed excellent reproducibility of the lab-based instrument signal, suitable for in situ evaluation of the demineralization status and solvent saturation control.

Keywords: Raman technology; recovery and resource utilization technology; process control; biogenic carbonate waste; demineralization process control; chitin; calcium acetate drug; carotenoids

1. Introduction

The demineralization of crustacean shells has wide applicability not only in materials science but also in aquatic research and environmental studies, especially in light of the increasing demands of the bioeconomy sector. The main objective of the blue bioeconomy is to obtain new and valuable products from aquatic waste, making it one of the rapidly expanding research fields in the context of current priorities [1].

The abundant aquatic waste material in focus here is the biogenic calcium carbonate originating from crustacean or gastropod shells. Among these, the Atlantic blue crab *Callinectes sapidus* (Rathbun, 1896) and whelk *Rapana venosa* (Valenciennes, 1846) are ranked as some of the most invasive species in the Mediterranean and Black Sea basins. There are many studies related to the increased awareness regarding the most invasive species in the Mediterranean Sea, such as the one conducted by Marchessaux et al., who advocate the resilient idea of turning the threat into new opportunities [2–4].

The Atlantic blue crab *C. sapidus* and the whelk *R. venosa* are both ranked among the 100 worst invasive species, with a negative impact on the ecology of invaded areas, seashore ecosystems, touristic areas, as well as the aquaculture exploitation of the local bivalve products. These species have garnered attention not only from environmental scientists but also from the blue bioeconomy research field due to their potential for multidisciplinary approaches. Aquatic resources can be sustainably exploited to produce value-added compounds or innovative products from aquatic waste.

We demonstrated in several recently published papers that the highly ordered 3D nanostructure of the *C. sapidus* shells, comprising mineral and organic components [5], possesses an intricate porosity, which could be exploited for various applications [6]. These include serving as an efficient material for solution loading and slow release [7,8], a drug carrier, an efficient absorbent of pollutants [9], and a new biostimulant [10], while being compliant with the regulations regarding their heavy metal content [11].

The complex scaffold of chitin-protein fibrils, which supports biomineralization, has never been considered with respect to its potential utility as a chitin-based polymer foil. This green product might be available as a subject of the intact raw material being demineralized, without powdering. As recently reviewed, numerous studies reporting chitin or chitosan production from crustaceans [12] have used powdering as a main step in the production process, where understanding dependencies for efficient extraction is crucial. When speeding up the process, the energy consumption and the workload for the multistep preparation of powders are high, while a slower process could eliminate all these steps. The demineralization process, however, can be considered without powdering, heating, and stirring of the demineralization bath, resulting in a material that can be further tailored according to the desired products.

Chitin, the second most abundant polymer after cellulose, is widely produced from the primary source of aquatic waste derived from crab and shrimp processing. Industrially, chitin is produced throughout acid treatment, with hydrochloric acid being the preferred demineralization agent, even though it may negatively impact the molecular mass and the degree of deacetylation of the resulted polymer. Therefore, it may impair the purified chitin's inherent qualities [13]. The demineralization process is followed by deproteinization and decolorization to obtain pure, colorless chitin, without any residues, reaching the required quality for specific applications in many fields, such as drug delivery and tissue engineering as well as agriculture, food industry, and others [12]. Due to the variability of the chitin source, the entire production process requires optimization.

To obtain chitin from various crustacean species, the demineralization process may depend on the structural and morphological characteristics of each species; thus, knowledge-based decisions on the most convenient process steps are required. Additionally, during the demineralization process, tools needed for informed decisions (to continue, to modify conditions, or to stop) are scant. Most of the studies rely on obtaining the "final product" under certain conditions and consider the necessary repeating of operations or improved conditions to achieve the optimal processing. This is done with the aim of ensuring compatibility with the transition to the industrial environment at the highest technology readiness level (TRL), while achieving a low-cost, high-quality final product, minimal workload, and an environmentally friendly chemical consumption [14].

When exposed to acid treatment (usually hydrochloric acid), the biogenic carbonate reacts to yield the secondary product, calcium acetate, in addition to CO_2 and water. Calcium acetate, approved by the regulatory bodies [15], is widely used as a food additive, an acidity regulator, a preservative and stabilizing agent for nutraceuticals, a calcium supplement, and a medication for patients with kidney disease undergoing dialysis, to control hyperphosphatemia. There are two widely used industrial methods to obtain calcium acetate. One involves the reaction between calcium carbonate and acetic acid, often using natural limestones or marble as the starting material, resulting in calcium acetate, carbon dioxide, and water. The other method employs using calcium hydroxide and acetic acid, yielding calcium acetate and water. The biogenic carbonate waste material, typically referring to calcium carbonate derived from crustaceans' waste, bivalve or mollusc shells, is not usually considered, as it is often blamed for the potential impurity residues that may alter the quality of the calcium acetate product.

To the best of our knowledge, there are neither reported studies on obtaining transparent polymeric foils of chitin from unground, unpowdered biogenic materials derived from *Callinectes sapidus*, the mantis shrimp *Squilla mantis* (Linnaeus, 1758), and the European spider crab *Maja squinado* (Herbst, 1788), nor reports on calcium acetate obtained from these crustaceans. Iftekhar et al. [16] noted the preparation of an optically transparent crab shell by removing non-chitin components (e.g., calcium carbonate, proteins, lipids, and pigments) to create transparent nanocomposites with improved properties. The species considered was *Chionoecetes opilio* (Fabricius, 1788), and the demineralization process occurred under hydrochloric acid treatment [16].

Previous studies have employed Raman spectroscopy as a tool for characterizing shells from various species. Polyene pigments (carotenoids) and inorganic calcium carbonate have been identified within the shells of several gastropod species through Raman spectroscopy [17], while investigations of the inorganic composition of bivalve [18] and crustacean [19] shells have also been conducted using Raman techniques. All these studies analyzed the efficiency of Raman spectroscopy in marine shell characterization. However, we use Raman spectroscopy in controlling the demineralization process, which has not been previously explored.

In this paper, we demonstrate the usefulness of the Raman spectroscopy techniques to assist in the demineralization process of the biogenic carbonates derived from three distinct crustacean species to obtain chitin or derived from *R. venosa* shells to valorize the abundant calcium content for calcium acetate production, as this gastropod shell does not comprise chitin. We compared the in-line process control results in terms of Raman spectroscopy signal, to check the appearance of the chitin signal in the resulting demineralized biological samples in a timely manner using both the lab-based Raman system and the hand-held Raman instrument. We further comparatively evaluate the calcium acetate by-products resulting from application of a 'green' method, using the acetic acid reaction on the biogenic waste at room temperature, without powdering or heating. Finally, we evaluate the quality of the calcium acetate resulting from the crustaceans and snail shells. Fourier-transform infrared spectroscopy (FT-IR), X-ray diffraction (XRD), and scanning electron microscopy combined with energy dispersive X-ray spectroscopy (SEM-EDX) were employed as cross-validation methods to confirm the identity and the morphology of the final bio-products.

Considering the time, cost, effort, and chemicals required for such a crucial industrial approach, here we propose the implementation of Raman technology as an effective tool to assist every step of this economically important activity.

2. Materials and Methods

2.1. Biogenic Material Selection and Processing

Biological samples from three crustacean species—*C. sapidus*, *S. mantis* and *M. squinado*—were acquired through a collaboration between the Babeș-Bolyai University and the University of Dubrovnik, originating from the Neretva River Delta (Southeasten Adriatic Sea). The specimens of *S. mantis* and *M. squinado* were caught and maintained in frozen conditions, while the shells from *C. sapidus* represented food waste from cooked crabs. One specimen from each of *S. mantis* and *M. squinado* were eviscerated. We considered waste biogenic material from cooked carapace fragments of *C. sapidus*, cuticle segments of the abdomen, the telson cuticle from *S. mantis*, and the whole, raw carapace of *M. squinado* for the experimental studies. Fresh specimens of the *R. venosa* snail were gathered from a cluster of individuals along the Romanian shores of the Black Sea, specifically Năvodari, at the geographical coordinates 44°18′07.4″ N, 28°37′38.4″ E. The *R. venosa* shells were randomly selected from a large stock comprising both the specimens with intensive pink-orange pigmentation and the specimens with pronounced blue pigmentation.

The selected crustacean shells were cleaned from adherent aquatic materials, degreased, washed abundantly with deionized pure water (resistivity 18.2 MΩ × cm at 22 °C) and immersed in pure glacial acetic acid. *R. venosa* specimens were immersed in a vinegar bath (acetic acid 9%) after undergoing a thorough cleaning process with the removal of soft tissue from the shells. Demineralization occurred at room temperature, with the process systematically monitored through periodic analysis of the biological samples by Raman techniques.

2.2. Chemicals

Glacial acetic acid was provided by Sigma Aldrich, St. Louis, MO, USA, while geogenic calcium carbonate was purchased from CHIMREACTIV S.R.L, with both substances being used without any further purification.

2.3. Demineralization By-Products and Reference Calcium Acetate Synthesis

The immersion bathing solutions of the biological samples were evaporated by exposure to controlled heat to fully investigate the demineralization process. Given that calcium carbonate is the main mineral of the biological samples considered, the reaction between this compound and acetic acid was observed to obtain geogenic calcium acetate as a reference material of the demineralization by-products. Geogenic calcium acetate was synthetized using 1 g of standard geogenic calcium carbonate dissolved in a mixture solution of 2 mL pure glacial acetic acid and 3 mL acetic acid aqueous solution 10%. The obtained mixture was prepared under magnetic stirring at a controlled temperature for 60 min, with the resulting solution being evaporated under controlled heat. All powders obtained were dried in an oven at 60 °C for 24 h and investigated through Raman spectroscopy and X-ray diffraction.

2.4. Instrumentation

As a lab-based instrument to validate the hand-held Raman system, we employed the Renishaw InVia Reflex Raman system (Renishaw, Gloucestershire, UK) with a Leica confocal microscope. For Raman excitation, a laser diode emitting at 785 nm was employed to characterize the starting materials and the ones during the demineralization progress. An additional Cobolt diode pumped solid-state laser emitting at 532 nm was employed to control the presence of native carotenoids in biogenic shells, exploiting their selective signal under resonance Raman conditions, and a He-Ne- laser providing the 632.8 nm excitation line was used for detecting the carotenoprotein resonant signal. The laser power and the exposure time were adjusted for each acquisition to achieve optimal signal-to-noise ratio, with the data presented in Figure S1. The laser power was selected within the range of 1–50 for 785 nm excitation and 0.5–10 for 532 nm and 633 nm. Due to the different thicknesses of the samples, the exposure times varied accordingly, from 1 s to 10 min in the case of extended acquisitions (Figure S1). The instrument calibration was achieved with the internal silicon providing the band centered at 520 cm^{-1}. WiRE™ 3.4 Software (Renishaw, UK) was used for data acquisition. The spectral resolution was 1 cm^{-1} in NIR and 0.5 cm^{-1} for the visible range excitation.

A hand-held TacticID® Mobile Raman system, model BWS493TSII (BWTEK, a Metrohm Group Company, Herisau, Switzerland), with a NIR-laser emitting at 1064 nm, 220 mW, with a TE-Cooled InGaAs Array detector, was used to record spectra during the process in the 176–2000 cm^{-1} spectral range, with a spectral resolution of 11 cm^{-1}. The system is equipped with a database of 1200 spectra of synthetic chemicals, narcotics, drugs, explosives, cutting agents, precursors, and solvents [20].

A Shimadzu FT-IR IRSPIRIT with an QATR-S accessory (Shimadzu Europa GmbH, Duisburg, Germany), holding a single-reflection integration-type ATR module and with a diamond prism, was employed to record the FT-IR spectra of the demineralized fragments in the 650–4000 cm^{-1} spectral range, setting 50 accumulations per spectrum, with 4 cm^{-1} spectral resolution selected in the LabSolutions IR software version 2. 31 (3 February 2023).

X-ray powder diffraction (XRD) analyses were achieved using a Bruker D8 Advance diffractometer (Bruker Corporation (Bruker AXS Advanced X-ray Solutions GmbH), Karlsruhe, Germany) in Bragg–Brentano geometry, possessing a Cu tube with λ_{K_α} = 0.15418 nm, a Ni filter, and a LynxEye detector. Corundum (NIST SRM1976a) was used as an internal standard. The data were collected in the 3.8–64° 2θ interval at a 0.02° 2θ step, measuring each step for 0.2 s. Acetate precipitates were ground in an agate mortar and placed in Bruker PMMA sample holders. In the case of the demineralized foils, surface XRD was performed on foil fragments, with the surface of the fragments aligned to the X-ray beam. The identification of mineral phases was performed with the Diffrac.Eva 2.1 software (Bruker AXS) using the PDF2 (2023) database from the ICDD (International Centre for Diffraction Data).

Scanning electron microscopy and energy-dispersive X-ray spectroscopy (SEM-EDX) analyses were achieved using a Hitachi SU8320 ultra-high resolution cold field emission scanning electron microscope (Hitachi, Ibaraki, Japan) with a Quorum Q150T gold sputtering coater of a controlled thickness of 11 nm at a rate of 14 nm/min and evaporating carbon for EDX analysis using an Oxford energy-dispersive X-ray module (Oxford, UK) for semiquantitative elemental analysis of the demineralized shell fragments.

The dataset underwent comprehensive processing and analyzing using OriginPro 2021b version 9.8.5.212, OriginLab Corporation, Northampton, MA, USA. The data processing steps of the Raman spectra include smoothing and background subtraction. Raw spectra together with the spectral acquisition parameters are presented in the supplementary materials (Figure S1).

3. Results and Discussion

3.1. Demineralization Process

The stock of the selected materials for exploring the green demineralization process, consisting of untreated fragments of crustacean shells, is shown in Figure 1A. All the anatomical fragments were subjected to the same demineralization treatment, described above. Following the applied treatment, transparent and flexible samples were obtained, as noted in Figure 1B. In addition, the disappearance of the stiffness indicates the dissolution of the mineral component of the crustaceans' shells (Supplementary Material, Video S1). The process was accompanied by the extraction of carotenoids, leading to the discovery of acetic acid's role in the depigmentation of crustacean shells. In the case of *R. venosa*, Figure 1C highlights the effects of shell demineralization after 14 days of vinegar immersion. The fully developed specimen showed, overall, different prominent signs of shell degradation from the appearance of through holes, shell apex withdrawal, and strong interior depigmentation. Furthermore, fragments of the exterior-side shell layer with brown pigmented lines started to detach from the still-solid body.

Figure 1. (**A**) Untreated anatomical parts: carapace fragments (a) and ventral fragment (b) of *C. sapidus*; cuticle segments of abdomen (c), uropod (d), and telson cuticle (e) from *S. mantis*; carapace of *M. squinado* (f); (**B**) crustacean anatomical shell fragments after exposure to acetic acid for 65 days: carapace fragment (a) and ventral fragment (b) of *C. sapidus*; cuticle segment of abdomen (c), uropod (d) and telson cuticle (e) from *S. mantis*; carapace of *M. squinado* (f); (**C**) adult *R. venosa* shell during vinegar demineralization (a) and after 14 days of treatment in ventral (b) and lateral (c) view.

3.2. Raman Spectral Analyses during the Demineralization Process

The Raman spectra of the starting materials were checked to comply with the already known [5–9] characteristics: under 785 nm excitation, the crab species shells exhibit the dominant, characteristic bands of calcite at 1085, 712, 281, and 156 cm^{-1}, weak bands of pigments (free and non-covalently bond astaxanthin showing the C=C stretching modes at 1516 and 1494 cm^{-1}, respectively), and several weak trace bands originating from the most intense modes of chitin above a typical background, even for NIR excitation (Supplementary Figure S1). Particularly, in the case of the *S. mantis* shell, besides the above bands typical for crabs, a strong band is observed for phosphate at 954 cm^{-1}, which is overlapped with the more prominent chitin bands. To track the demineralization process under acetic acid exposure, we analyzed the biological samples at different moments of time, marked in Figure 2 for each individual crustacean specimen.

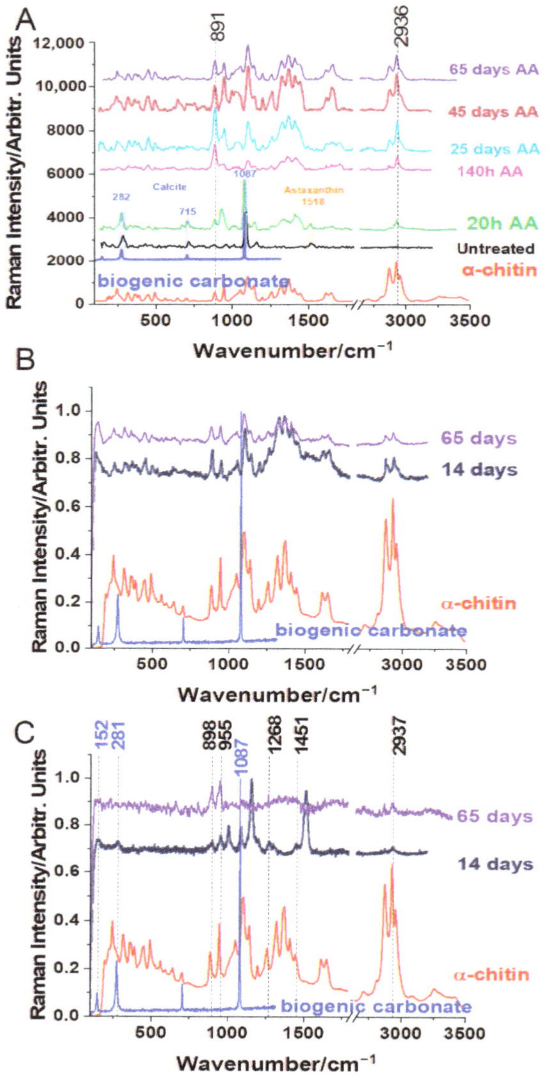

Figure 2. Raman spectra evolution of the biogenic material from *C. Sapidus* (**A**), *S. mantis* (**B**), and *M. squinado* (**C**) at various times (indicated as hours or days) during the acetic acid demineralization process, using a 785 nm laser line, compared to the reference spectrum of α-chitin, as indicated. The blue spectra show the reference signal of calcium carbonate to highlight its disappearance in the final products.

By monitoring the evolution of the Raman spectrum of the crustacean shells, from the typically raw biogenic calcium carbonate presence to the appearance of the characteristic signal of chitin and the disappearance of the calcium carbonate Raman bands at 1085, 712, 281, and 156 cm^{-1}, the spectra taken during the process control promptly guided decision making regarding whether to stop or to continue the acetic acid exposure of the crustacean shells. It turned out that the different species required varying durations for complete demineralization, with the thick carapace of *M. squinado* requiring a longer exposure compared to the shell fragments of *C. sapidus* or *S. mantis*.

The Raman spectra of the biogenic material from *C. sapidus* after acetic acid exposure are shown in Figure 2A. The presence of calcium carbonate in the polymorphic form of calcite is revealed by the Raman spectra acquired from the *C. sapidus* untreated shell and the one exposed to acetic acid for 20 h. Additionally, the strongest band in the astaxanthin Raman spectrum is also present in the spectra of the two discussed samples. The α-chitin specific signals start to appear only after exposure to acetic acid, being covered by the background in the case of the untreated fragment shell. After 140 h of immersion in acetic acid, the *C. sapidus* shell fragment was completely demineralized. Thus, through exposure to acetic acid, markedly weaker acid than the hydrochloric acid, which was commonly employed in many previous studies [12], the process of dissolving the mineral component in crustacean shells yielded promising outcomes. Following this success, we monitored the evolution of the chitin Raman bands through an extended acetic acid treatment. Once the exposure time increased, the Raman spectrum acquired on the shell fragments began to show an improved spectral resolution of the α-chitin signal. After 65 days of immersion in acetic acid, the spectrum of the shell fragment from the *C. sapidus* presented the bands of the reference α-chitin with highest accuracy. Additional vibrational bands express the presence of lipids and proteins in fragments of *C. sapidus* shells exposed to acetic acid.

The different time needed for chitin clear band appearance along the demineralization process could be explained by the fact that the various parts of the crab shell have different compositions and thicknesses [21]. Considering the biological samples, the whole carapace of *M. squinado* is more mineralized and thicker than the *S. mantis* cuticles; thus, the latter experienced faster demineralization and clear observation of chitin Raman band occurrence. The earliest chitin bands were observed after 18 h of acetic acid bath treatment (Figure 2B). The calcium carbonate Raman signal was still present after two weeks of treatment in the carapace of *M. squinado*, indicating an intricate structure and low demineralization process (Figure 2C).

Time-dependent Raman spectra acquired under 785 nm excitation revealed different chitin bands in the case of *S. mantis* and *M. squinado*. The first chitin bands observed after 18 h of treatment in *S. mantis* are present at other points in time. In the case of *M. squinado*, the first chitin bands were visible two weeks after the treatment. At the same time, calcium carbonate bands at 152 cm^{-1}, 281 cm^{-1}, and 1087 cm^{-1} were still present. Besides chitin and calcium carbonate, the peaks with the highest intensity were those from the carotenoids and the carotenoproteins (Figure 3). The excitation of the carapace at the end of the treatment (after 65 days) revealed a high background from the remaining proteins (Figure 2C). Due to the high background, the calcium carbonate bands were often not seen. Carotenoids seemed not to be present at the time, or the point of the carapace did not contain carotenoids. The color originating from the carotenoids was still visible with the naked eye at the end of the treatment (Figure 1B).

The crab cuticles contain carotenoids and carotenoproteins, as Figure 3 shows. The carotenoids were still present after 14 days of acetic acid bath solution. In the case of *C. sapidus* and *M. squinado*, the carotenoid profile did not experience shifts in the wavenumber position but appeared with lower relative intensity (Figure 3A,C). *S. mantis* carotenoids exhibited shifts in the wavenumber position, and the relative intensities of the vibrational bands were lower after acetic acid treatment (Figure 3B).

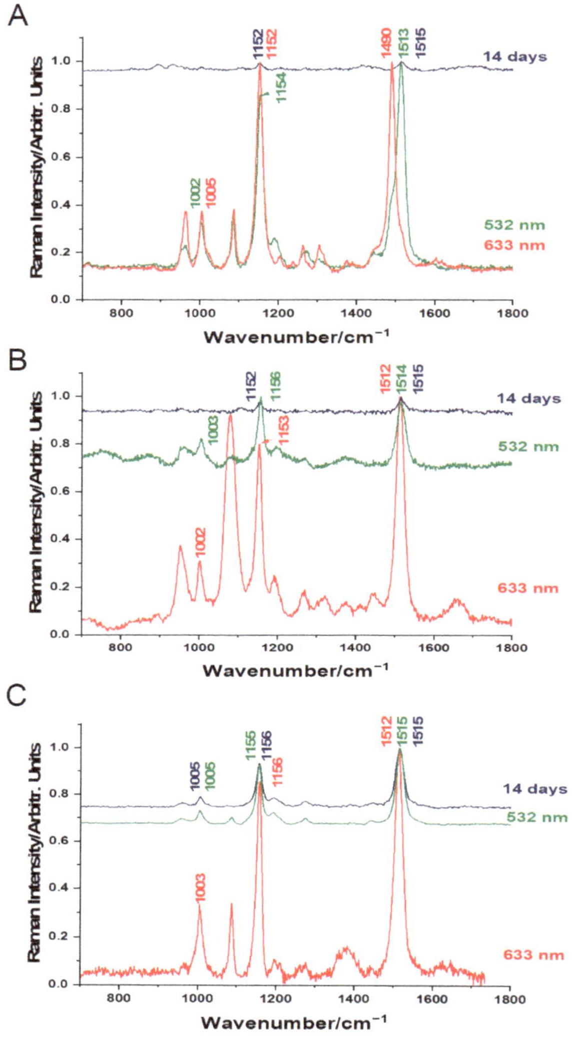

Figure 3. Pigments detected in the raw crustacean shells of *C. sapidus* (**A**), *S. mantis* (**B**), and *M. squinado* (**C**) under resonant Raman excitation with 532 nm for carotenoids (green line) and 633 nm for carotenoproteins (red line) before demineralization, and after 14 days of exposure to the acetic acid bath solution with 532 nm (top spectra, navy blue line in each case).

3.3. Validation Results Using FT-IR Spectral Analyses of Demineralized Foil Products Derived from Crustaceans

The FT-IR spectra of the demineralized crustacean foils are shown in Figure 4. In the case of all three crustacean species (*C. sapidus*, *S. mantis*, and *M. squinado*), some new vibrational bands were identified at the end of the demineralization process (65 days) in addition to the specific vibrational signals previously reported for raw crustacean fragments [9]. Among these, chitin-specific FT-IR bands were identified in all specimen fragments, as follows: 891 cm^{-1} (ring stretching), 1411 cm^{-1} (CH_2 bending and CH_3 deformation), 1556 and 1315 cm^{-1} (amide II (N-H bending and amide III (C-N stretching) and 1628 cm^{-1} (stretching of amide I) in agreement with other FT-IR data reported on chitin [22,23]. *S. mantis* and *M. squinado* demineralized shell fragments also revealed chitin bands at 1113 cm^{-1} (asymmetric in phase ring stretching) and 1377 cm^{-1}, while higher CO-stretching mode was identified in the case of the *C. sapidus*-treated shell fragment at 1022 cm^{-1}, [22,23].

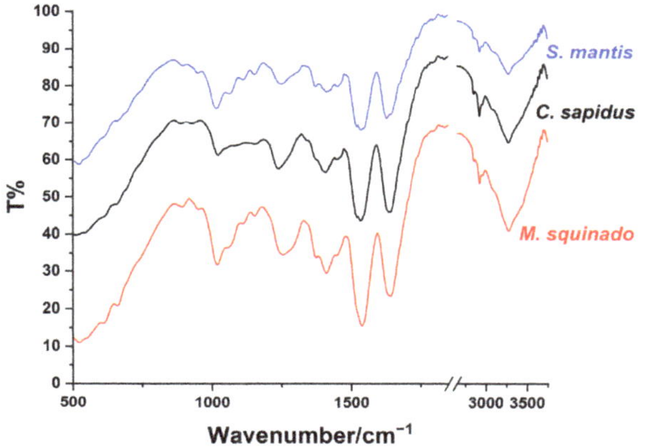

Figure 4. FT-IR spectra of the intact, transparent, demineralized foil products resulting from the three crustacean species after 65 days of acetic acid treatment: *S. mantis* (blue line), *C. sapidus* (black line), and *M. squinado* (red line).

3.4. Validation of the Demineralized Intact Foil Product Content Using X-ray Diffraction

The X-ray diffraction patterns of the demineralized foils from each of the three studied crustacean species are shown in Figure 5. The presence of the characteristic peaks corresponding to the poly-glucosamine functional group, which are indicative of the structural framework of the chitin biopolymer, confirmed the demineralized foils' chitin content. The diffraction peaks were recorded around the following 2θ values: 9.2°, 19.3°, 23.4°, and 39°. These findings suggest that all three biological demineralized specimen fragments exhibit the same chitin semi-crystalline structure signal, consistent with previously reported characteristics of chitin extracted from insects and cuttlebone [24].

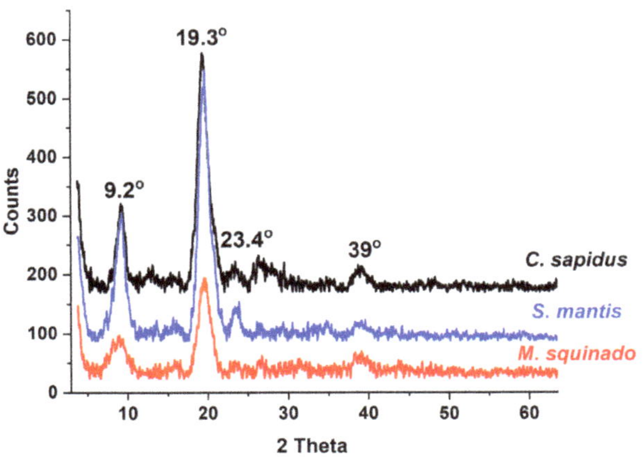

Figure 5. XRD diffraction patterns of the intact, transparent, demineralized foil products resulting from the three crustacean species after 65 days of acetic acid treatment: *C. sapidus* (black line), *S. mantis* (blue line), and *M. squinado* (red line).

3.5. Validation of the Calcium Acetate By-Product Using X-ray Diffraction Patterns of the Crystallized Compound from Acetate Bath Solutions

Aside from the production of demineralized biological shells, acetic acid treatment of the crustaceans' cuticle and vinegar treatment of the gastropod resulted in a calcium acetate by-product, a chemical of high interest in the medical field.

According to the XRD data of the calcium acetate resulting from the evaporated demineralization bathing liquid of the crustacean shells, the saturated solution indicates a mixture of at least three distinct compounds: calcium acetate monohydrate, calcium acetate hemihydrate (calcium acetate half-hydrate) [25], and two additional peaks at 9.42° and 9.74° (Figure 6), which can be possibly assigned to calcium magnesium acetate hydrate, as previously noted [26].

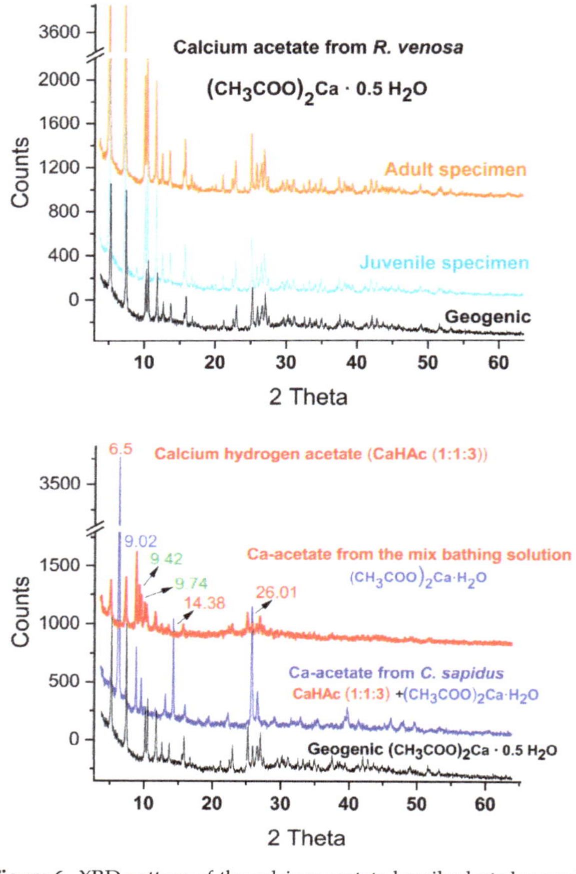

Figure 6. XRD pattern of the calcium acetate hemihydrate by-products from *R. venosa* snail (top panel), from an adult specimen featuring pink shell color (orange spectrum) and from the blue pigmented shells (light-blue plotted diffractogram). Calcium acetate by-products from crustaceans (bottom panel) show different hydrated forms of Ca-acetate from the bathing solution of mixed crustacean shells (dark-red line), Ca-acetate from *C. sapidus* as a mixture of Ca-hydrogen acetate and hydrate (navy-blue line), and geogenic Ca-acetate hemihydrate (black line), as indicated on each diffractogram. Additional peaks in the Ca-acetate from the bathing solution of mixed crustacean shells (dark-red line) are plotted in green, indicating other contributions.

The calcium acetate produced from the blue crab (*C. sapidus*) is in fact a mixture of calcium hydrogen acetate (CaHAc 1:1:3) and calcium acetate monohydrate, while the geogenic form described above, and produced as a reference, is composed of the half-hydrate form ($(CH_3COO)_2Ca \cdot 0.5\ H_2O$), as the X-ray diffraction patterns show. This form was the only one resulting from the comparative demineralization solutions of the two randomly selected *R. venosa* specimens, appearing as a more pink or blue pigmented shell color. From these comparative data, we can conclude that using *R. venosa* shell waste for calcium acetate production might be more effective and deserves more elaborate study (under development); these findings may be explained by the lower magnesium content of the *R. venosa* mineral shell, containing a mixture of aragonite and calcite polymorphs of calcium carbonate. In the case of the crustacean shells, the chitin flexible foils resulting from whole-cuticle macro-fragments as a compact polymer could be recovered, preserving their shape integrity, while the organic component of the demineralized *R. venosa* shells, where chitin is not present, was dispersed in the acetic acid solution once the biomineral component was dissolved.

3.6. Validation of the Calcium Acetate By-Product Using Raman spectra of the Crystallized Compound from Acetate Bath Solutions

To comprehend the structural variations between the obtained calcium acetate hydrates, Raman spectroscopic analyses of the powdered Ca-acetate by-products were conducted in addition to X-ray diffraction. As shown in Figure 7, calcium acetate derived from *R. venosa*, both from adult, orange pigmented specimen and the one from a blue pigmented specimen exhibited Raman bands similar to those from the geogenic compound [27]. Regarding the calcium acetate forms obtained from crustaceans, both mixed and only from *C. sapidus*, the Raman spectra showed the presence of specific vibrational bands of the geogenic form, but their profile is slightly different. Notable differences are observed in the case of *C. sapidus* calcium acetate by the new vibrational bands at 1697 cm^{-1} and 903 cm^{-1} assigned to calcium hydrogen acetate as well as the shift of the 1467 cm^{-1} signal. These marked differences can be attributed to the saturation of the bathingsolution, and the co-existence of two acetate forms, also confirmed by the XRD analyses.

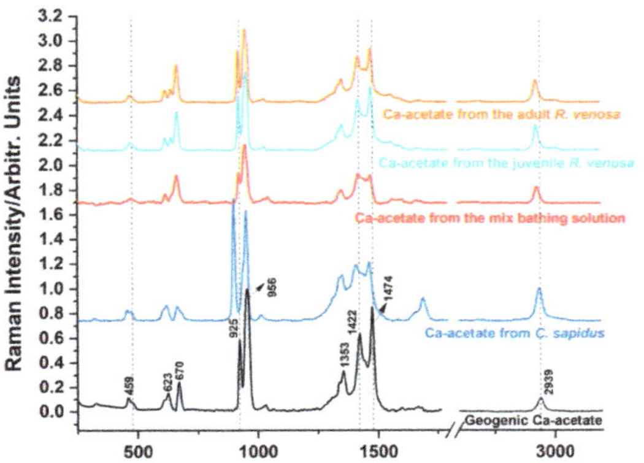

Figure 7. Raman Spectra of the calcium acetate hydrates formed in the demineralization bath solutions of tested pink-orange pigmented (orange line) and blue pigmented *R. venosa* (light-blue line) specimens, mixed crustacean specimens (dark-red line), and *C. sapidus* (navy-blue line), compared to the geogenic calcium acetate (black line).

3.7. Demineralization Process Tracked with a Handheld TacticID Raman System

Using the hand-held Raman instrument, we were able to record the signal of the acetic acid bath solution comprising intact crustacean cuticle fragments exposed to the demineralization process, as well as the minerals in various stages of demineralization. Additionally, three isolated and crystalized calcium acetate by-products could be tracked through the recipient or plastic bag sample. The advantage of the hand-held Raman system relies on the possibility of accurately recording the Raman spectrum of the bathing solution through the glass container. Thus, it is effective for controlling the bathing solution in terms of reagent consumption, the occurrence of the dissolved compounds during reaction, and the status of the intact fragments regarding their calcium carbonate dissolution, as illustrated in Figure 8A. During the process, the TacticID Raman instrument was used for detection of the chitin bands in the treated *C. sapidus* shell fragment after one week of the acetic acid treatment (Figure 8B). The occurrence of the chitin bands in comparison with the reference α-chitin shows linear correlation (R^2 = 0.998 and Pearson's r = 0.998), demonstrating the value of the employment of the system for optimizing the demineralization process (Figure 8C).

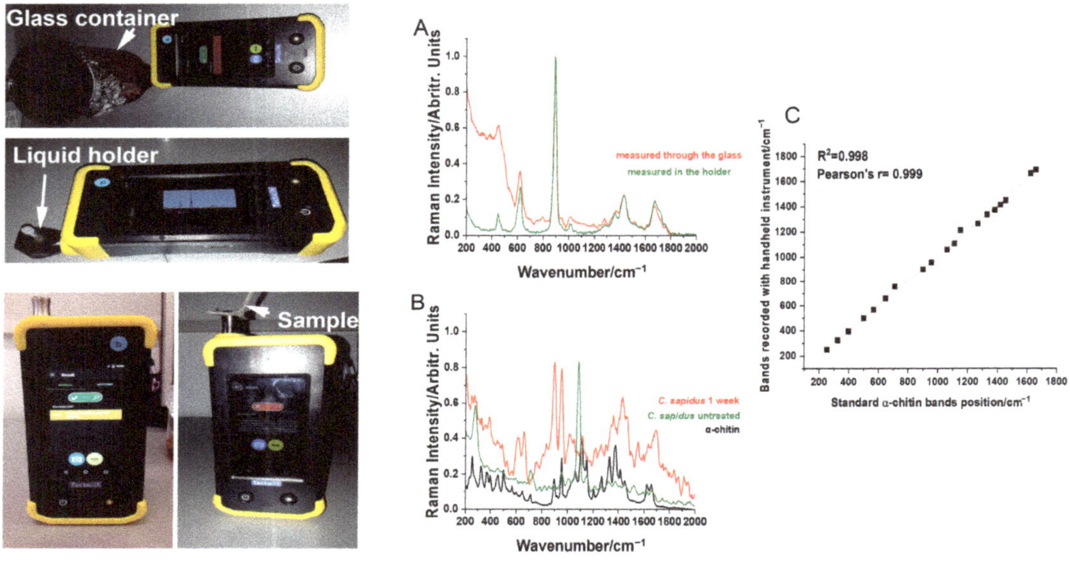

Figure 8. Using a TacticID Raman hand-held instrument to control the demineralization process: images of the instrument at work and the displayed Raman spectra of the acetic acid bath solution (**A**), measured through the demineralization glass container (upper image, red line in the graph) and through its holder accessory for liquids (lower image, green line in the graph). (**B**) Setup of measurement of *C. sapidus* fragments: untreated (down-left image, green line in the graph), showing the strong calcium carbonate signal of raw shells and their disappearance after acetic acid treatment (down-right image, red line in the graph). The chitin reference signal is shown for comparison. Excitation: 1064 nm. (**C**) Linear correlation (R^2 = 0.998, Pearson's r = 0.998) of the chitin bands recorded with TacticID Raman hand-held instrument from the final foil product, with the standard α-chitin bands recorded with a lab-based instrument.

Compared to the lab-based Raman system, where excitation with 785 nm of the demineralized shells in various stages results in a highly fluorescent background, with several steps needed to process multiple data (Supplementary Figure S1), the hand-held system incorporated technology to instantly "see" the processed signal without additional work-loading. The observed Raman bands collected from the resulting intact demineralized shell fragments of the three crustacean species are summarized in Table 1, together with their assignment. In addition to the chitin bands, clearly observed either with the lab-based or with the portable handheld instrument, several additional bands were observed, which might be attributed to proteins or lipids [28]. The by-products (crystalline calcium acetate) were identified through the plastic bag, as samples used in other validation techniques, and confirmed the identity of the product.

Table 1. Summarized Raman bands observed in spectra of the demineralized, chitin-based foil products derived from the three crustacean species, compared to the pure α-chitin [29–32].

Chitin-Based Foil from *C. sapidus*	**Waste Shell of *C. sapidus***	**Waste Fragment of *M. squinado***	**Waste Shell *S. mantis* (Abdomen Cuticle)**	**α-Chitin Raman Bands/cm^{-1}**	**Assignments**
Hand-held TacticID Raman Instrument, 1064 nm	Renishaw InVia Reflex Raman system, 785 nm	Renishaw InVia Reflex Raman system, 785 nm	Renishaw InVia Reflex Raman system, 785 nm		
250	253		254	253	$\delta(C-NH-C)$, $\gamma(OH)$
325	325		325	325	$\gamma(OH)$, $\gamma(\phi)$
				366	
			369	369	
	373			373	
395	395			397	
457	456		457	458	$d(C-C-C)$ *ring*
499	501		501	499	$(C-C)$ *skeletal backbone*
	527			530	
567	565			566	$d(C-C)$, $\delta_{op}(O-H)$
	599			599	$\tau(C-C)$, $\delta(C-O)$, $\delta(C-H)$, $\nu\left(PO_4^{3-}\right)$
658	650			649	$\tau(C-C)$, $\delta(C-O)$, $\delta(C-H)$
755	709			710	$\delta_{op}(C-O)$, $\delta_{op}(C-H)$, $d(N-H)$
899	894	898	894	899	$d(CH_x)$
955	953	955	952	955	$\delta(CH_3)$, $\delta(C-O-H)$
1059	1059			1059	$\nu(C-O)$, $\nu(C-C)$, *ring*
1109	1109		1108	1109	$\nu_s(C-O-C)$, *ring*
	1146		1147	1149	$\nu_{as}(C-O-C)$
1266	1263	1268	1265	1266	Amide III, $\nu(C-H)$ $\delta_{ip}(N-H)$ $\delta_{ip}(C=O)$
1337	1328		1330	1328	Amide III, $\delta_s(CH_3)$

Table 1. *Cont.*

Chitin-Based Foil from *C. sapidus*	Waste Shell of *C. sapidus*	Waste Fragment of *M. squinado*	Waste Shell *S. mantis* (Abdomen Cuticle)	α-Chitin Raman Bands/cm^{-1}	Assignments
1373	1372		1374	1378	$\rho(C-CH_2)$ $\delta(C-CH_3)$
1416	1414		1415	1415	$\omega(CH_2), \nu_s(COO^-)$
1451	1448	1451		1451	$\delta(CH_2)$, $\delta_{as}(CH_3)$
1629	1620		1621	1622	Amide I, $\delta(N-H)$
1663	1657		1658	1657	Amide I, $\nu(C-O)$
Out of the instrument range	2880		2882	2881	$\nu_s(CH_2)$
	2913			2909	$\nu_s(CH_3)$
	2937	2937	2937	2936	$\nu_{as}(CH_2)$
	2958			2963	$\nu_{as}(CH_3)$

ν_s = symmetrical stretching, ν_{as} = asymmetrical stretching, δ = bending, δ_s = symmetrical bending, δ_{as} = asymmetrical bending, δ_{ip} = in-plane bending, δ_{op} = out-of-plane bending, ρ = rocking, τ = twisting, ω = wagging, d = deformation.

3.8. Surface Morphology of the Demineralized Crustacean Foils Determined with SEM

SEM images show surface morphology of the treated cuticle shells at the end of the acetic acid treatment (65 days). The surface morphology of the transparent polymeric foils derived from *C. sapidus* fragment shells shows the long, unbroken fiber texture of chitin (Figure 9). Additionl images and semi-quantitative analyses of the demineralized crustacean foils determined with SEM-EDX are given in the supplementary material (Figure S2).

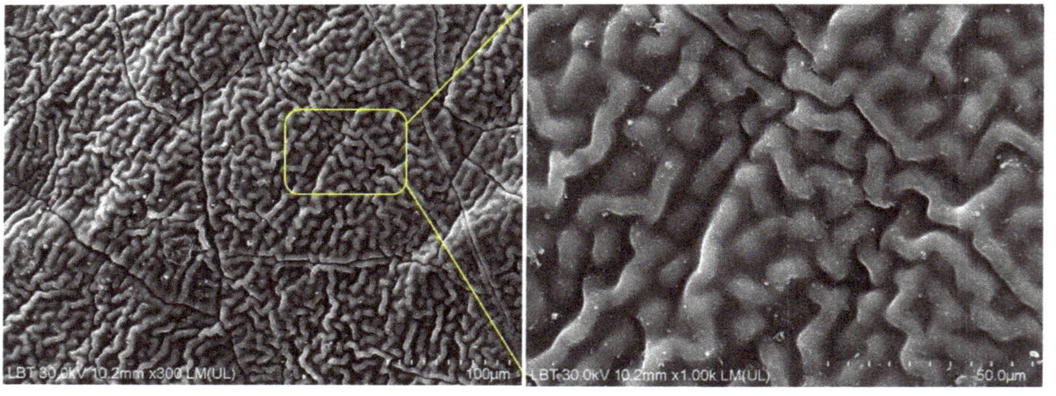

Figure 9. SEM images of the surface morphology of the chitin foils derived from the *C. sapidus* cuticle. Scale bar: 100 µm (**left**); 50 µm (**right**).

4. Conclusions and Outlook

The demineralization process of the waste biogenic carbonates from crustacean and sea snail shells can be optimized and controlled by using Raman techniques and technologies. Optimization could involve monitoring the rate of waste shell demineralization in a time-dependent manner. Such process control is non-destructive and does not compromise the shells. Besides using Raman spectroscopy in laboratory settings, hand-held Raman instruments are promising and more convenient tools for monitoring the demineralization process. Moreover, obtaining chitin-based polymeric materials from the carapace or fragments of the crustacean shells, along with the recovery of calcium acetate as a by-product of interest, which can be crystalized from the treatment bath solution, exemplifies a bioeconomic process. Biogenic calcium acetate was also produced through the environmentally friendly demineralization process of *R. venosa* shells. Overall, the implementation ofRaman techniques and technologies can be a relevant and convenient tool for successful demineralization processes, increasing the reutilization of marine waste shells for obtaining compounds of interest, such as chitin and calcium acetate, with various applications.

The chitin-based polymeric materials are of increasing interest in the light of proposing a future circular carbon and plastic economy [33] centered on several targets, such as reducing and eliminating 50% of all plastic materials and products, replacing all fossil-fuel-based plastics with those sourced from alternative bio-based waste, accelerating carbon recirculation through the use of biomass and CO_2, minimizing the environmental footprint, and maximizing recycling. We expect to use such tools to achieve these goals in finding technical solutions, from waste to new-generation polymeric materials, through sustainable approaches to convert aquatic waste into recyclable polymers.

The conventional demineralization process is conducted using aggressive chemicals, such as hydrochloric acid, and the progressive decomposition of calcium carbonate into water-soluble calcium salts is rarely investigated, unless the final products are taken, filtered, washed and examined. The process is eventually repeated with one or more optimized parameters. After separation by filtration to recover the chitin residue, the residual bath that is left over is eventually discarded, while the solid residue, after extensive washing, is the subject of acidimetric titration to prove its identity (chitin in the case of biogenic carbonates from crustaceans). Titration, a conventional, well-established method, involves chemicals, pH control, and additional workload to complete the description of the demineralization process products. Such processes involve dependencies, and their understanding allows repeated and optimized new processes. These dependencies include the dissolution of respective minerals, which is species-specific; reaction time, temperature, and milling procedures to transform shell fragments in powders; and eventual calibration by milling to control particle size, distribution, acid concentration, and solute-to-solvent ratio, just to name a few. In the approach described here, a significant number of these steps are surpassed.

Regarding the implementation of an in-line tool for demineralization process control, seven major steps and considerations are highlighted:

1. Understanding the process: based on the previous studies within our group [5–9], robust knowledge on the structure, morphology, and distribution of organic and inorganic components of the shell waste as well as the dependencies of the Raman signal on the material status (waste shells derived from fresh or cooked seafood products) is needed to control the process. Thus, the parameters to be controlled are the presence of the carbonate Raman bands on shells under process, the persistence of the carotenoid as organic components in shells or their release in solution, the solution status (acetic acid Raman bands decreasing while acetate bands increasing), and the critical quality attributes of the final product.
2. Characterization of raw waste material to check the initial mineral content, composition, and properties. This baseline data served as a reference for assessing the effectiveness of the demineralization process of the given raw material stock.
3. Selecting the suitability of the Raman monitoring techniques, bearing in mind the classical demineralization approach, reporting only the final product characteristics and ideas to optimize a new approach with several improved parameters. This involves many chemicals and operations, to read out the properties of the final product, without considering the intermediary steps, which can be tailored according to the indicated Raman signal of the intermediary status of product.

4. The efficiency of the in-line monitoring techniques, to provide meaningful information during the demineralization process (for example, the presence of a carbonate Raman signal on the material and the presence of the acetic acid Raman band in solution). Depending on the signal, a decision can be made, such as to remove the material if a carbonate band is absent or to supply more acetic acid if the acetate solution is saturated and the process is incomplete. Classically, this issue may include multiple techniques and workloads, such as conductivity measurement, pH monitoring, specific ion analysis, or other relevant metrics specific to shell demineralization.
   Here, factors such as sensitivity, accuracy, reliability, compatibility with shell material, and ease of integration into the production line could be evaluated by comparing the hand-held Raman instrument performance with the lab-based instrument performance on the same materials.
5. For an upscaled approach, comparable with the industrial environment, integration with the production line can be performed, including monitoring Raman equipment requesting of minimal integration solutions without burdening the existing process equipment and control systems, since the minimal parameters to control are defined (carbonate, chitin, and acetate Raman signal monitoring in optimized acquisition conditions, after understanding dependencies).
6. For calibration and validation of the in-line monitoring equipment, the manufacturer specifications are easy to calibrate, with minimal training of operators (or AI-integrated solutions), and validate, in terms of the performance under actual operating conditions, using representative samples of shell material and lab-based techniques as well as complementary techniques to evaluate the Raman output information.
7. Additionally, decision-making tools during the process allow the implementation of a feedback control loop, using Raman data from the in-line monitoring tool to adjust process parameters in real time, which may include (but are not limited to) the use of sorted raw materials or mixtures, milled or raw grinded materials, temperature, bath solutions, or other factors to optimize demineralization efficiency according to the specific purpose regarding final product composition, quality and morphology, and continuous improvement.

According to the present results, these tailored steps can be effectively implemented as an in-line tool for shell demineralization process control, ensuring consistent quality and performance of the demineralized shell material for various applications or industrial demand.

Supplementary Materials: The following supporting information can be downloaded at: https://www.mdpi.com/article/10.3390/pr12040832/s1, Figure S1: Recorded Raman spectra without any processing, with the spectral acquisition parameters. Parameters are shown in this manner: exposure time, number of acquisitions, laser power, magnification objective for lab-based instrument, laser line/nm; Figure S2: SEM images of the surface morphology of the acetic acid bath treated cuticle of three species with the corresponding EDX graphs; *C. sapidus* (A), *S. mantis* (B), *M. squinado* (C); Video S1: Demonstration of the flexibility of the intact, transparent, demineralized foil products resulted from the three crustacean species after 65 days of acetic acid treatment.

Author Contributions: Conceptualization, S.C.P.; methodology, S.C.P., I.-C.P., K.M. and D.-A.D.; validation, S.C.P., I.-C.P., K.M., T.-L.T., D.-A.D., and B.N.; formal analysis, S.C.P., K.M. and I.-C.P.; investigation, S.C.P., I.-C.P., K.M., D.-A.D., F.N.,T.-L.T., L.B.-T. and B.N.; resources, S.C.P., F.N., D.-A.D. and B.N.; data curation, S.C.P., K.M., I.-C.P., L.B.-T. and B.N.; writing—original draft preparation, I.-C.P. and K.M.; writing—review and editing, S.C.P., I.-C.P., K.M., D.-A.D., T.-L.T. and B.N.; visualization, S.C.P., I.-C.P., K.M., D.-A.D., L.B.-T., T.-L.T., F.N. and B.N.; supervision, S.C.P. All authors have read and agreed to the published version of the manuscript.

Funding: This research received no external funding.

Data Availability Statement: Data supporting the reported results are included in the manuscript and supplementary information.

Conflicts of Interest: Author Neculai Bogdan was employed by the company Metrohm Analytics Romania SRL. The remaining authors declare that the research was conducted in the absence of any commercial or financial relationships that could be construed as a potential conflict of interest. The [company Metrohm Analytics Romania SRL in affiliation] had no role in the design of the study; in the collection, analyses, or interpretation of data; in the writing of the manuscript, or in the decision to publish the results.

References

1. Bioeconomy Strategy—European Comission, Research-and-Innovation.ec.europa.eu. Available online: https://research-and-innovation.ec.europa.eu/research-area/environment/bioeconomy/bioeconomy-strategy_en (accessed on 20 February 2024).
2. Glamuzina, B.; Vilizzi, L.; Piria, M.; Žuljević, A.; Cetinić, A.B.; Pešić, A.; Dragičević, B.; Lipej, L.; Pećarević, M.; Bartulović, V.; et al. Global warming scenarios for the Eastern Adriatic Sea indicate a higher risk of invasiveness of non-native marine organisms relative to current climate conditions. *Mar. Life Sci. Technol.* **2023**, *6*, 143–154. [CrossRef] [PubMed]
3. Marchessaux, G.; Gjoni, V.; Sarà, G. Environmental drivers of size-based population structure, sexual maturity and fecundity: A study of the invasive blue crab Callinectes sapidus (Rathbun, 1896) in the Mediterranean Sea. *PLoS ONE* **2023**, *18*, e0289611. [CrossRef] [PubMed]
4. EC—Commission of the European Communities. Green Paper on the Management of Bio-Waste in the European Union. COM(2008) 811 Final. Available online: https://eur-lex.europa.eu/legal-content/EN/TXT/PDF/?uri=CELEX:52008DC0811&from=EN (accessed on 19 December 2022).
5. Nekvapil, F.; Pinzaru, S.C.; Barbu–Tudoran, L.; Suciu, M.; Glamuzina, B.; Tamaș, T.; Chiș, V. Color-specific porosity in double pigmented natural 3d-nanoarchitectures of blue crab shell. *Sci. Rep.* **2020**, *10*, 3019. [CrossRef] [PubMed]
6. Nekvapil, F.; Aluas, M.; Barbu-Tudoran, L.; Suciu, M.; Bortnic, R.-A.; Glamuzina, B.; Pinzaru, S.C. From Blue Bioeconomy toward Circular Economy through High-Sensitivity Analytical Research on Waste Blue Crab Shells. *ACS Sustain. Chem. Eng.* **2019**, *7*, 16820–16827. [CrossRef]
7. Lazar, G.; Nekvapil, F.; Hirian, R.; Glamuzina, B.; Tamas, T.; Barbu-Tudoran, L.; Pinzaru, S.C. Novel Drug Carrier: 5-Fluorouracil Formulation in Nanoporous Biogenic Mg-calcite from Blue Crab Shells—Proof of Concept. *ACS Omega* **2021**, *6*, 27781–27790. [CrossRef] [PubMed]
8. Lazar, G.; Nekvapil, F.; Glamuzina, B.; Tamaș, T.; Barbu-Tudoran, L.; Suciu, M.; Cinta Pinzaru, S. pH-Dependent Behavior of Novel 5-FU Delivery System in Environmental Conditions Comparable to the Gastro-Intestinal Tract. *Pharmaceutics* **2023**, *15*, 1011. [CrossRef] [PubMed]
9. Nekvapil, F.; Mihet, M.; Lazar, G.; Pinzaru, S.C.; Gavrilović, A.; Ciorîță, A.; Levei, E.; Tamaș, T.; Soran, M.-L. Comparative Analysis of Composition and Porosity of the Biogenic Powder Obtained from Wasted Crustacean Exoskeletonsafter Carotenoids Extraction for the Blue Bioeconomy. *Water* **2023**, *15*, 2591. [CrossRef]
10. Nekvapil, F.; Ganea, I.-V.; Ciorîță, A.; Hirian, R.; Tomšić, S.; Martonos, I.M.; Cintă Pinzaru, S. A New Biofertilizer Formulation with Enriched Nutrients Content from Wasted Algal Biomass Extracts Incorporated in Biogenic Powders. *Sustainability* **2021**, *13*, 8777. [CrossRef]
11. Nekvapil, F.; Ganea, I.-V.; Ciorîță, A.; Hirian, R.; Ogresta, L.; Glamuzina, B.; Roba, C.; Cintă Pinzaru, S. Wasted Biomaterials from Crustaceans as a Compliant Natural Product Regarding Microbiological, Antibacterial Properties and Heavy Metal Content for Reuse in Blue Bioeconomy: A Preliminary Study. *Materials* **2021**, *14*, 4558. [CrossRef]
12. Younes, I.; Rinaudo, M. Chitin and Chitosan Preparation from Marine Sources. Structure, Properties and Applications. *Mar. Drugs* **2015**, *13*, 1133–1174. [CrossRef]
13. Gadgey, K.K.; Bahekar, A. Studies on extraction methods of chitin from crab shell and investigation of its mechanical properties. *IJMET* **2017**, *8*, 220–231.
14. Gortari, M.C.; Hours, R.A. Biotechnological processes for chitin recovery out of crustacean waste: A mini-review. *Electron. J. Biotechnol.* **2013**, *16*, 14. [CrossRef]
15. GRAS Notice (GRN) No. 712. Available online: https://www.fda.gov/Food/IngredientsPackagingLabeling/GRAS/NoticeInventory/default.htm (accessed on 3 March 2024).
16. Iftekhar Shams, M.d.; Nogi, M.; Berglund, L.A.; Yano, H. The transparent crab: Preparation and nanostructural implications for bioinspired optically transparent nanocomposites. *Soft Matter* **2012**, *8*, 1369–1373. [CrossRef]
17. De Oliveira, L.N.; De Oliveira, V.E.; D'ávila, S.; Edwards, H.G.M.; De Oliveira, L.F.C. Raman spectroscopy as a tool for polyunsaturated compound characterization in gastropod and limnic terrestrial shell specimens. *Spectrochim. Acta Part. A Mol. Biomol. Spectrosc.* **2013**, *114*, 541–546. [CrossRef]
18. Wehrmeister, U.; Jacob, D.E.; Soldati, A.L.; Loges, N.; Häger, T.; Hofmeister, W. Amorphous, nanocrystalline and crystalline calcium carbonates in biological materials. *J. Raman Spectrosc.* **2011**, *42*, 926–935. [CrossRef]
19. Hegna, T.A.; Czaja, A.D.; Rogers, D.C. Raman spectroscopic analysis of the composition of the clam-shrimp carapace (Branchiopoda: Laevicaudata, Spinicaudata, Cyclestherida): A dual calcium phosphate-calcium carbonate composition. *J. Crustac. Biol.* **2020**, *40*, 756–760. [CrossRef]
20. Metrohm. Available online: https://www.metrohm.com/ro_ro/products/b/wt-8/bwt-840000920.html (accessed on 21 February 2024).

21. Watling, L.; Thiel, M. (Eds.) . *The Natural History of the Crustacea;* Oxford University Press: Oxford, UK; New York, NY, USA, 2013; pp. 141–148.
22. Dahmane, E.M.; Taourirte, M.; Eladlani, N.; Rhazi, M. Extraction and Characterization of Chitin and Chitosan from *Parapenaeus longirostris* from Moroccan Local Sources. *Int. J. Polym. Anal. Charact.* **2014**, *19*, 342–351. [CrossRef]
23. Vino, A.B.; Ramasamy, P.; Shanmugam, V.; Shanmugam, A. Extraction, characterization and in vitro antioxidative potential of chitosan and sulfated chitosan from Cuttlebone of Sepia aculeata Orbigny, 1848. *Asian Pac. J. Trop. Biomed.* **2012**, *2*, S334–S341. [CrossRef]
24. Kaya, M.; Sargin, I.; Aylanc, V.; Tomruk, M.N.; Gevrek, S.; Karatoprak, I.; Colak, N.; Sak, Y.G.; Bulut, E. Comparison of bovine serum albumin adsorption capacities of α-chitin isolated from an insect and β-chitin from cuttlebone. *J. Ind. Eng. Chem.* **2016**, *38*, 146–156. [CrossRef]
25. Musumeci, A.W.; Frost, R.L.; Waclawik, E.R. A spectroscopic study of the mineral paceite (calcium acetate). *Spectrochim. Acta Part. A Mol. Biomol. Spectrosc.* **2007**, *67*, 649–661. [CrossRef]
26. Miller, J.R.; LaLama, M.J.; Kusnic, R.L.; Wilson, D.E.; Kiraly, P.M.; Dickson, S.W.; Zeller, M. On the nature of calcium magnesium acetate road deicer. *J. Solid. State Chem.* **2019**, *270*, 1–10. [CrossRef]
27. Koleva, V. Vibrational Behavior of Calcium Hydrogen Triacetate Monohydrate, $CaH(CH_3COO)_3H_2O$. *Croat. Chem. Acta* **2005**, *78*, 581–591.
28. Khoushab, F.; Yamabhai, M. Chitin Research Revisited. *Mar. Drugs* **2010**, *8*, 1988–2012. [CrossRef]
29. Socrates, G. *Infrared and Raman Characteristic Group Frequencies: Tables and Charts*, 3rd ed.; Wiley: Chichester, UK; New York, NY, USA; Weinheim, Germany, 2001; pp. 51–52.
30. Zając, A.; Hanuza, J.; Wandas, M.; Dymińska, L. Determination of N-acetylation degree in chitosan using Raman spectroscopy. *Spectrochim. Acta Part. A Mol. Biomol. Spectrosc.* **2015**, *134*, 114–120. [CrossRef]
31. Agbaje, O.B.A.; Brock, G.A.; Zhang, Z.; Duru, K.C.; Liang, Y.; George, S.C.; Holmer, L.E. Biomacromolecules in recent phosphate-shelled brachiopods: Identification and characterization of chitin matrix. *J. Mater. Sci.* **2021**, *56*, 19884–19898. [CrossRef]
32. Nekvapil, F.; Glamuzina, B.; Barbu-Tudoran, L.; Suciu, M.; Tămaş, T.; Pinzaru, S.C. Promoting hidden natural design templates in wasted shells of the mantis shrimp into valuable biogenic composite. *Spectrochim. Acta Part. A Mol. Biomol. Spectrosc.* **2021**, *250*, 119223. [CrossRef]
33. Vidal, F.; Van Der Marel, E.R.; Kerr, R.W.F.; McElroy, C.; Schroeder, N.; Mitchell, C.; Rosetto, G.; Chen, T.T.D.; Bailey, R.M.; Hepburn, C.; et al. Designing a circular carbon and plastics economy for a sustainable future. *Nature* **2024**, *626*, 45–57. [CrossRef]

processes

MDPI

Article

Co-Product of Pracaxi Seeds: Quantification of Epicatechin by HPLC-DAD and Microencapsulation of the Extract by Spray Drying

Raimundo Lopes da Silva [1], Lindalva Maria de Meneses Costa Ferreira [1], José Otávio Carréra Silva-Júnior [2], Attilio Converti [3,*] and Roseane Maria Ribeiro-Costa [1]

[1] Laboratory of Pharmaceutical Nanotechnology, College of Pharmacy, Federal University of Pará, Belém 66075-110, PA, Brazil; raimundo.lopes13@yahoo.com.br (R.L.d.S.); lindalva.costa.ferreira@ics.ufpa.br (L.M.d.M.C.F.); rmrc@ufpa.br (R.M.R.-C.)
[2] Laboratory of Pharmaceutical and Cosmetic R&D, College of Pharmacy, Federal University of Pará, Belém 66075-110, PA, Brazil; carrera@ufpa.br
[3] Department of Civil, Chemical and Environmental Engineering, University of Genoa, Pole of Chemical Engineering, Via Opera Pia 15, 16145 Genova, Italy
* Correspondence: converti@unige.it

Abstract: In the industrial processing of fruits, co-products are generated, which are often not used. The pracaxi co-product, obtained by cold pressing its seeds, contains phenolic compounds with antioxidant activity, which in this work were extracted and microencapsulated by spray drying. The pracaxi extract was characterized by Fourier-transform infrared spectroscopy (FTIR) and high-performance liquid chromatography (HPLC-DAD), and its antioxidant activity was quantified by the ABTS and DPPH assays. Total polyphenol and flavonoid contents in the extract and microparticles were determined by UV-Vis spectrophotometry. Microparticles were then characterized regarding their moisture content, morphology (by scanning electron microscopy), size, polydispersity index and zeta potential. The FTIR spectra revealed functional groups that may be related to phenolic compounds. The extract showed good antioxidant activity according to both selected assays, while the HPLC-DAD analysis evidenced epicatechin as the main compound, whose content was quantified and validated according to the guidelines of recognized national and international agencies. The total polyphenol contents were 20.61 ± 0.20 mg gallic acid equivalent (GAE)/g in the extract and 18.48 ± 0.10 mg GAE/g in the microparticles, while the total flavonoid contents were 28.29 ± 0.70 mg quercetin equivalent (QE)/g and 13.73 ± 0.10 mg QE/g, respectively. Microparticles had a low moisture content, spherical shape, size less than 1 μm and negative zeta potential. Furthermore, they were able to satisfactorily retain phenolic compounds, although in a smaller amount compared to the extract due to small losses resulting from degradation during the drying process. These results, taken as a whole, demonstrate that the pracaxi co-product can be a promising candidate in obtaining products of interest for the cosmetic and food sectors by aiming to exploit its antioxidant activity.

Keywords: Amazonian fruits; flavonoids; validation; antioxidant activity; spray drying

1. Introduction

Pentaclethra macroloba (Willd.) Kuntze (pracaxi) is an oilseed plant from the Amazon region that is widely used in the cosmetic industry due to the light-yellow liquid oil that results from pressing its seeds at room temperature [1,2]. However, the secondary product generated by pressing, called pracaxi co-product, is little explored, which suggests investigating its possible exploitation.

The liquid and solid co-products generated by the industrial processing of natural raw materials, despite containing biodegradable organic matter, accumulate in the environment and cause imbalances in the ecosystem. However, the increase in ecological awareness, which began at the end of the 20th century, has led to the realization that one of the humanity's biggest challenges for the coming decades is to balance the production of goods and services, economic growth, social equality and environmental sustainability [3,4]. Given this, the Brazilian agroindustry increasingly requires the appropriate techniques for treating such co-products, whose applicability, however, is often limited by the lack of knowledge of their nutritional and economic values, for example, in the production of animal feed [5,6].

The vast majority of the co-products from vegetable raw materials are considered to have no economic value, despite their bioactive potential due to their high contents of enzymes, proteins and secondary metabolites (polyphenols, flavonoids, carotenoids, etc.). For this reason, they have become the target of research aimed at developing new products [7,8]. These bioactive compounds, such as phenolic compounds, have in fact antioxidant potential that may be exploited to treat various pathologies [9,10], thus contributing to a better quality of life for the population.

Among the phenolic compounds found in co-products, flavonoids stand out, especially epicatechin, which has high antioxidant activity and the potential to reduce the risks of cardiovascular diseases and tumors as well as the level of low-density lipoprotein (LDL) [11,12].

Antioxidants play an important role in the human body's defense system, as they are able to delay or prevent cell oxidation, which is responsible for cell aging and many degenerative processes in humans [12]. Furthermore, antioxidants have a wide application in the pharmaceutical, cosmetic and nutritional sectors, as they exhibit a wide range of biological properties, such as anti-allergic, anti-inflammatory and antimicrobial activities [13]. Despite these advantages, bioactive compounds are unstable and susceptible to oxidation, polymerization and condensation when exposed to variations in light, oxygen, temperature and pH, among others, which can reduce their biological potential [14–19].

Microencapsulation, which consists of coating small particles or droplets of the extract using one or more encapsulating agents, has therefore become an alternative to minimize these losses. Among these agents, maltodextrin, a carbohydrate formed by the partial hydrolysis of starch thanks to the action of acids and/or enzymes [20–22], has emulsifying properties and low hygroscopicity, viscosity and cost and allows good film formation, in addition to improving the taste and odor of products [20]. In this sense, the spray drying of plant extracts, despite exposure to high temperatures, is used to promote the protection and stability of their active components, in addition to increasing their application potential [23–25] thanks to spray drying's microencapsulation capacity [26].

In this context, exploring the potential of the pracaxi co-product by evaluating its antioxidant activity becomes promising, in addition to contributing to reducing waste, minimizing the environmental impact and adding value to the production chain. Therefore, in this work, the epicatechin-rich pracaxi co-product extract was microencapsulated by spray drying, with the aim of preserving its stability and bioactivity in order to prepare new products with possible applications in the pharmaceutical, cosmetic and food sectors.

2. Materials and Methods

2.1. Chemicals, Reagents and Standards

Sodium carbonate, sodium hydroxide, aluminum chloride and potassium bromide were purchased from Synth (São Paulo, Brazil), while potassium persulfate was from Impex (São Paulo, Brazil). Gallic acid, caffeic acid, 2-hydroxycinnamic acid, *trans*-cinnamic acid, catechin, epicatechin, quercetin, rutin, 2,2′-azino-bis [3-ethylbenzothiazoline-6-sulfonic acid] (ABTS) (≥98%), 2,2-diphenyl-1-picrylhydrazyl (DPPH), (±)-6 hydroxy-2,5,7,8-tetramethylchroman-2-carboxylic acid (Trolox) and Folin–Ciocalteu reagent were acquired from Sigma-Aldrich (São Paulo, Brazil), while maltodextrin with dextrose equivalent (DE) 10 was from Corn Products (São Paulo, Brazil).

2.2. Obtaining Raw Material from Pracaxi Co-product

The pracaxi seed co-product obtained from the Amazon oil company underwent a cooking process at 65 °C for 45 min and was then subjected to cold pressing to extract the oil, which generated the co-product, the raw material used in this study. The co-product was collected shortly after oil extraction and transferred to the Pharmaceutical Nanotechnology Laboratory of the Faculty of Pharmacy (Nanofarm) of the Federal University of Pará (UFPA). The material was dehydrated for seven days in an oven (model SL-102 SOLAB®, Piracicaba, Brazil) with air circulation at a temperature of 40 ± 2 °C until a constant weight was obtained. Next, the sample was crushed in a knife mill to reduce the size of the particles, placed in polyethylene bags and stored under refrigeration (−18 °C) to guarantee material stability.

2.3. Preparation of Pracaxi Co-product Extract

The hydroethanolic extract of the pracaxi co-product was obtained by percolation using 70:30 (*v*/*v*) ethanol:water as the extraction solvent. In the percolator, which contained filter paper with a diameter of 12.5 cm, 1 kg of pracaxi seed co-product was added and 10 L of solvent was dripped intensely (with solvent renewal) until exhaustion, according to the Phytotherapy Form of the *Brazilian Pharmacopoeia* (2nd Edition) with adaptations [27]. After percolation, the extract was concentrated in a rotary evaporator (model 840, Fisatom®, São Paulo, Brazil) at a temperature of 40 °C under low pressure until complete ethanol evaporation. The obtained crude extract was stored in a freezer (−18 °C).

2.4. Determination of Total Polyphenols

To a 10 mL volumetric flask containing 100 µL of the sample, 500 µL of the Folin–Ciocalteu reagent and 6 mL of distilled water were added. After stirring for 1 min, 2 mL of 10% (*w*/*v*) sodium carbonate solution was added and the mixture was stirred for another 30 s before the volume was completed with distilled water. The solution was left to rest for 2 h, and readings were taken on a UV-Vis spectrophotometer (model UV-1800, Shimadzu®, Kyoto, Japan) at a wavelength of 760 nm [28]. The standard curve was constructed using gallic acid solutions at concentrations of 6 to 20 mg/mL. The results were expressed in milligrams of gallic acid equivalent per gram of crude extract (mg GAE/g). The blank consisted of a solution containing all reagents except the sample.

2.5. Determination of Total Flavonoids

To a 10 mL volumetric flask containing 800 µL of the sample, 400 µL of 2.5% (*v*/*v*) aluminum chloride solution was added, and the volume was completed with distilled water. The solutions remained at rest for 30 min in the absence of light. Then, readings were taken on the same spectrophotometer mentioned above at a wavelength of 425 nm. The quercetin standard curve was constructed using concentrations of 5 to 35 mg/mL. The results were expressed as milligram of quercetin equivalent per gram of crude extract (mg QE/g) [29]. The blank consisted of a solution containing all reagents except the sample.

2.6. Identification and Quantification of Epicatechin by High-Performance Liquid Chromatography (HPLC-DAD)

2.6.1. HPLC-DAD Analyses

HPLC-DAD analyses were performed using a high-performance liquid chromatograph (model 1260 Infinity®, Agilent, Santa Clara, CA, USA). The standards used were gallic acid, caffeic acid, 2-hydroxycinnamic acid, *trans*-cinnamic acid, catechin, epicatechin, quercetin and rutin. The standard solutions at a concentration of 0.5 mg/mL and the extract at a concentration of 1 mg/mL were filtered through membranes with a 0.22 µm pore diameter (MF Millipore®, Bedford, MA, USA). After injecting 10 µL of the sample or standard, the analysis was performed on an Xbridge C18 column (250 × 4.6 mm, 5 µm particle size) maintained at 30 °C. The mobile phase, composed of acidified water formic acid pH 3 (solvent A) and methanol (solvent B), was fed at a flow rate of 0.8 mL/min according to the following elution program: 10 min, 5% solvent B and 95% solvent A; 15 min, 25% solvent B and 75% solvent A; 24 min, 35% solvent B and 65% solvent A; 33 min, 80% solvent B and 20% solvent A; return to the initial condition within 36 min. Peaks were recorded in the wavelength range of 190 to 400 nm and detection at wavelengths of 254, 280, 325 and 365 nm. Peaks were confirmed by comparing the retention times and UV spectra with those of reference standards. To locate the peak corresponding to the compounds in the analyzed sample, co-elution was carried out using a 100 μL aliquot of each standard in 1 mL of the sample, and injection was performed under the same conditions as for the extract [30].

2.6.2. Validation of the Epicatechin Quantification Method

The epicatechin quantification method was validated according to the parameters linearity, precision, accuracy, robustness, limits of detection and quantification of ICH Q2 (R1) [31] and Resolution RE 166/17 of the Brazilian National Health Surveillance Agency (ANVISA) [32].

Linearity

To evaluate linearity, a standard solution of 1 mg/mL epicatechin in methanol was used to prepare diluted solutions at different concentrations (31.25, 62.50, 125, 250 and 500 μg/mL). An analytical curve was then constructed and used under the chromatographic conditions described in Section 2.6.1. The results were plotted on a graph of concentration versus area, from which the coefficient of determination (R^2) and the straight-line equation were obtained.

Precision

Precision was assessed at two levels, repeatability (intra-run) and intermediate precision (inter-run). Repeatability was evaluated based on 6 sample determinations at a concentration of 125 μg/mL, injected on the same day by the same analyst under the same chromatographic conditions. On the other hand, intermediate precision, also known as reproducibility, was evaluated by two different analysts, on different days, carrying out 6 analyses of the standard sample at the same concentration (125 μg/mL). It was possible to establish these parameters through the coefficient of variation (CV%) defined according to Equation (1):

$$\text{CV\%} = \frac{\textit{Standard deviation}}{\textit{Mean concentration}} \times 100 \tag{1}$$

Accuracy

Accuracy was assessed by the recovery assay, using three concentrations (31.25, 125 and 500 μg/mL), all in triplicate. The recovery value (R%) was obtained according to Equation (2):

$$\text{R\%} = \frac{\textit{Mean experimental concentration}}{\textit{Theoretical concentration}} \times 100 \tag{2}$$

Robustness

Using the same concentration (125 μg/mL), two variations were carried out in triplicate at an oven temperature of 30 ± 2 °C and a mobile phase flow rate of 0.8 mL/min ± 0.1 mL/min.

Limits of Detection and Quantification

The limits of detection and quantification were calculated based on the standard deviation under the slope of the calibration curve, multiplied by 3 and 10 for detection and quantification, respectively.

2.7. Determination of Total Antioxidant Activity

2.7.1. Antioxidant Activity by the ABTS Radical Cation Assay

The antioxidant activity of the crude extract of the pracaxi co-product was evaluated via the capture of the ABTS radical cation ($ABTS^{+\cdot}$) using the same UV-Vis spectrophotometer mentioned above (model UV-1800, Shimadzu®, Kyoto, Japan). For analysis, a Trolox standard curve was prepared with concentrations varying from 100 to 2000 μM. To obtain the $ABTS^{+\cdot}$ solution, 5 mL of the ABTS stock solution (7 mM) was allowed to react with 88 μL of the potassium persulfate solution (140 Mm) at room temperature (25 ± 0.5 °C) in the absence of light for 16 h. After that, 1 mL of the mixture was diluted in ethanol until reaching an absorbance of 0.7 ± 0.05 at a wavelength of 734 nm. Then, 30 μL of the crude extract of the pracaxi co-product was added to 3.0 mL of the $ABTS^{+\cdot}$ solution and, 6 min after vortex homogenization, readings were carried out at the same wavelength [33]. Only ethanol was used in the blank. The analysis was performed in triplicate, and the results were expressed in μMTrolox/g of sample.

2.7.2. Antioxidant Activity by the DPPH Radical Assay

The antioxidant activity of the crude extract was also assessed by the DPPH radical capture method (DPPH·), as described by Brand-Williams et al. [34], with adaptations from Silva et al. [35]. For this purpose, 75 μL of crude pracaxi extract and 2925 μL of methanolic DPPH· Solution (25 mg/mL) were used. The reaction was performed, protected from light, at room temperature, for an incubation period of 30 min, with the absorbance measured in the same spectrophotometer mentioned above at a wavelength of 515 nm. Methanolic DPPH· solution only was used as the blank. A calibration curve was prepared with Trolox solutions at concentrations of 25 to 1000 μM. The assay was performed in triplicate, and the results were expressed in μMTrolox/g of sample.

2.7.3. Antioxidant Activity by the Ferric Reducing Antioxidant Power (FRAP) Assay

Antioxidant activity measurements via the iron reduction method were performed on the same spectrophotometer mentioned above at a wavelength of 595 nm. The FRAP reagent was prepared by mixing 100 mL of 0.3 M acetate buffer, 10 mL of a 10 mM 2,4,6-tripyridyl-s-triazine (TPTZ) solution and 10 mL of a 20 mM aqueous ferric chloride solution. In a dark environment, an aliquot of 90 μL of the pracaxi co-product crude extract was transferred into 270 μL of distilled water and 2.7 mL of FRAP reagent, then the mixture was homogenized by vortexing and was kept in a water bath at 37 °C for 30 min. Readings were then carried out at 595 nm. FRAP reagent was used as a blank to calibrate the spectrophotometer. Antioxidant activity was calculated based on a Trolox standard curve (160–1600 μM). The final concentration was expressed in μM Trolox/g of sample [36].

2.8. Preparation of Microparticles from the Pracaxi Co-product Extract

We prepared a 5% (*w*/*v*) maltodextrin solution as the wall material by adding 20 g of maltodextrin into 400 mL of distilled water and a 10% (*w*/*v*) solution of crude pracaxi extract by homogenizing 5 g of the extract in 50 mL of distilled water. Then, 20 mL of the extract solution was added to the maltodextrin solution and the resulting mixture was stirred for 30 min and fed to a spray dryer (model LM-MSD 1.0, Labmaq Brasil®, São Paulo, Brazil) operated at a drying air inlet temperature of 160 °C and a feed flow rate of 10.0 mL/min. Finally, the dried material was weighed in a container protected by aluminum foil and stored in a desiccator at room temperature [28].

2.8.1. Drying Yield

The drying yield (*DY*), expressed as a percentage, was calculated according to Equation (3) [37]:

$$DY\ (\%) = \left[\left(\frac{\frac{W-M}{S_c}}{100}\right) \times Q_e\right] + \text{Adj} \times 100 \quad (3)$$

where W is the mass recovered in the spray dryer collector (g), M is the mass of moisture in the dry extract (g), S_c is the solids content of the dry extract (g/100 g of DE), Q_e is the total mass placed for spray drying (g) and Adj is the total mass of maltodextrin used for drying (g).

2.8.2. Extraction of Phenolic Compounds from Microparticles

To 0.4 g of microparticles produced in the spray dryer, 2 mL of a 50:8:42 (*v*/*v*/*v*) ethanol/acetic acid/water solution was added. The suspension was homogenized by vortexing for 1 min, placed in an ultrasonic bath (model CD-4820, Cleaner Kondentech®, São Paulo, Brazil) for 20 min and centrifuged (model 5804 R, Eppendorf®, São Paulo, Brazil) at 7500 rpm for 15 min. The supernatant was filtered through membranes with a 0.45 µm pore diameter [11] and used to determine the polyphenol and flavonoid contents of according to Sections 2.4 and 2.5.

2.8.3. Microencapsulation Yield

The microencapsulation yield was calculated, as a percentage, according to Equation (4) [28]:

$$Y_{TP}\ (\text{or } Y_{TF}) = \left(\frac{\text{TP (or TF)}\ \textit{in the microparticle}}{\text{TP (or TF)}\ \textit{in the crude extract before microencapsulation}} \right) \times 100 \quad (4)$$

where TP and TF are the contents of total polyphenols or total flavonoids and Y_{TP} and Y_{TF} are their respective microencapsulation yields.

2.9. *Characterization of Microparticles*

2.9.1. Content of Total Polyphenol (TP)

A 100 µL aliquot of the solution containing the phenolic compounds extracted from the microparticles was transferred to a 10 mL volumetric flask along with 500 µL of Folin–Ciocalteu reagent and 6 mL of distilled water. After resting for 1 min, 2 mL of 10% (*w*/*v*) sodium carbonate solution was added and the mixture stirred for 30 s, completing the flask volume with distilled water. After resting for 2 h, readings were taken in the same UV-Vis spectrophotometer mentioned above at 760 nm. A calibration curve with gallic acid concentrations ranging from 6 to 20 mg/mL was used to interpolate the results of the total polyphenol content expressed in mg GAE/g [38]. The analysis was carried out on both the microparticles and the crude extract before drying.

2.9.2. Content of Total Flavonoids (TF)

In a 10 mL volumetric flask, 800 µL of the solution containing the phenolic compounds extracted from microparticles and 400 µL of 2.5% (*w*/*v*) aluminum chloride solution were added together; the volume was completed with distilled water. After leaving the system to rest for 30 min in the absence of light, readings were taken using the same UV-Vis spectrophotometer mentioned above at a wavelength of 425 nm. A calibration curve of quercetin at concentrations between 5 and 35 mg/mL was prepared to determine the flavonoid content expressed in mg QE/g [29]. The analysis was also performed on the extract before drying.

2.9.3. Moisture Content

The moisture content was calculated based on the weight loss of particles before and after drying. To this end, the particles were subjected to a temperature of 105 °C in an oven until constant weight [14].

2.9.4. Size, Polydispersity Index and Zeta Potential of Microparticles

Particle size, polydispersity index and zeta potential were determined via dynamic light scattering using a Zetasizer (model 2000, Malvern Instruments®, Malvern, UK). For this purpose, before reading, the microparticles were diluted in ultrapure water in a ratio of 1:100 (v/v) [39].

2.9.5. Particle Morphology by Scanning Electron Microscopy

Particle morphology was examined with a scanning electron microscope (VEGA® 3, Tescan, Brno, Czech Republic). The microparticles were placed on stubs with carbon tape, metallized with gold/palladium to allow the electrical conductivity necessary for image formation (SC7620 sputter coater, Quorum Technologies, Kent, UK) and observed in secondary electron mode with magnitudes of 250×, 1000× and 5000× [14].

3. Results

3.1. Identification and Quantification of Epicatechin

Identification of the epicatechin peak in the HPLC analysis of crude extract was based on the comparison of the retention time and DAD spectrum with the epicatechin standard and confirmed by co-injection. The chromatogram of the crude extract shows an intense peak at a retention time of 21.05 min (280 nm) for epicatechin at a concentration of 69 µg/mL. Figure 1 shows the chromatographic profiles of the extract, the epicatechin standard and the extract plus epicatechin (co-elution), respectively.

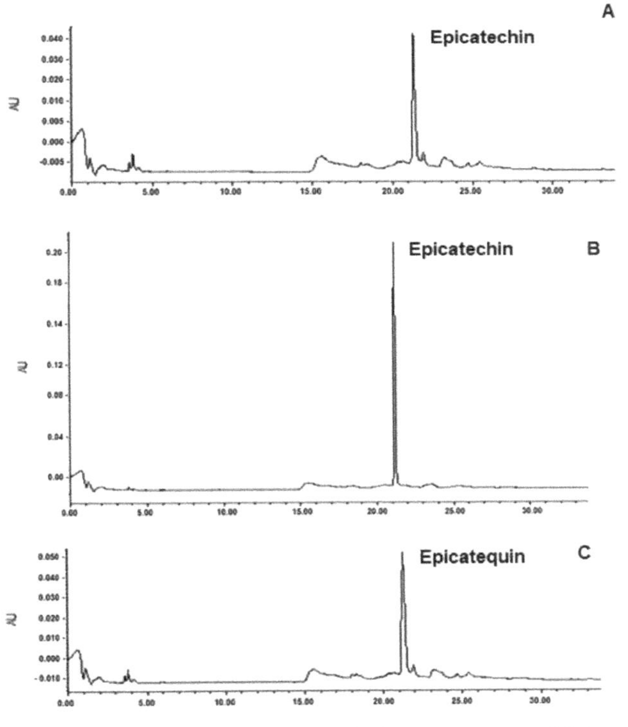

Figure 1. HPLC-DAD chromatograms of (**A**) pracaxi extract and (**B**) epicatechin standard. (**C**) Co-elution of the standard with the extract (280 nm). AU = arbitrary units.

3.2. Validation

The method used to quantify epicatechin was linear, as seen in Figure 2.

The value of the coefficient of variation of repeatability (intra-run) (Table 1) was lower than 5%, i.e., the limit established by ANVISA [32]. Even the coefficient of variation of reproducibility, also called intermediate precision (inter-run) (Table 1), whose measurement was made by two different analysts on distinct days, remained within the limit established by legislation and showed very little difference compared to that of repeatability. Therefore, it can be concluded that the method proposed in this work to quantify epicatechin in the pracaxi extract proved to be accurate.

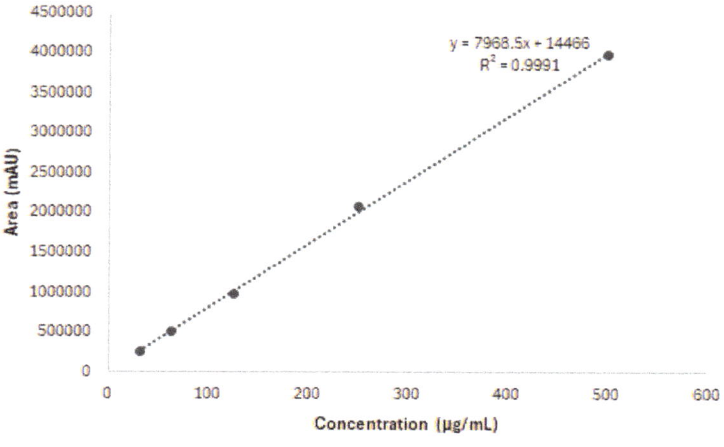

Figure 2. Epicatechin calibration curve used for its quantification in the extract for the HPLC method (31.25 a 500 µg/mL).

Table 1. Precision by repeatability (intra-run) and reproducibility (inter-run) of the epicatechin standard at a concentration of 125 µg/mL.

Precision	Average (µg/mL)	[a] CV%
Repeatability	130.60 ± 3.75	2.87
Reproducibility	130.50 ± 3.60	2.76

[a] CV = Coefficient of variation.

Furthermore, according to the results obtained in this work, the method proved to be accurate, as the recovery value remained in the range of 80 to 120% (Table 2), as recommended by ANVISA [32].

Table 2. Accuracy results of the chromatographic method for low, medium and high concentrations, all performed in triplicate.

Standard (µg/mL)	31.25	125	500
Average (µg/mL)	31.61	125.43	551.20
Recovery (%)	101.17	100.35	110.24

In the robustness assessment (Table 3), it was observed that even when varying the oven temperature (reduction to 28 °C or increase to 32 °C) and the mobile phase flow rate (reduction to 0.7 mL/min or increase to 0.9 mL/min) there was no significant influence on the values of peak area and retention time. Furthermore, the area values were similar to those obtained in the precision and accuracy assessments (concentration of 125 µg/mL), with the retention time (21 min) corresponding to that of the epicatechin standard obtained in the chromatographic analysis. The coefficient of variation for both parameters remained below the 5% limit established by ANVISA [32], which shows that the method was robust in relation to the changes made.

A detection limit of 3.79 µg/mL and a quantification limit of 12.64 µg/mL were obtained by mathematical extrapolation from the calibration curve in accordance with the analytical methods validation guide [32].

Table 3. Robustness results for parameters subjected to variations, i.e., the oven temperature and the mobile phase flow rate.

Parameters	Variation	Area (mAU)	Retention Time (min)	[a] CV %
Oven temperature–Mobile phase flow rate	28 °C–0.7 mL/min	1,008,325 1,030,253 1,013,049	21.68 21.68 21.64	1.15
	32 °C–0.9 mL/min	1,053,434 1,061,086 1,078,367	21.23 21.24 21.33	1.21

[a] CV = Coefficient of variation.

3.3. *Antioxidant Activity*

The antioxidant activities of the pracaxi co-product extract and the standard used (ascorbic acid) by the ABTS, DPPH and FRAP methods are listed in Table 4.

Table 4. Antioxidant activity of the pracaxi co-product crude extract by the ABTS, DPPH and FRAP methods.

Sample	ABTS (μMTrolox/g)	DPPH (μMTrolox/g)	FRAP (μMTrolox/g)
Standard (ascorbic acid)	2380.04 ± 0.0000	1132.68 ± 0.005	2408.57 ± 0.058
Co-product extract	910.82 ± 7.33	906.68 ± 1.20	1549.89 ± 0.020

3.4. *Characterization of Microparticles*

3.4.1. Contents of Total Polyphenols, Total Flavonoids and Microencapsulation Yield

As there are still no data in the literature on the presence of phenolic compounds in the co-product of pracaxi seeds; however, the determination of is of great significance due to the fact that these compounds are considered to be responsible for the majority of biological activities found in plant species. So, the contents of total polyphenols and flavonoids were assessed in both the crude extract and microparticles (Table 5), and the microencapsulation yield was also determined for the microparticles.

Table 5. Total polyphenol and flavonoid contents in the crude extract and microparticles and respective microencapsulation yields.

	Phenolic Compounds		Microencapsulation Yield	
	Total Polyphenols (mg GAE/g)	Total Flavonoids (mg QE/g)	Total Polyphenols (%)	Total Flavonoids (%)
Crude extract	20.61 ± 0.20	28.29 ± 0.70	-	-
Microparticles	18.48 ± 0.10	13.73 ± 0.10	89.66 ± 0.20	48.53 ± 0.30

3.4.2. Drying Yield and Moisture Content

The drying yield was 20.28 ± 1.20%, and the microparticles were found to have a low moisture content (3.46 ± 0.06%).

3.4.3. Size, Polydispersity Index and Zeta Potential of Microparticles

The microparticles were smaller than 1 μm in size, with a polydispersity index (PDI) value revealing polydisperse and heterogeneous characteristics, and had a relatively large negative zeta potential (Table 6).

Table 6. Physical characteristics of microparticles prepared by spray drying.

Particle Size (µm)	Polydispersity Index (PDI)	Zeta Potential (mV)
0.38 ± 0.11	0.43 ± 0.12	−42.80 ± 0.10

Results expressed as means and standard deviations (n = 3).

3.4.4. Scanning Electron Microscopy Examination

The photomicrographs of microparticles illustrated in Figure 3 show their spherical shape and smooth surface, in addition to confirming their heterogeneous size. Moreover, there was little agglomeration among the microparticles and no evidence of the presence of ruptures, which guarantees the protection of the active ingredients inside them.

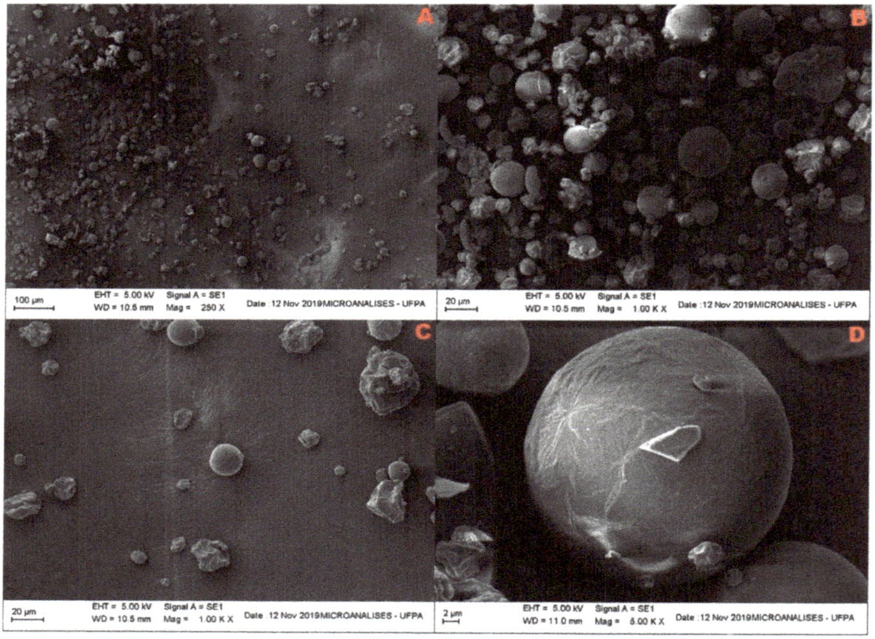

Figure 3. Photomicrographs of pracaxi co-product microparticles obtained by spray drying. Magnitude: (**A**) (250×), (**B,C**) (100×), (**D**) (5000×).

4. Discussion

The physicochemical characterization of agroindustrial co-products is an important step in the identification of their chemical constituents, which can lead to their better exploitation through valorization [40]. The extract chromatogram displays several other peaks that may be related to the presence of phenolic acids and flavonoids. However, since it is a complex mixture, a good separation of these peaks is necessary to obtain spectra that are more characteristic [28].

The presence of epicatechin in the pracaxi co-product extract is promising because of the antioxidant activity displayed by this flavonoid, which is considered capable of reducing free radicals and consequently cell damage [11]. Gabbay Alves et al. [37] identified epicatechin as one of the main phenolic compounds present in cocoa co-product extract. Da Costa et al. [41] also reported the presence of epicatechin in the crude extract of the cupuaçu seed co-product. Epicatechin acts as a vasodilator, being able to dilate blood vessels throughout the body, increase blood flow toward the extremities and brain and reduce blood pressure [42]. Furthermore, evidence demonstrates its ability to modulate immunological and inflammatory functions as well as its use in malaria prophylaxis [43–45].

The method for identifying epicatechin by HPLC-DAD in the pracaxi co-product was validated to ensure that it provides reliable information, which is a fundamental step in determining any metabolite of interest using new analytical methods [46]. In validating the analytical method, the peak areas were directly proportional to the concentration of the epicatechin standard, therefore being considered linear. In fact, coefficient of determination values very close to 1.000, as seen in this work (R^2 = 0.9991), indicate a minimum dispersion for the set of experimental points, corroborating the RE 166/17 resolution of the Brazilian National Surveillance Agency (ANVISA) [32], for which R^2, to be acceptable, must be at least 0.99.

The high levels of epicatechin, detected by the HPLC-DAD method, and of total polyphenols and flavonoids in the pracaxi co-product extract can justify its good antioxidant activity determined by both ABTS and DPPH methods. It is worth mentioning that at least two methods are needed to evaluate antioxidant activity in order to an obtain acceptable reliability for the results [47,48].

Antioxidant activity is linked to the number of hydroxyls present mainly in the structure of the extracted flavonoids, which means that, depending on the type of flavonoid present in the extract, the antioxidant capacity may be greater or lesser in different samples [49]. Compounds with antioxidant activity are mainly concentrated in plant peels and seeds. As the pracaxi co-product was obtained from pressing the seeds, this justifies its good antioxidant capacity [50], which opens up the possibility of its use in the development of new products.

The pracaxi co-product extract showed good antioxidant activity via the use of the ABTS, DPPH and FRAP methods. The values found were close to that for ascorbic acid, which is used as a standard, thus suggesting that the extract, mainly because it is an agroindustrial co-product, could be used as a promising natural antioxidant.

From this perspective, spray drying stands out among the alternatives to microencapsulate bioactive compounds due to its ability to effectively remove water from the product, thus avoiding microbiological attack as well as degradation and oxidation reactions. After spray drying, parameters such as the microencapsulation yield must be evaluated to check the efficiency of the process [11].

The high microencapsulation yield of the total polyphenols found in the microparticles containing the pracaxi co-product indicates that, in general, these phenolic compounds were little degraded in relation to the crude extract. However, the microencapsulation yield of total flavonoids was less than 50%, which suggests that these compounds were particularly sensitive to the high temperature of the spray drying process, despite the short exposure in the drying chamber [23].

The high temperatures in the drying chamber of the spray dryer during microencapsulation may have led to the degradation of a portion of the epicatechin present in the pracaxi co-product extract, thus significantly reducing the total flavonoid content in the microparticles compared to the crude extract. Furthermore, the use of only maltodextrin as a coating agent may have limited the protection of epicatechin, which could be solved by combining it with other wall materials to improve process efficiency [51]. Although the value for the total flavonoid microencapsulation yield suggested that the drying process was responsible for the degradation of about one half of these compounds, it still preserved a significant fraction of the other bioactive polyphenols in the dried sample.

The lower the moisture content of the microparticles, the greater their physical, chemical and microbiological stability [52,53]. The result of this work was an indicator of the quality of the powder, which showed low moisture content, probably due to the high temperature used for spray drying (160 °C).

High spray drying inlet temperatures are known to favor water evaporation due to heat transfer from hot air to water droplets [14,54]. The low moisture content found in the microparticles of the pracaxi co-product is an important finding, as it can avoid the chemical degradation of polyphenols and flavonoids that often occurs when their water content is high [55].

The drying yield is another important parameter in the evaluation of powder quality. Several factors may have contributed to the low value for this parameter observed in this study (20.28 ± 1.20%), such as the loss of powder due to microparticle adhesion onto the wall of the drying chamber, the transport of a portion of fine particles directly to the drying filter, thus bypassing the collector [56], and the presence of sugars and lipids in the extract [57]. Ferreira et al. [28], who microencapsulated tucumã co-product extract by spray drying, observed that the higher the concentration of maltodextrin, the higher the drying yield, probably due to the increase in the content of solids. The relatively low maltodextrin concentration used in this study (5%) may also have been responsible for the low drying yield.

Particle size is one of the most important parameters to be considered in microparticle evaluations, as it influences some important characteristics such as their optical properties and viscosity, among others. It is related to the concentration of the wall material, i.e., the lower the concentration of the wall material, the smaller the particle size [58]; consequently, the larger the surface area, the more improved the bioavailability and solubility of the encapsulated bioactive compounds [57]. From this point of view, therefore, the low concentration of maltodextrin used in this study, despite having impaired the drying yield, may have favored the formation of small microparticles (0.38 ± 0.11 μm). Da Costa et al. [41] found an even smaller size (0.27 μm) for microparticles containing cupuaçu co-product using 5% maltodextrin as an encapsulating agent.

It is known that the lower the value of the polydispersity index (PDI), the greater the homogeneity of microparticles and vice versa [59]. PDI values < 0.2 indicate low polydispersity or monodispersity, while values above 0.4, such as the one observed in this study (0.43 ± 0.12), indicate polydispersity and heterogeneity [60].

As for the zeta potential, its relatively large and negative value (−42.80 ± 0.10 mV) indicates microparticles that are stable in colloidal dispersions, in which repulsive interactions tend to minimize their agglomeration [41,60]. The low tendency for the microparticles to agglomerate was confirmed in scanning electron microscopy photomicrographs.

Morphological examination of the microparticles revealed no cracked walls, which suggests nucleus protection, as air permeability was likely to be minimal [37,53]. The use of high drying air temperatures in spray drying, like the one used in this study, makes the particles more morphologically defined [53]. Da Costa et al. [41] reported photomicrographs where microparticles containing cupuaçu co-product had a spherical and regular structure without agglomeration. On the other hand, Gabbay Alves et al. [37] observed microparticles containing cocoa co-product that had a rounded shape, without roughness and without any rupture, like those observed in this study.

5. Conclusions

Chemical characterization by UV-Vis spectrophotometry revealed high levels of total polyphenols and total flavonoids in the pracaxi co-product extract, and high-performance liquid chromatography revealed the presence of flavonoids. Epicatechin was the main flavonoid identified, which suggests that it may be largely responsible for the good antioxidant activity of the extract. In relation to the microparticles obtained by spray drying, the process led to degradation of a portion of total polyphenols and flavonoids, which impaired the microencapsulation yield. The greatest losses were observed for total flavonoids. The microparticles showed a low moisture content and small size, in addition to good morphological characteristics, as a result of the conditions used in the drying process. The results obtained open up the possibility of their use in different biotechnological processes, as a potential natural source of antioxidants in foods and pharmaceutical and cosmetic products or as a raw material to obtain new functional ingredients for the food industry. This would lead to better exploitation of the pracaxi co-product compared to its current disposal as a waste, which would also avoid environmental problems, in addition to leading to better environmental sustainability of natural products from the Amazon rainforest.

Author Contributions: Conceptualization, R.M.R.-C., A.C. and J.O.C.S.-J.; Methodology, R.L.d.S. and L.M.d.M.C.F.; Investigation, R.L.d.S. and L.M.d.M.C.F.; Writing—original draft, R.L.d.S. and R.M.R.-C.; Writing—review and editing, R.M.R.-C., A.C. and J.O.C.S.-J.; Supervision, R.M.R.-C. and J.O.C.S.-J.; Project administration, R.M.R.-C.; Funding acquisition, R.M.R.-C., A.C. and J.O.C.S.-J. All authors have read and agreed to the published version of the manuscript.

Funding: This research was funded by Coordination for the Improvement of Higher Education Personnel (CAPES, R.L.d.S postgraduate scholarship), finance code 001. National Council for Scientific and Technological Development (CNPq), process: 401816/2022-2. Financing Agency of Studies and Projects (FINEP), process: 0058/124, for the implementation of Nanofarma/UFPA.

Data Availability Statement: All the data are presented in this work.

Acknowledgments: The authors thank the Dean of Research and Graduate Studies (PROPESP) of Federal University of Pará (UFPA) for providing the infrastructure for developing this research.

Conflicts of Interest: The authors declare no conflicts of interest.

References

1. Dantas, A.R.; Marangon, M.C.G.; Feliciano, A.L.P.; Lira-Guedes, A.C. Spatial distribution of a population of *Pentaclethra macroloba* (willd) Kuntze in a floodplain forest of the amazon estuary. *Rev. Árvore* **2017**, *41*, e410406. [CrossRef]
2. Costa, M.N.F.S.; Muniz, M.A.P.; Negrão, C.A.B.; Da Costa, C.E.F.; Lamarão, M.L.N.; Morais, L.; Silva Junior, J.O.C.; Costa, R.M.R. Characterization of *Pentaclethra macroloba* oil. *J. Therm. Anal. Calorim.* **2014**, *115*, 2269–2275. [CrossRef]
3. Sousa, M.S.B.; Vieira, L.M.; Lima, A. Total phenolics and in vitro antioxidant capacity of tropical fruit pulp wastes. *Braz. J. Food Technol.* **2011**, *14*, 202–210. [CrossRef]
4. Costa, A.P.G.C.; Cruz, F.G.G.; Rufino, J.P.F.; Feijó, J.C.; Melo, R.D. Economic viability of tucumã residue flour in broiler chicken feed. *Rev. Agropecuária Técnica* **2017**, *38*, 225–233. [CrossRef]
5. Meneghetti, C.C.; Domingues, J.L. Nutritional characteristics and use of agroindustry by-products in cattle feeding. *Nutr. Rev. Eletrônica* **2008**, *5*, 512–536.
6. Viana, F.J.C.; Araujo, D.C.; Souza, A.R.; Reis, A.L.A.; Camargo, P.H.G.; Santana, T.S.B.; Biagiotti, D. Alternative feeds for semi-arid rabbit breeding. *Rev. Bras. Cunicult.* **2018**, *14*, 1–13.
7. Matias, M.F.O.; Oliveira, E.L.; Gertrudes, E.; Magalhães, M.M.A. Use of fibres obtained from the cashew (*Anacardium ocidentale* L) and guava (*Psidium guayava*) fruits for enrichment of food products. *Braz. Arch. Biol. Technol.* **2005**, *48*, 143–150. [CrossRef]
8. Menezes, C.R.; Durrant, L.R. Xylooligosaccharides: Production, applications and effects on human health. *Cienc. Rural* **2008**, *38*, 587–592. [CrossRef]
9. Hu, Z.; Chen, H.; Yuan, J.S. Removal of Congo red from aqueous solution by cattail root. *J. Hazard. Mater.* **2010**, *173*, 292–297. [CrossRef]
10. Kallel, F.; Driss, D.; Chaari, F.; Belghith, L.; Bouaziz, F.; Ghorbel, R.; Chaabouni, S.E. Garlic (*Allium sativum* L.) husk waste as a potential source of phenolic compounds: Influence of extracting solvents on its antimicrobial and antioxidant properties. *Ind. Crops Prod.* **2014**, *62*, 34–41. [CrossRef]
11. Da Costa, R.S.; Santos, O.V.; Lanne, S.C.S.; Casazza, A.A.; Aliakbarian, B.; Perego, P.; Ribeiro-Costa, R.M.; Converti, A.; Silva Junior, J.O.C. Bioactive compounds and value-added applications of cupuassu (*Theobroma grandiflorum* Schum.) agroindustrial by-product. *Food Sci. Technol.* **2020**, *40*, 401–407. [CrossRef]
12. Moraes, G.V.M.; Jorge, G.M.; Gonzaga, R.V.; Dos Santos, D.A. Antioxidant potential of flavonoids and therapeutic applications. *Res. Soc. Dev.* **2022**, *11*, e238111436225. [CrossRef]
13. Martínez, R.; Torres, P.; Meneses, M.A.; Figueroa, J.G.; Pérez-Alvaréz, J.A.; Viudas-Martos, M. Chemical, technological and in vitro antioxidant properties of mango, guava, pineapple and passion fruit dietary fibre concentrate. *Food Chem.* **2012**, *135*, 1520–1526. [CrossRef]
14. Paini, M.; Aliakbarian, B.; Casazza, A.A.; Lagazzo, A.; Botter, R.; Perego, P. Microencapsulation of phenolic compounds from olive pomace using spray drying: A study of operative parameters. *LWT-Food Sci. Technol.* **2015**, *62*, 177–186. [CrossRef]
15. Liu, Y.; She, X.R.; Huang, J.B.; Liu, M.C.; Zhan, M.E. Ultrasonic-extraction of phenolic compounds from *Phyllanthus urinaria*: Optimization model and antioxidant activity. *Food Sci. Technol.* **2018**, *38*, 286–293. [CrossRef]
16. Manochai, B.; Ingkasupart, P.; Lee, S.H.; Hong, J.H. Evaluation of antioxidant activities, total phenolic content (TPC), and total catechin content (TCC) of 10 sugar apple (*Annona squamosa* L.) cultivar peels grown in Thailand. *Food Sci. Technol.* **2018**, *38*, 294–300. [CrossRef]
17. Rezaei, A.; Fathi, M.; Jafari, S.M. Nanoencapsulation of hydrophobic and low-soluble food bioactive compounds within different nanocarriers. *Food Hydrocoll.* **2019**, *88*, 146–162. [CrossRef]

18. Ordoñez-Quintana, E.; Salmeron, I.; Chavez-Flores, D.; Ramos, V.; Gutierrez, N.; Morales-Oyervides, L.; Delgado, E.; Kwofie, E.; Ngadi, M.; Perez-Vega, S.B. Supercritical and subcritical extraction of ursolic acid and polyphenols from apple pomace: Effect of variables on composition and antioxidant capacity. *J. Food Process. Preserv.* **2020**, *44*, e14296. [CrossRef]
19. Silva, V.; Singh, R.K.; Gomes, N.; Soares, B.G.; Silva, A.A.; Falco, V.; Capita, R.; Alonso-Calleja, C.; Pereira, J.E.; Amaral, J.S.; et al. Comparative insight upon chitosan solution and chitosan nanoparticles application on the phenolic content, antioxidant and antimicrobial activities of individual grape components of Sousão variety. *Antioxidants* **2020**, *9*, 178. [CrossRef]
20. Li, Y.; Tang, B.; Chen, J.; Lai, P. Microencapsulation of plum (*Prunus salicina* lindl.) phenolics by spray drying technology and storage stability. *Food Sci. Technol.* **2018**, *38*, 530–536. [CrossRef]
21. Rezende, Y.R.R.S.; Nogueira, J.P.; Narain, N. Microencapsulation of extracts of bioactive compounds obtained from acerola (*Malpighia emarginata* DC) pulp and residue by spray and freeze drying: Chemical, morphological and chemometric characterization. *Food Chem.* **2018**, *124*, 281–291. [CrossRef]
22. Ferreira, L.M.M.C.; Pereira, R.R.; Carvalho, F.B.; Santos, A.S.; Ribeiro-Costa, R.M.; Junior, J.O.C.S. Green extraction by ultrasound, microencapsulation by spray drying and antioxidant activity of the tucumã coproduct (*Astrocaryum vulgare* Mart.) almonds. *Biomolecules* **2021**, *11*, 545. [CrossRef]
23. Çam, M.; Içyer, N.C.; Erdoğan, F.F. Pomegranate peel phenolics: Microencapsulation, storage stability and potential ingredient for functional food development. *LWT-Food Sci. Technol.* **2014**, *55*, 117–123. [CrossRef]
24. Engel, B.; Baccar, N.M.; Marquardt, L.; Rohlfes, A.L.B. Use of the spray dryer in the food industry: A brief review. *Rev. Jovens Pesqui.* **2017**, *7*, 2–11. [CrossRef]
25. Archaina, D.; Vasile, F.; Jiménez-Guzmán, J.; Alamilla-Beltrán, L.; Schebor, C. Physical and functional properties of roselle (*Hibiscus sabdariffa* L.) extract spray dried with maltodextrin-gum arabic mixtures. *J. Food Process. Preserv.* **2019**, *43*, e14065. [CrossRef]
26. Pereira, K.C.; Ferreira, D.C.M.; Alvarenga, G.F.; Pereira, M.S.S.; Barcelos, M.C.S.; Da Costa, J.M.G. Microencapsulation and release controlled by the diffusion of food ingredients produced by spray drying: A review. *Braz. J. Food Technol.* **2018**, *21*, e2017083. [CrossRef]
27. *Brazilian Pharmacopoeia Phytotherapy Form*, 2nd ed.; Agência Nacional de Vigilância Sanitária: Brasilia, Brazil, 2021; p. 126. Available online: https://www.gov.br/anvisa/pt-br/assuntos/farmacopeia/formulario-fitoterapico/2024-fffb2-1-er-3-atual-final-versao-com-capa-em-word-2-jan-2024.pdf (accessed on 10 May 2024).
28. Ferreira, L.M.M.C.; Pereira, R.R.; Carvalho-Guimarães, F.B.; Remígio, M.S.N.; Barbosa, W.L.R.; Ribeiro-Costa, R.M.; Silva-Junior, J.O.C. Microencapsulation by spray drying and antioxidant activity of phenolic compounds from tucuma coproduct (*Astrocaryum vulgare* Mart.) almonds. *Polymers* **2022**, *14*, 2905. [CrossRef]
29. Funari, C.S.; Ferro, V.O. Propolis analysis. *Food Sci. Technol.* **2006**, *26*, 171–178. [CrossRef]
30. Iacopini, P.; Baldi, M.; Storchi, P.; Sebastiani, L. Catechin, epicatechin, quercetin, rutin and resveratrol in red grape: Content, in vitro antioxidant activity and interactions. *J. Food Compos. Anal.* **2008**, *21*, 589–598. [CrossRef]
31. ICH–Harmonised Tripartite Guidance. Validation of Analytical Procedures: Text and Methodology Q2(R1). In International Conference on Harmonisation of Technical Requirements for Registration of Pharmaceuticals for Human Use. 2005. Available online: https://database.ich.org/sites/default/files/Q2(R1)%20Guideline.pdf (accessed on 10 May 2024).
32. ANVISA (National Health Surveillance Agency). *Resolution RE 166. Validation of Analytical Methods*; ANVISA (National Health Surveillance Agency): Brasília, Brazil, 2017. Available online: https://antigo.anvisa.gov.br/documents/10181/2721567/RDC_166_2017_COMP.pdf/d5fb92b3-6c6b-4130-8670-4e3263763401 (accessed on 10 May 2024).
33. Re, R.; Pellegrini, N.; Proteggenete, A.; Pannala, A.; Yang, M.; Rice-Evans, C. Antioxidant activity applying an improved ABTS radical cation decolorization assay. *Free Radic. Biol. Med.* **1999**, *26*, 1231–1237. [CrossRef]
34. Brand-Williams, W.; Cuvelier, M.E.; Berset, C. Use of a free radical method to evaluate antioxidant activity. *LWT-Food Sci. Technol.* **1995**, *28*, 25–30. [CrossRef]
35. Silva, L.P.; Stringheta, P.C.; Teófilo, R.F.; De Oliveira, I.R.N. Parameter optimization for spray-drying microencapsulation of jaboticaba (*Myrciaria jaboticaba*) peel extracts using simultaneous analysis of responses. *J. Food Eng.* **2013**, *117*, 538–544. [CrossRef]
36. Benzie, F.F.; Strain, J. The ferric reducing ability of plasma (FRAP) as a measure of "antioxidante power": The FRAP assay. *Anal. Biochem.* **1996**, *239*, 70–76. [CrossRef]
37. Gabbay Alves, T.V.; Da Costa, R.S.; Aliakbarian, B.; Casazza, A.A.; Perego, P.; Carréra Silva Júnior, J.O.; Ribeiro Costa, R.M.; Converti, A. Microencapsulation of *Theobroma cacao* L. waste extract: Optimization using response surface methodology. *J. Microencapsul.* **2017**, *34*, 111–120. [CrossRef]
38. Aliakbarian, B.; Casazza, A.A.; Perego, P. Valorization of olive oil solid waste using high pressure-high temperature reactor. *Food Chem.* **2011**, *128*, 704–710. [CrossRef]
39. Butstraen, C.; Salaün, F. Preparation of microcapsules by complex coacervation of gum Arabic and chitosan. *Carbohydr. Polym.* **2014**, *99*, 608–616. [CrossRef]
40. Rockenbach, I.I.; da Silva, G.L.; Rodrigues, E.; Kuskoski, E.M.; Feet, R. Solvent influence on total polyphenol content, anthocyanins, and antioxidant activity of grape (*Vitis vinifera*) bagasse extracts from Tannat and Ancelota-different varieties of *Vitis vinifera* varieties. *Food Sci. Technol.* **2008**, *28*, 238–244. [CrossRef]

41. Da Costa, R.S.; Teixeira, C.B.; Gabbay Alves, T.V.; Ribeiro-Costa, R.M.; Casazza, A.A.; Aliakbarian, B.; Converti, A.; Silva Júnior, J.O.C.; Perego, P. Optimization of spray drying conditions to microencapsulate cupuassu (*Theobroma grandiflorum*) seed by-product extract. *Nat. Prod. Res.* **2019**, *33*, 2600–2608. [CrossRef]
42. Corti, R.; Lammer, A.; Hollenberg, N.K.; Luscher, T.F. Cocoa and cardiovascular Health. *Circulation* **2009**, *119*, 1433–1441. [CrossRef]
43. Addai, F.K. Natural cocoa as diet-mediated antimalarial prophylaxis. *Med. Hypotheses* **2010**, *74*, 825–830. [CrossRef]
44. Ellam, S.; Williamson, G. Cocoa and human health. *Annu. Rev. Nutr.* **2013**, *33*, 105–128. [CrossRef]
45. Panneerselvam, M.; Ali, S.S.; Finley, J.C.; Kellerhals, S.E.; Migita, M.Y.; Head, B.P.; Patel, P.M.; Roth, D.M.; Patel, H.H. Epicatechin regulation of mitochondrial structure and function is opioid receptor dependent. *Mol. Nutr. Food Res.* **2013**, *57*, 1007–1014. [CrossRef]
46. Ribani, M.; Bottoli, C.H.; Collins, I.C.S.; Jadim, I.C.S.F.; Melo, L.F.C. Validation for chromatographic and electrophoretic methods. *New Chem.* **2004**, *27*, 771–780. [CrossRef]
47. Agourram, A.; Ghirardello, D.; Rantsiou, K.; Zeppa, G.; Belviso, S.; Romane, A. Phenolic content, antioxidant potential, and antimicrobial activities of fruit and vegetable by-product extracts. *Int. J. Food Prop.* **2013**, *16*, 1092–1104. [CrossRef]
48. Copra-Janicijevic, A.; Culum, D.; Vidic, D.; Tahirović, A.; Klepo, L.; Basic, N. Chemical composition and antioxidant activity of the endemic *Crataegus microphylla* Koch subsp. *malyana* KI Chr. & Janjić from Bosnia. *Ind. Crops Prod.* **2018**, *113*, 75–79. [CrossRef]
49. Tittona, N.F.; Schumacher, A.B.; Dani, C. Comparative study of the amount of total polyphenols and antioxidant activity in different chocolates: Milk, bittersweet, bittersweet and soy. *Ciência Em Mov.* **2014**, *2*, 77–84. [CrossRef]
50. Abdille, M.H.; Singh, R.P.; Jayaprakasha, G.K.; Jena, B.S. Antioxidant activity of the extracts from *Dillenia indica* fruits. *Food Chem.* **2005**, *90*, 891–896. [CrossRef]
51. Meng, Y.; Cloutier, S. Gelatin and other proteins for microencapsulation. In *Microencapsulation in the Food Industry: A Practical Implementation Guide*; Gaonkar, A.G., Vasisht, N., Khare, A.R., Sobel, R., Eds.; Academic Press: Cambridge, MA, USA, 2014; pp. 227–236.
52. Kang, Y.R.; Lee, Y.K.; Kim, Y.J.; Chang, Y.H. Characterization and storage stability of chlorophylls microencapsulated in different combination of gum Arabic and maltodextrin. *Food Chem.* **2019**, *272*, 337–346. [CrossRef]
53. Cortés-Rojas, D.F.; Souza, C.R.F.; Oliveira, W.P. Optimization of spray drying conditions for production of *Bidens pilosa* L. dried extract. *Chem. Eng. Res. Des.* **2015**, *93*, 366–376. [CrossRef]
54. Sampaio, R.C.A.; Da Costa, R.S.; De Souza, C.R.F.; Junior, A.P.D.; Ribeiro-Costa, R.M.; Da Costa, C.E.F.; De Oliveira, W.P.; Converti, A. Thermal characterization of *Arrabidaea chica* (Humb. & Bonpl.) B. Verl. dry extracts obtained by spray dryer. *J. Therm. Anal. Calorim.* **2016**, *123*, 2469–2475. [CrossRef]
55. Tupuna, D.S.; Paese, K.; Guterres, S.S.; Jablonski, A.A.; Flôres, S.H.; de Oliveira Rios, A. Encapsulation efficiency and thermal stability of norbixin microencapsulated by spray-drying using different combinations of wall materials. *Ind. Crops Prod.* **2018**, *111*, 846–855. [CrossRef]
56. Álvarez-Henao, M.V.; Saavedra, N.; Medina, S.; Cartagena, C.J.; Alzate, L.M.; Londonõ-Londonõ, L. Microencapsulation of lutein by spray-drying: Characterization and stability analyses to promote its use as a functional ingredient. *Food Chem.* **2018**, *256*, 181–187. [CrossRef]
57. Oliveira, O.W.; Petrovick, P.R. Spray drying of plant extracts: Basic remarks and application. *Rev. Bras. Pharmacogn.* **2010**, *20*, 641–650. [CrossRef]
58. Liu, C.H.; Wu, C.T. Optimization of nanostructured lipid carriers for lutein delivery. *Colloids Surf. A Physicochem. Eng. Asp.* **2010**, *353*, 149–156. [CrossRef]
59. Fernández, K.; Aburto, J.; Plessing, C.V.; Rockel, M.; Aspé, E. Factorial design optimization and characterization of poly-lactic acid (PLA) nanoparticle formation for the delivery of grape extracts. *Food Chem.* **2016**, *207*, 75–85. [CrossRef]
60. Santiago-Adame, R.; Medina-Torres, L.; Gallegos-Infante, J.A.; Calderas, F.; González-Laredo, R.F.; Rocha-Guzmán, N.E.; Ochoa-Martínez, L.A.; Bernad-Bernad, M.J. Spray drying-microencapsulation of cinnamon infusions (*Cinna-momum zeylanicum*) with maltodextrin. *LWT-Food Sci. Technol.* **2015**, *64*, 571–577. [CrossRef]

processes

MDPI

Article

Evaluation of Heavy Metal Contamination in Black Soil at Sanjiang Plain: From Source Analysis to Health Risk Assessment

Zijie Gao [1], Jie Jiang [1,*] and Guoxin Sun [2,3,*]

[1] College of Environmental Science and Engineering, Beijing Forestry University, No. 35 Qinghua East Road, Beijing 100083, China; gaaaoooooo@163.com
[2] State Key Lab of Urban and Regional Ecology, Research Center for Eco-Environmental Sciences, Chinese Academy of Sciences, No. 18 Shuangqing Road, Beijing 100083, China
[3] College of Resources and Environment, University of Chinese Academy of Sciences, No. 1 Yanqihu East Road, Beijing 101408, China
* Correspondence: jiangjie@bjfu.edu.cn (J.J.); gxsun@rcees.ac.cn (G.S.)

Abstract: Heavy metals were discharged into the agricultural soil through coal mining, transportation, etc., posing a threat to human health through the food chain. In order to investigate the sources of heavy metals and potential risk to the population, we collected 298 surface soil samples in the black soil area of Sanjiang Plain in Heilongjiang province and tested the concentrations of seven heavy metals. Toxic element contamination in the soil was evaluated by combining the potential ecological risk index and environmental capacity, and pollution sources are identified through positive matrix factorization. The results indicate that the concentrations of Cd and As exceed background values by 1.74 and 1.51 times, respectively, and the ecological risk of Cd is significantly higher than those of other toxic elements. The comprehensive ecological risk level is a moderate comprehensive ecological risk level at 78.5% and a low comprehensive ecological risk level at 21.5%. The sources of heavy metal elements include pesticide spraying (36.5%), input fertilizer and transport activities (20.5%), and mining and metallurgy-related activities (43.1%). When linking the PMF to the Human Health Risk Assessment model, it was found that about 56% of the samples pose a carcinogenic risk to children. Knowledge of soil pollution can certainly help to understand the sources of toxic elements and the health risks to people in the black soil area and provide a scientific basis for the prevention of heavy metal pollution.

Keywords: pollution risk evaluation; potential ecological risk index (RI); environmental capacity; positive matrix factorization model (PMF); health risk assessment

1. Introduction

The rapid development of resource-based cities leads to extensive mining and industrial activities, leaving abandoned mines and resulting in land collapse and water pollution, further leading to the continued deterioration of the soil environment [1]. Heavy metal contamination of crops is widespread; previous studies have shown that in the conventional agricultural areas of southern India, heavy metals pose potential health hazards [2], and in the town of Matehuala in northern San Luis Potosí, Mexico, which is located in a mining and agricultural practice zone, the main contaminants found in the soil are As, Cu, Fe, Mn, Pb, and Zn, with concentrations far exceeding the permissible limits [3]. In the farmlands of Ilokun, Ekiti State, in southwestern Nigeria, Mn, Cu, Pb, and Zn are the main contaminating metals [4]. According to previous studies, it has been found that the types of soil contaminated with HMs in China are concentrated in agricultural soils [5,6] or industrial soils [7], among others. Studies have shown that the ecological risk of heavy metals in the soil of the Sanjiang Plain is mostly low, but the risks of Hg and Cd are prominent [8]. However, the literature in the past had fewer sampling points and failed to provide feedback on the more serious Cd pollution. In 2014, the report on the national general survey of soil contamination found the exceedance rate of total soil point was 16.1%, and the exceeding rates of seven toxic elements for Cd, As, Cu, Pb, Cr, Zn, and Ni were 7.0%, 2.7%, 2.1%, 1.5%, 1.1%, 0.9%, and 4.8%, respectively. Therefore, Pb, Cd, Cr, and As are listed as major pollutants in China [9]. Furthermore, heavy metals have accumulative, biotoxic, and nondegradable properties; they can accumulate in soil and crops, posing a threat to human health through the food chain [10–12]. Prolonged human exposure to As can lead to skin lesions. Pb can impair the respiration and the gastrointestinal tract system [13–15], and Cd exposure leads to kidney disease and bone fragility [13,16]. In addition, an excessive intake of Zn affects the nervous system and disrupts body balance [17,18]. Thus, it is necessary to identify the region of soil heavy metal pollution, investigate the source of metals, and carry out a source-oriented health risk analysis.

The ecological risk factor (ER), combined with the potential ecological risk index (RI), has been used to systematically assess soil metal contamination [19]. In addition, environmental capacity could also be analyzed for soil metal contamination, including the static environmental carrying capacity and dynamic environmental capacity. The environmental capacity of a heavy metal is the maximum load of a heavy metal that the soil can sustain while meeting environmental criteria, sustaining agricultural output, and maintaining biological quality [20,21]. As a result, the dynamic environmental capacity of soil is used to estimate its self-purification ability and predict future changes, while the static environmental carrying capacity calculates particular local pollutant elements. Static environmental carrying capacity was utilized to assess soil pollution levels, while the potential ecological risk index (RI) evaluates the ecological impact of soil heavy metals, accounting for both their toxicity and concentration [22], reducing the influence of human factors, and resulting in more objective and accurate evaluation results.

Heavy metal sources are extensive. In addition to the parent material, the input of pollution from external sources is also severe [23]. Previous studies demonstrated that soil heavy metals mainly come from transportation, industrial activities [24], and mineral resource exploration [25,26]. Recent studies have found that mining and smelting activities are significant to promoting the accumulation of heavy metals in soil [27]. Considering its widespread use in source apportionment, the positive matrix factorization (PMF) model is chosen as the fundamental receptor model [28]. It can quantify sources, optimize the data's standard deviation, and deal with missing and incorrect data [29]. Moreover, a combination of the PMF model and human health risk assessment (HHRA) can quantify the hazards to human health posed by soil heavy metals and the sources of those risks [19]. Chandra [30] calculated cancer risk values based on the PMF–HHRA model and found that agricultural activities had a high risk of cancer in adults and children, while traffic emissions were high risk in women. Dai [12] applied the PMF model to determine human health risks principally from agriculture and found that As is the primary toxic element that causes health risks. Therefore, the heavy metal sources from PMF analysis can be used as an orientation to evaluate health risks and systematically estimate risk sources.

In this study, soil samples were collected in Sanjiang Plain, eastern Heilongjiang Province, China, which is the primary black soil with abundant mineral resources. However, the high organic matter concentration of black soils contributes to loose soil structure [31], while intensive coal mining and the application of chemical fertilizers and pesticides have had a negative impact on soil quality [32,33]. This research developed an integrated framework for investigating polluted locations, distinguishing likely sources, and assessing source-oriented health hazards caused by heavy metals. The major objectives are (1) to identify heavy metal pollution by combining the potential ecological risk index and environmental capacity; (2) to statistically identify underlying sources of soil heavy metals by utilizing the PMF model paired with correlation analysis as well as spatial distribution; and (3) to provide a concentration-oriented and source-oriented perspective to assessing health hazards. The findings in this study can provide theoretical and practical references for identifying sources, assessing health risks, and synthesizing and managing heavy metal contamination in black soil areas.

2. Materials and Methods

2.1. Study Locations and Soil Sampling

Sanjiang Plain is located in the eastern part of Heilongjiang Province, China, including the cities of Shuangyashan, Hegang, Jixi, Jiamusi, and Qitaihe (45°–48°9′ N, 129°40′–132°31′ E). Local soil types include dark-brown earths, meadow soils, Chernozems, black soils, Albic soils, and brown earths [34]. The soil thickness is 120–160 cm, the annual average temperature is 0–5 °C, and the annual mean precipitation is 400–600 mm, which belongs to the cold-temperate and temperate continental monsoon climate with four distinct seasons. Expressways such as G1011, G11, and other national and provincial highways pass through, providing convenient transportation. The predominant crops in the sampling sites are spring wheat, corn, sorghum, and soybeans.

A total of 298 surface layer (0–20 cm) soil samples were collected in August 2023, based on the standard method of HJ/T 166-2004 [35]. Through thorough mixing and quartering, approximately 1 kg of soil was selected as a representative mixed sample for the area. Air-dried soil samples were passed through a 100-mesh soil sieve. GPS was utilized to accurately locate each sampling point (Figure 1).

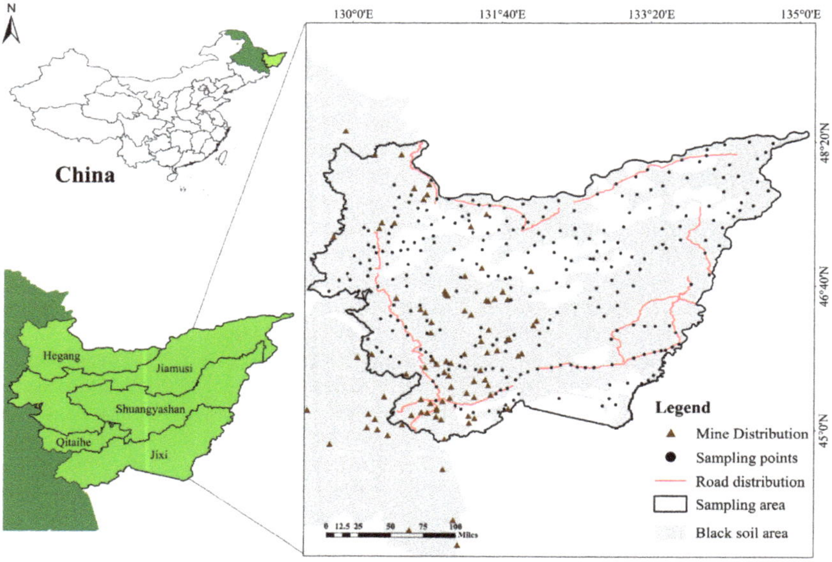

Figure 1. Distribution map of soil sampling in Sanjiang Plain black soil region.

2.2. Sample Analyses

Subsamples of soil were digested based on the technique of "HJ 832-2017", and after that Cr, Ni, Cu, As, Pb, Zn, and Cd were digested by the HNO_3-HCl-$HClO_4$ method. The Zn concentration was determined by an inductively coupled plasma emission spectrometer (ICP-OES, PerkinElmer Optima 8300, Prodigy7, Waltham, MA, USA), and the concentrations of other elements were detected by ICP-MS (NexION 300 Platinum Elmer Instruments Shanghai Co. Waltham, MA, USA). The available phosphorus (AP) was extracted by a sodium fluoride–hydrochloric acid solution and determined by the molybdenum anti colorimetric method (NY/T 1121.7-2014, Manufacturer: Pulse Analyzer; Beijing, China). The available potassium (AK) was extracted with neutral ammonium acetate and determined by a flame photometer (NY/T 889-2004, Manufacturer: PerkinElmer; Waltham, MA, USA). The calculation of organic matter (OM) values used the amount of soil organic carbon (NY/T 1121.6-2006, Manufacturer: Shenzhen Ding Xinyi Experimental Equipment Co., Ltd.; Shenzhen, China). The detection limits for Cr, Ni, Cu, As, Pb, Zn, and Cd are 2 mg/kg, 2 mg/kg, 0.7 mg/kg, 0.2 mg/kg, 1 mg/kg, 5 mg/kg, and 0.03 mg/kg. The respective detection limits for OM, AK, and AP are 0.6 g/kg, 0.5 mg/kg, and 0.1 mg/kg. Duplicate samples, blanks, and standard reference soils (GSS-5a and GSS-6a) were utilized for quality assurance. With relative standard deviations (RSDs) of the duplicates less than 5%, the recoveries of the standards were 85% to 123%. Table S1 shows the national soil nutrient grading standards to evaluate the fertility of agricultural soils.

2.3. Characterization of Soil Heavy Metal Pollution

2.3.1. Potential Ecological Risk Index

The ecological risk factor (E_r) is a parameter used to assess the potential adverse effects of pollutants on ecosystems and their associated organisms [3]. The potential ecological risk was determined based on the local element background values of soil heavy metals [36]. The formula is as follows:

$$RI = \sum_{i}^{n} E_r^i = \sum_{i=1}^{n} \left(T_r^i \times \frac{W_i}{B_i} \right) \quad (1)$$

where RI is the potential ecological risk index; E_r^i is the ecological risk index factor of heavy metal element i; T_r^i i is the toxic response factor of heavy metal i; W_i is the measured value for heavy metal i (mg/kg); and B_i is the local background value of heavy metal i (mg/kg). The toxic response factors for heavy metals Cr, Ni, Cu, As, Pb, Zn, and Cd are 2, 5, 5, 10, 30, 5, and 1, respectively [36]. Evaluation criteria for potential ecological risk indices are contained in Table S2.

2.3.2. Static Environmental Carrying Capacity

The static environmental carrying capacity (QI) is a type of environmental unit that adheres to environmental quality standards at a certain time, ensures agricultural product grain yield and quality, and does not generate environmental contamination that is beyond the maximum load that the soil can withstand [37,38]. It is considered to combine soil heavy metal critical and background concentration values [39], but also the quality of natural soil environments and sustainable use of resources [21]. In addition, QI can specifically identify elements of pollution and determine the degree of pollution.

$$QI{-}i = \frac{Q_i}{Q_{ib}} \quad (2)$$

where $QI - i$ denotes the metal i single environmental carrying index; Q_i and Q_{ib} indicate the metal's existing environmental capacity as well as static environmental capacity. They are determined as follows:

$$Q_i = 10^{-6} \times M \times (C_{is} - C_{ip}) \quad (3)$$

$$Q_{ib} = 10^{-6} \times M \times (C_{is} - C_{ib}) \quad (4)$$

where M denotes the soil weight per hectare of arable land; C_{ip} and C_{ib} are the detected concentrations and background values of metal i; and C_{is} indicates metal i's estimation standard for the selected value of the soil-hazard-control criterion (GB15618-2018, China). The following equations were made in order to assess the environmental carrying capacity of individual sites for different metals [39]:

$$QI = \frac{1}{n} \sum_{i=1}^{n} QI - i \quad (5)$$

In this equation, n denotes the overall amount of the chosen heavy metals. QI values are categorized into 5 categories (Table S3).

2.3.3. Dynamic Environmental Capacity

Dynamic environmental capacity, as a significant warning indicator of soil heavy metal contamination, can be used to assess whether the soil could withstand the maximum load of heavy metal pollution [40]. The formula of calculating the annual dynamic soil environmental capacity is as follows:

$$W_n = M \times \left(C_{in} - C_{ip} \times K^n\right) \times \frac{1-K}{K \times (1-K^n)} \times 10^{-6} \quad (6)$$

where W_n is the annual dynamic environmental capacity of toxic element i in soil (kg/hm^2); M represents the weight of the tillage layer (0–20 cm) of land per hectare, taking an empirical value of 2.25×10^6 kg/hm^2; C_{in} is the critical value of toxic element i in soil after n years (mg/kg); C_{ip} is the measured value of soil toxic element i at the monitoring site (mg/kg); K is the residual rate of toxic metals in soil, which takes the value of 0.9 [37]; and n is the number of control years.

2.4. PMF Model

The data are divided into two matrices of factor assessment (G) and factor profile (F) by the PMF model. PMF software uses data concentration and uncertainty data files to run the model without requiring source configuration [11]. In this study, PMF is implemented through the EPA-PMF 5.0 software [41]. The specific calculation parameters and formulas refer to Shao [42]. The uncertainties were estimated according to Chen et al. [40].

$$X_{ij} = \sum_{k=1}^{P} g_{ik} f_{ki} + e_{ij} \quad (7)$$

where X_{ij} denotes the j heavy metals content in sample i; f_{ki} expresses the j heavy metals content in source k; g_{ik} indicates the source k to sample i contribution; and e_{ij} indicates the residual matrix.

$$Q = \sum_{i=1}^{n} \sum_{j=1}^{m} \left(\frac{e_{ij}}{u_{ij}}\right)^2 \quad (8)$$

According to Equation (8), through PMF modeling, the factor contributions and profiles are derived by minimizing the objective function Q [42]. Here, u_{ij} expresses uncertainty about the j-th heavy metal in the i-th sample [43].

2.5. PMF-Based Human Health Risk Evaluation Models (HHRA)

Based on the health risk evaluation model (HHRA) offered by the U.S. Environmental Protection Agency (USEPA), we assessed the carcinogenic and non-carcinogenic risks in the study area for two age groups, children and adults, through three exposure pathways: direct ingestion (ing), inhalation (inh), and dermal contact (dermal) [44]. Previous studies showed that the direct ingestion of heavy metals poses the greatest health risk to the population, followed by dermal exposure and inhalation. Analyzed from the perspective of heavy metals, chronic non-carcinogenic health concerns include Cr, Ni, Cu, As, Pb, Zn, and Cd, while Cr, Ni, As, and Cd are carcinogenic [45]. With relevant parameters referring to Liu [46], the detailed calculations are as follows:

$$ADD_{ing} = C_S \times \frac{R_{ing} \times EF \times ED}{BW \times AT} \times 10^{-6} \quad (9)$$

$$ADD_{Inh} = C_S \times \frac{R_{inh} \times EF \times ED}{PEF \times BW \times AT} \quad (10)$$

$$ADD_{der} = C_S \times \frac{AF \times SA \times ABS \times EF \times ED}{BW \times AT} \times 10^{-6} \quad (11)$$

$$HI_i = \sum \frac{ADD_i}{RFD_i} \quad (12)$$

$$TCR = \sum CR_i \quad (13)$$

$$CR_i = \sum (ADD_i \times SF) \quad (14)$$

$$THI = \sum HI_i = \sum (HI_{ing} + HI_{inh} + HI_{der}) \quad (15)$$

where C_S is the measured soil heavy metals (mg/kg), and the other parameters are shown in Tables S4 and S5.

Under the three exposure pathways, single-metal non-carcinogenic health risk values (HI) and total non-carcinogenic health risk values (THI) are calculated. If HI or THI > 1, it indicates that heavy metals pose a non-carcinogenic risk to humans; otherwise, it means that soil heavy metals pose no health risk to humans [46]. Carcinogenic risk (CR) is utilized to evaluate the probability of individual exposure to carcinogenic risks [11,47]. The i in CR_i represents different pathways; TCR indicates heavy metals as integrated carcinogenic risk values. If CR or TCR > 1×10^{-4}, it is considered that heavy metals pose a carcinogenic risk to humans; CR or TCR < 10^{-6} suggests no carcinogenic risk [48]. The range of 1×10^{-6} to 1×10^{-4} indicates that the soil heavy metals pose an acceptable carcinogenic risk to the population [49].

This study coupled typical health risk assessment with the PMF model, which can quantitatively distribute the health risk contribution of various heavy metal sources [50]. The specific calculation is to multiply the health risk value of heavy metals by the contribution of the identified source, obtaining the identified source-induced health risk [51].

2.6. Data Processing Analysis

IBM SPSS Statistics 26.0, EPA PMF 5.0, and Excel 2021 applications were used to complete statistical analysis; Arc GIS 10.8.1 and Origin 2022 applications were used to draw charts.

3. Results and Discussion

3.1. Heavy Metal Concentration and Fertility Characteristics of Soil

Descriptive statistics of soil samples as well as soil background values (BGVs) [52] of Sanjiang Plain in Heilongjiang Province are shown in Table 1. The average concentrations of Cr, Ni, Cu, As, Cd, Pb, and Zn were 60.6 mg kg^{-1}, 27.7 mg kg^{-1}, 19.5 mg kg^{-1}, 11.0 mg kg^{-1}, 0.2 mg kg^{-1}, 14.5 mg kg^{-1}, and 81.2 mg kg^{-1}, respectively. The levels of Cr, Ni, As, Cd, and Zn were 1.03, 1.17, 1.51, 1.74, and 1.15 times higher than the BGV [52], respectively. In contrast, Cu and Pb elements had lower average concentrations than the BGV. The exceeding rates of Cr, Ni, As, Cd, and Zn were 52.68%, 73.83%, 91.28%, 53.36%, and 62.42%, respectively. Only 1% of Cr and Ni exceeded the risk screening values (RSVs), but 7.38% of Cd and 1.34% of Zn showed as higher than the RSVs, respectively. This indicated that As and Cd were the major heavy metal contaminants, which is supported by previous research [53]. The coefficient of variation (CV) reflects the data dispersion between sample points [54]. The CV values of Cd were strongly differentiated (CV > 0.5), while other elements were moderately differentiated ($0.2 < CV \leq 0.5$). Anomalously high values further suggested that Cd may be affected by external factors of interference [55].

Figure S1 illustrates the black soil zone nutrient profile. Some studies had shown that AP and AK can be directly utilized by plants and are considered soil fertility and plant growth effective indicators [56]; OM can help to improve soil physical properties [57,58]. According to the data in Table S1, it can be concluded that the soil was plentiful in AP (Figure S1a), AK (Figure S1b), and OM (Figure S1c).

Table 1. Statistical description of soil heavy metals in Heilongjiang Province.

Metal	Unit	Max	Min	Average	SD	CV	BGV [52]	Average /BGV
Cr	mg/kg	255.79	0.87	60.62	23.22	0.32	58.6	1.03
Ni	mg/kg	69.39	2.58	27.72	7.36	0.27	23.65	1.17
Cu	mg/kg	38.18	6.24	19.52	4.80	0.25	20	0.98
As	mg/kg	20.89	0.60	11.01	3.00	0.27	7.3	1.51
Pb	mg/kg	31.34	5.47	14.48	4.77	0.33	24.2	0.60
Zn	mg/kg	270.85	11.47	81.15	34.59	0.43	70.7	1.15
Cd	mg/kg	0.62	0.03	0.15	0.14	0.94	0.086	1.74
OM	g/kg	187.48	6.73	43.30	16.87	0.39	-	-
AP	mg/kg	135.68	7.89	32.72	20.81	0.64	-	-
AK	mg/kg	1088.3	41.21	222.16	137.75	0.62	-	-

3.2. Identification of Soil Heavy Metal Contamination Areas

3.2.1. Potential Ecological Risk Index Result

The ecological risk index (E_r^i) for seven heavy metals in the soil is shown in Figure 2. In the study area, the average E_r^i is as follows: Cd (53.71) > As (15.03) > Ni (5.86) > Cu (4.88) > Pb (2.99) > Cr (2.06) > Zn (1.15). Compared to other toxic elements, the Cd data are very prominent, indicating Cd is highly ecologically hazardous. In Shuangyashan City (Figure 2a), 64.15% of the locations are at considerable risk and 35.85% are low risk. Jixi (Figure 2c) shows that the high risk, considerable risk, moderate risk, and low risk percentages of the total study area are 7.14%, 32.14%, 8.93%, and 51.79%, respectively, with low risk and considerable risk predominantly. Qitaihe (Figure 2e) has a moderate risk percentage of total study area of 47.83%, with a considerable risk percentage of 13.04%. Jiamusi (Figure 2d) has a moderate risk proportion of 8.26%, a considerable risk proportion of 3.67%, and only one point is at the high risk; the pollution levels are lower. Element Cd in the majority of points in Hegang (Figure 2b) is low risk. Some studies [59–61] on heavy metal pollution in coal mining areas showed that soil Cd exhibits severe ecological risks, which are highly consistent with the results of this study.

From the perspective of the integrated ecological risk index (RI) (Figure 3), the minimum RI of heavy metals in the soil around the study area is 32.66, the maximum is 243.71, and the average is 85.68. The moderate risk level is predominant, accounting for 78.5%. Shuangyashan, Jixi, and Qitaihe had high contributions of moderate risk, accounting for 56.6%, 37.5%, and 52.2%, respectively. Based on the average values, the comprehensive ecological risk levels classified by RI are considerable, implying that ecological risk prevention and control efforts should be increased in the research area.

3.2.2. Environmental Capacity Evaluation

Table 2 shows the soil environmental quality evaluation results of heavy metals in the region. Elements Cu and Pb were below the pollution quality level; Cr, Ni, As, Cd, and Zn were found to have some polluted points. Polluted areas with Cd (Figure S2) were localized in the east and south-central regions of the research area, accounting for 42.95%, and the moderate-heavy pollution areas were located in the center of the region, accounting for 22.82%. Therefore, Cd pollution in Sanjiang Plain is serious and needs attention.

Because static environmental carrying capacity does not consider soil yield and self-purification [37], soil environmental capacity studies need further analysis. Predicting the future capacity of the soil environment is essential for avoiding soil quality decrease. In this investigation, 10 years, 20 years, 30 years, and 50 years were chosen as control years to calculate the dynamic environmental capacity of soil heavy metals in 2034, 2044, 2054, and 2074 (Table 3). The results showed that the data in 2034, 2044, 2054, and 2074 will decrease year after year, all below the existing environmental capacity. The environmental capacity of Cr, Ni, Cu, As, Zn, Pb, and Cd in the study area is significantly decreased to 65.7, 19.1, 15.4, 7.23, 26.9, 53.1, and 0.11, respectively, up to 2074. In the next 10 years, predictions indicated that the largest decrease is Pb with 81.15%, and the smallest decrease is Zn (75.96%). In the next 20–50 years, Pb is the largest decrease among heavy metals. This finding suggests that if no action is taken to improve the soil environment in the research area, the soil capacity for heavy metal pollutants will steadily decline, and the soil self-purification capacity for pollutants will become weaker [46].

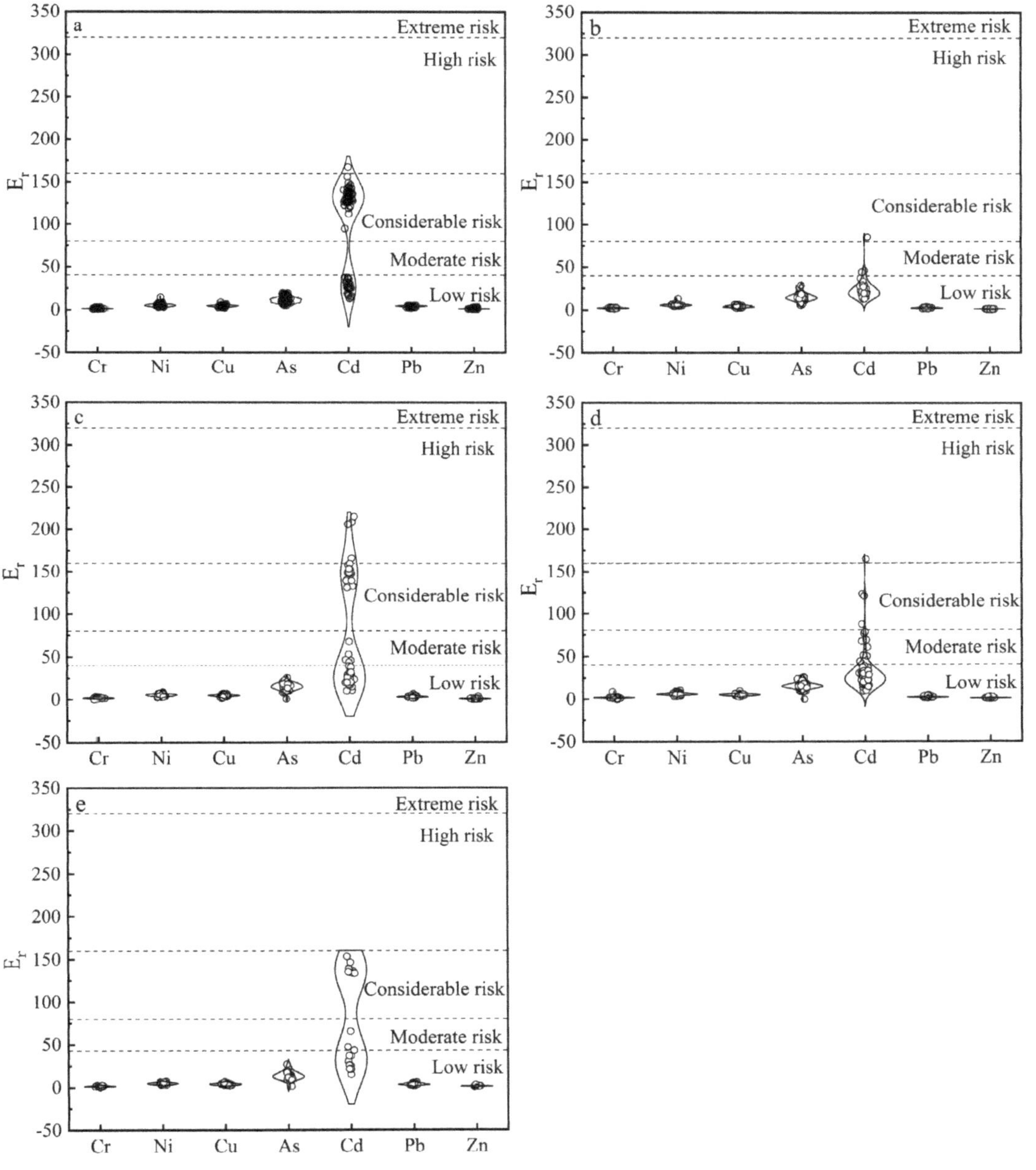

Figure 2. E_r^i of soil heavy metals: (**a**) Shuangyashan, (**b**) Hegang, (**c**) Jixi, (**d**) Jiamusi, and (**e**) Qitaihe.

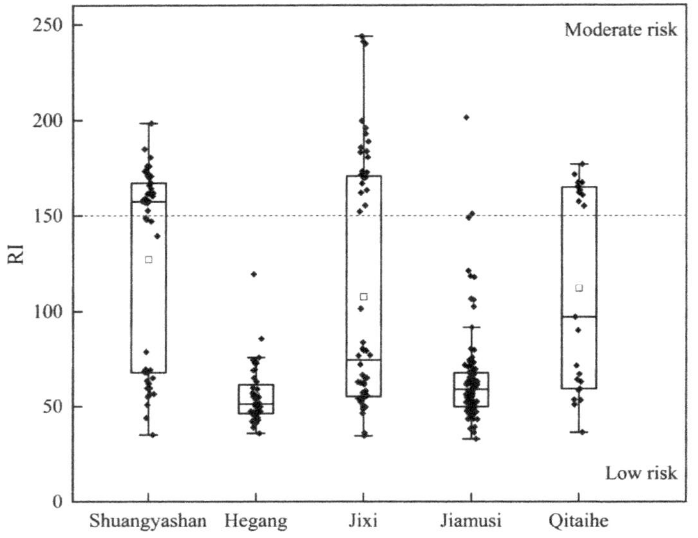

Figure 3. Box plot of RI of soil heavy metals.

Table 2. Evaluation table of heavy metal soil environmental quality.

Area	Cr	Ni	Cu	As	Pb	Zn	Cd
Shuangyashan	1.04	0.92	1.01	0.89	1.06	0.69	0.40
Hegang	0.96	0.90	1.02	0.83	1.16	1.04	1.05
Jixi	0.99	0.95	1.01	0.82	1.10	0.99	0.59
Jiamusi	0.98	0.87	0.99	0.81	1.16	0.93	1.00
Qitaihe	1.05	1.02	1.03	0.88	1.10	0.98	0.51

Table 3. Prediction table for soil heavy metal dynamic environmental capacity (kg/hm^2).

Metal	Present	10 Years	Amplitude [1]	20 Years	Amplitude [2]	30 Years	Amplitude [3]	50 Years	Amplitude [4]
Cr	452.047	92.443	−79.55%	72.410	−21.67%	67.660	−6.56%	65.687	−2.92%
Ni	108.559	25.529	−76.48%	20.720	−18.84%	19.579	−5.51%	19.105	−2.42%
Cu	93.138	20.933	−77.52%	16.784	−19.82%	15.800	−5.86%	15.391	−2.59%
As	40.287	9.594	−76.19%	7.824	−18.44%	7.405	−5.36%	7.231	−2.35%
Pb	207.968	39.202	−81.15%	30.013	−23.44%	27.833	−7.26%	26.928	−3.25%
Zn	292.747	70.382	−75.96%	57.439	−18.39%	54.369	−5.34%	53.094	−2.34%
Cd	0.626	0.146	−76.75%	0.118	−18.88%	0.112	−5.52%	0.109	−2.43%

Amplitude [1]: the magnitude of change compared to the present; Amplitude [2]: change compared to 10 years; Amplitude [3]: change compared to 20 years; and Amplitude [4]: change compared to 30 years.

All variables, including soil type, natural environment, and biological variables, have an impact on environmental capacity. Soil type was the most important factor because different soil types have different environmental capacities [34]. Nevertheless, the different sources of heavy metals lead to inputs of different range and intensity into the soil, which in turn affects the extent of changes in dynamic environmental capacity. In addition, the mobility and residual properties of heavy metals influence their dynamic environmental capacity [55,56]. The migration capacities and bioavailabilities of Ni, Zn, As, and Cd are relatively low with a mean decrease of 76.4% [55,57]. In contrast, the mobility of Pb has been shown to be slightly higher than that of Zn [58], with a higher migration capacity, so that the environmental capacity has a greater influence.

3.3. Spatial Distribution of Soil Heavy Metals

In agricultural soils, the heavy metal contamination distribution shows a pattern of "clear geographic variations as well as complicated development" [62]. Using inverse distance weight interpolation (IDW) to map heavy metal spatial distribution in the study area (Figure 4) can make the pollution distribution clearer. On the research area's center Cu, Ni, and Zn levels were high. The distribution of Ni was highly fragmented and patchy (Figure 4b), and Cu high-numerical areas demonstrated a "T" shape (Figure 4c). High concentrations of Zn were densely distributed in Shuangyashan, which further confirms static environmental carrying capacity calculations (Figure 4f). In contrast, As (Figure 4d) and Cr (Figure 4a) were focused on the north-west and south-east, possibly related to the same geologic influences. The spatial distribution of Cd (Figure 4g) and Pb (Figure 4e) demonstrated strong consistency, with high content in the southern central research area. The survey found that the high spots belong to the four major coal cities in Heilongjiang. Furthermore, the principal causes of heavy metal pollution in soil are coal mining, industrial production, and transportation emissions [56]. The study area's transportation is convenient, automobile emissions from the usage of leaded gasoline [56], as well as tire wear, contribute to Cd and Pb enrichment [63,64]. According to the heavy metal distribution level features, a group of potentially sourced elements can be briefly identified [65].

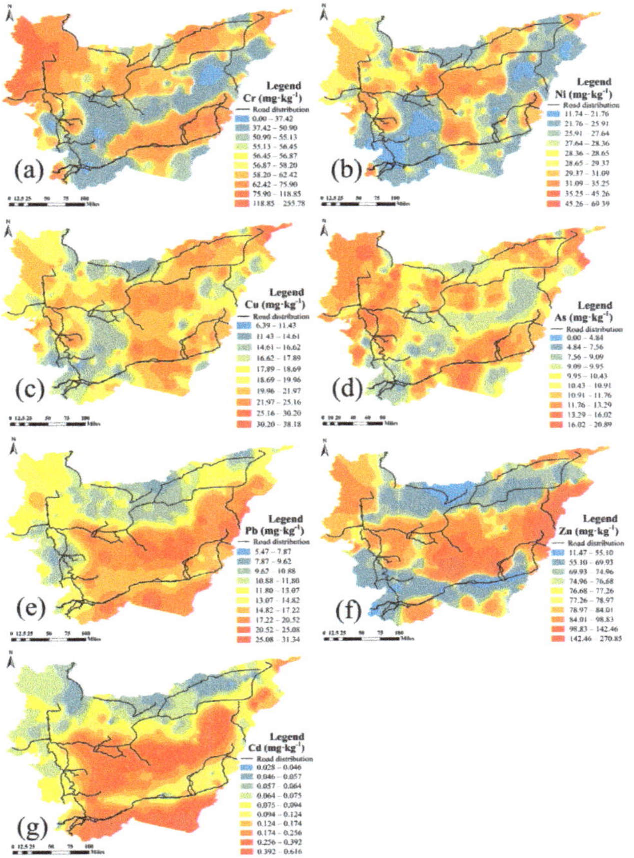

Figure 4. Spatial distribution characteristics of seven toxic elements in soil. (Figures (**a**–**g**) show the heavy metal distribution of Cr, Ni, Cu, As, Pb, Zn and Cd in the study area, respectively).

3.4. Source Analysis of Soil Heavy Metals

3.4.1. Pearson Correlation Analysis of Soil Heavy Metals

Pearson correlation analysis was utilized to identify the connection among heavy metals (Figure 5), and then the PMF model was used to further distinguish and quantify the sources in the study region. The results indicate that the correlation coefficient between Ni and Cu was greater than 0.7, and Cr–Ni, Cr–Cu, Cr–As, and Cd–Pb also had high correlation coefficients ($p > 0.5$). In metals, strong and vital association suggests that they could have the same origin and geochemical features [65,66]. There is a strong and significant correlation between Cd and Pb, but not a remarkable interrelation with other elements, which implied that Cd is not homologous with the other elements [67] and may be derived from different sources [68]. Hence, the PMF model needs to be utilized to explore sources.

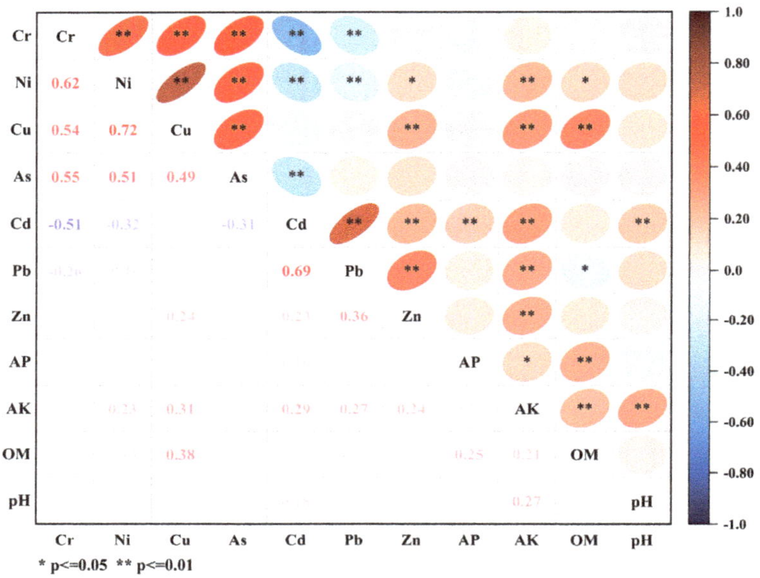

Figure 5. Correlation analysis of AP, AK, OM, pH, and seven toxic elements in soil.

3.4.2. IDW-Based PMF Analysis

After excluding anomalous data, the soil in the research area was modeled using software EPA PMF 5.0, and all of the elemental fits (R^2) were >0.7. The elemental fit (R^2) values for the metal elements are as follows for Cr, Ni, Cu, As, Cd, Pb, and Zn: 0.81, 0.78, 0.7, 0.73, 0.98, 0.82, and 0.96 (Figure S3). This demonstrated that the findings derived by the model were reasonable and dependable. The results show (Figure 6) that Factor 1 explained 36.5% of the weights (Figure 6b). This is mainly influenced by As, Cr, Ni, and Cu, accounting for 57.97%, 57.27%, 49.52%, and 47.95% of the total contribution, respectively (Figure 6a). Factor 2 contributed 20.4% (Figure 6b) and was mainly driven by Cd and Pb with contributions of 82.8% and 40.3% (Figure 6a). Factor 3 contribution, with 43.1% (Figure 6b), including Zn (62.4%) and Ni (34.6%), was mainly influenced by Zn (Figure 6a). Compared to the outcomes of the QI, the results of the PMF are consistent with the actual situation. The parameters are detailed in Table S6.

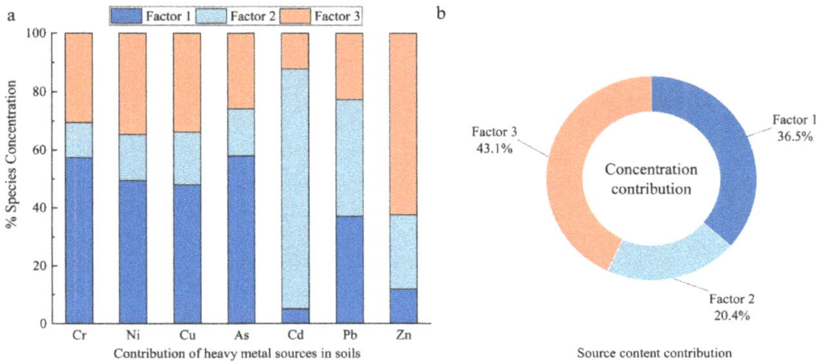

Figure 6. Heavy metal content and factor contribution (**a**), Contribution of heavy metal sources (**b**).

Factor 1 consisted mostly of As, Cr, Ni, and Cu, which were moderately differentiated, suggesting that they were less influenced by anthropogenic interference; these elements might be soil natural ingredients. The elements of As, Cr, Ni, and Cu showed similar distribution trends (Figure 4) and a strong correlation ($p \leq 0.01$) (Figure 5), indicating that they could have a similar source. Previous research has indicated that Cr and Ni in soil are frequently associated with the parent material [69]. It has been reported that As [68,70,71] and Cu [24,72,73] are the key ingredients in pesticides, and they generally enter the soil with the spraying of pesticides. Mainly due to low temperatures, the enzyme activity in biodegradation decreases, and the low volatility of pesticides leads to slow dissipation in the northern regions [74]. The lower soil dissipation rate, combined with high application rates, results in severe pollution in northern China, especially in Heilongjiang, Jilin, and Liaoning [75]. As a consequence, F1 was assumed to originate primarily from pesticide spraying.

Factor 2 was primarily composed of Cd and Pb, which were strongly correlated. Element Cd was concentrated in the central part of the region, which is a production site for commercial grains. Long-term fertilization [56,76] and sewage irrigation [11] have been demonstrated to cause Cd accumulation. Animal manure or manure-based fertilizers are considered one of the main sources of heavy metal contamination in agricultural soils [77]. Minerals added to livestock feed may also contain trace amounts of As and Cd, which subsequently migrate to peri-urban agricultural soils through the application of livestock manure [78]. Synthetic phosphate fertilizer is the primary source of Cd in north-eastern agriculture [79–81]. Notably, a similar distribution is observed for Pb, around transportation roads and mining areas. Lead is usually considered as a transport representative element. For example, vehicle exhaust [60,82] and tire wear produce Pb-containing particulate matter [61]. As a result, it was assumed that F2 predominantly originated from fertilizer input and transport.

Factor 3 was predominantly characterized by Zn and Pb, with the main component being Zn. Zinc is concentrated in the central region, where mineral resources are abundant; mining activities are significant sources of Zn accumulation in the soil [83]. The results indicated that the mining industries accounted for 16.1% of the total industrial output and play a vital role in the development of Heilongjiang Province [84]. Moreover, the Harbin-Tongjiang Expressway runs through the study area, providing convenient transportation. Previous research has indicated that Zn has become a typical component of transportation emissions [24,85]. During transport, tire wear and automotive exhaust fumes can cause the Zn accumulation in surface soils [48,86,87]. Therefore, it can be concluded that Zn enrichment is mostly due to coal mining and processing activities. Factor 3 was closely related to mining and metallurgy-related activities.

3.5. Human Health Hazard Assessment

As shown in Tables 4 and S7, the three routes of exposure, HI and CR, in descending order, were direct ingestion > dermal contact > inhalation, which is analogous to Shi's findings [88]. This shows that direct ingestion is the primary factor resulting in the health risk of the population.

Table 4. Evaluation table of carcinogenic and non-carcinogenic health risks of soil heavy metals.

		Non-Carcinogenic Risks				Carcinogenic Risks			
		HI_{ing}	HI_{inh}	HI_{dermal}	HI	CR_{ing}	CR_{inh}	CR_{dermal}	CR
Cr	Adults	3.43×10^{-2}	3.83×10^{-4}	6.09×10^{-3}	4.07×10^{-2}	1.76×10^{-5}	1.58×10^{-7}	2.50×10^{-6}	2.03×10^{-5}
	Children	2.45×10^{-1}	7.08×10^{-4}	7.58×10^{-11}	2.45×10^{-1}	3.15×10^{-5}	7.29×10^{-8}	3.08×10^{-6}	3.46×10^{-5}
Ni	Adults	2.38×10^{-3}	5.65×10^{-5}	3.14×10^{-5}	2.47×10^{-3}	2.78×10^{-5}	1.46×10^{-9}	2.47×10^{-6}	3.03×10^{-5}
	Children	1.70×10^{-2}	1.04×10^{-4}	3.91×10^{-13}	1.71×10^{-2}	4.96×10^{-5}	$6.76 \times 10-10$	3.04×10^{-6}	5.27×10^{-5}
Cu	Adults	7.87×10^{-4}	8.39×10^{-8}	9.32×10^{-6}	7.97×10^{-4}	—	—	—	—
	Children	5.62×10^{-3}	1.55×10^{-7}	1.16×10^{-13}	5.62×10^{-3}	—	—	—	—
As	Adults	6.40×10^{-2}	—	5.55×10^{-4}	6.46×10^{-2}	9.87×10^{-6}	1.06×10^{-8}	8.56×10^{-8}	9.97×10^{-6}
	Children	4.57×10^{-1}	—	6.91×10^{-12}	4.57×10^{-1}	1.76×10^{-5}	4.90×10^{-9}	1.05×10^{-7}	1.77×10^{-5}
Cd	Adults	2.16×10^{-4}	8.04×10^{-7}	3.07×10^{-5}	2.47×10^{-4}	—	4.97×10^{-11}	—	4.97×10^{-11}
	Children	1.54×10^{-3}	1.49×10^{-6}	3.82×10^{-13}	1.54×10^{-3}	—	2.29×10^{-11}	—	2.29×10^{-11}
Pb	Adults	7.77×10^{-3}	7.06×10^{-7}	1.58×10^{-4}	7.93×10^{-3}	6.80×10^{-8}	3.58×10^{-11}	—	6.80×10^{-8}
	Children	5.55×10^{-2}	1.31×10^{-6}	1.97×10^{-12}	5.55×10^{-2}	1.21×10^{-7}	1.65×10^{-11}	—	1.21×10^{-7}
Zn	Adults	4.66×10^{-4}	4.97×10^{-8}	8.27×10^{-6}	4.74×10^{-4}	—	—	—	—
	Children	3.33×10^{-3}	9.18×10^{-8}	1.03×10^{-13}	3.33×10^{-3}	—	—	—	—

3.5.1. Concentration-Oriented Health Risk

The adult median HI data, in descending order, are as follows: As (3.23×10^{-2}) > Cr (1.36×10^{-2}) > Pb (2.64×10^{-3}) > Zn (1.58×10^{-4}) > Cu (2.66×10^{-4}) > Ni (8.24×10^{-4}) > Cd (8.24×10^{-5}). The children's average HI values are shown below, and are as follows: As (2.29×10^{-1}) > Cr (8.18×10^{-2}) > Pb (1.85×10^{-2}) > Ni (5.71×10^{-3}) > Cu (1.87×10^{-3}) > Zn (1.11×10^{-3}) > Cd (5.14×10^{-4}). Hence, elements As, Cr, and Pb pose a high health risk to the public, and they are inevitably prioritized as non-carcinogenic risks [89].

Between adults and children, the effects of several heavy metals on CR were as follows: Pb > Ni > Cr > Cd > As. More importantly, compared with other elements, Pb makes the highest contribution to CR. Both children and adults displayed average CR values in the range of 1×10^{-11} to 1×10^{-5}, demonstrating a range that is within reasonable bounds. Notably, both direct ingestion and inhalation pose a more non-carcinogenic hazard to children than to adults; in different routes of cancer risk, direct ingestion and dermal exposure are higher in children than adults. As a result, children face higher health risks from heavy metals.

Individual heavy metal risks are insufficient to represent the integral human health hazards of heavy metals [90,91]. The population health risk contribution rates from different pollution sources are shown in Figure 7. Specifically, THIs in the study region were lower than one, suggesting that the risk was within an acceptable level for adults. The THI values for heavy metals were above one in children, mainly focused on the south-central and north-western sections of the research zone, suggesting that heavy metals constitute a non-carcinogenic risk.

In adults, the TCR data were higher than 1×10^{-5} in most areas; in contrast, only three data points were higher than 1×10^{-4}. As a result, the overall carcinogenic risk of heavy metals to people in the research region may be considered minimal. However, for children, the TCR data were higher than 1×10^{-4} in most areas, indicating that heavy metals pose a carcinogenic risk to children. The high value distribution is concentrated in the north-western and south-central parts of the area, specifically in Hulin City and Luobei County (Figure 7). Moreover, the trends in the distribution of non-carcinogenic and carcinogenic risks are similar in both adults and children, which is similar to the results of Jiang's study [50]. Collectively, in the research area, heavy metals might constitute a considerable threat to children's health.

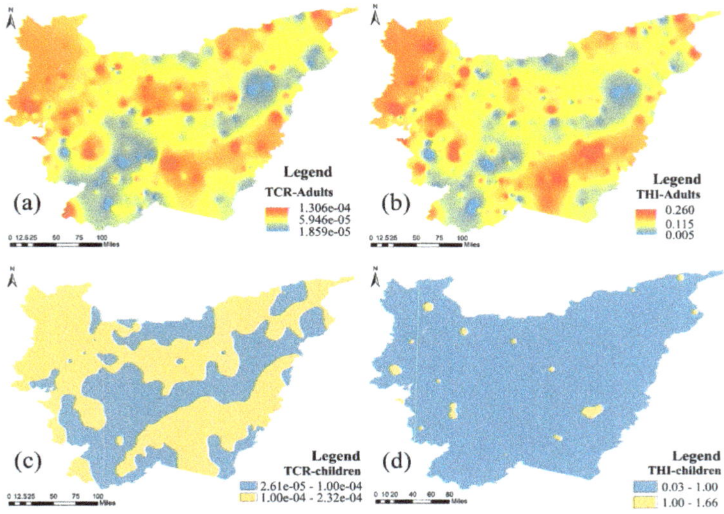

Figure 7. Spatial distribution of THI and TCR indexes for adults and children.(Figures (**a–d**) show adult carcinogenic risk, adult non-carcinogenic risk, children's carcinogenic risk, and children's non-carcinogenic risk, respectively).

In particular, compared with adults, children might face a more severe risk. The reason children face a greater risk than adults is attributed to their greater ingestion rate and lower body weight. As a result, children are exposed to a higher dose than adults [92]. Additional precautionary measures are offered to decrease children's exposure to soil.

3.5.2. Source-Oriented Health Risks

Under the source allocation results, the health hazards posed by various sources of heavy metal exposure were studied, and the THI and TCR were calculated utilizing the different source contributions. Figure S4 expresses the connection between heavy metals, pollution sources, and human health hazards.

THI-assessed sources showed that adults were dominated by Factor 1 (pesticide spraying) with 55.7%, followed by mining and metallurgy-related activities (27.6%), fertilizer input, and transport (16.8%). For children, pesticide spraying was the primary source, accounting for 55.6%. TCR sources suggested that pesticide spraying is a major risk source for both adults and children, explaining 53.51% and 53.50%, respectively. At the same time, in children (31.77%) and adults (31.79%), mining and metallurgy-related activities also play a role that cannot be ignored. Factor 1 is the primary factor contributing to health risk, which is consistent with another report [85]; mining and metallurgy-related activities are secondary. Figure S4 shows that diverse pollution sources contribute similarly to health hazards for adults and children, which is consistent with the results of Huang et al. [93].

Overall, in the study area, pesticide spraying should be considered as a priority source of risk. It can be concluded that human health may not be adequately safeguarded if pollution sources are regulated just by emissions. In order to prevent and reduce health risks, actions should be taken to control agricultural production methods in the study area, such as decreasing the use of fertilizers and herbicides [50].

4. Conclusions

This study investigates the ecological risk from toxic metals and the environmental capacity, and systematically assesses source-oriented and concentration-oriented health risks. The average concentrations of Cr, Ni, As, Cd, and Zn are significantly higher than the local background levels. The results of the environmental impact assessment show that the overall level of integrated ecological risk in the study area is moderate, with Cd making the largest contribution. Soil in the study area is heavily contaminated by toxic elements, with Cd being the most contaminating element. Sources include pesticide spraying, fertilizer application and transport, and mining and metallurgy-related activities, with the largest contribution coming from mining and metallurgy-related activities. Although heavy metals pose a low carcinogenic and non-carcinogenic risk to adults, the risk of carcinogenic and non-carcinogenic substances to children is not negligible, and Pb from natural source rock is highly carcinogenic. In summary, the exceedance rate of As is 91.28%, and it has non-carcinogenic effects on the population. As intake from rice accounts for 60% of total arsenic intake [94], and the high As content in rice poses a serious threat to the health of people who rely on it as a staple food. Cd mainly comes from fertilizer application, with an exceedance rate of 53.36%. The issue of excessive heavy metals in the soil of Chinese vegetable fields has become quite serious, with 70% of Cd in the human body coming from vegetables in food [95]. This is enough to warrant our attention. Overall, the results revealed significant ecological and human health risks. In the future, the reasonable use of organic fertilizer, timely shutting down of abandoned mining areas, and strengthening of protective measures in coal mines to prevent coal dust from continuously entering the soil environment should be implemented. It is recommended that local residents enhance their own protective measures.

Supplementary Materials: The following supporting information can be downloaded at: https://www.mdpi.com/article/10.3390/pr12122829/s1, Figure S1: The proportion of AP, AK, OM in the study area; Figure S2: Static environmental carrying capacity of Cd in the study area; Figure S3: Information on the curve of each element at the sampling sites in the study area; Figure S4: Comparison of population health risks from various pollution sources; Table S1: Classification standard table of soil nutrient content; Table S2: Potential ecological risk index (E_r) standard; Table S3: Classification of QI value of static environment carrying capacity; Table S4: Relevant input parameters for characterizing the daily exposure dose of heavy metals under various exposure routes; Table S5: Reference dose of heavy metals (RfD; mg/kg/day) and slope factor (SF; per mg/kg/day); Table S6: PMF model run parameters for 7 heavy metal concentration data; Table S7: Carcinogenic and non-carcinogenic health risks regarding metals. References [96–100] are cited in the Supplementary Materials.

Author Contributions: Z.G.: Writing—review and editing, Writing—original draft, Visualization, Methodology, Formal analysis, Data curation, Conceptualization. J.J.: Writing—review and editing, Supervision, Funding acquisition, Conceptualization. G.S.: Writing—review and editing, Supervision, Funding acquisition, Conceptualization. All authors have read and agreed to the published version of the manuscript.

Funding: This work was financially supported by the Strategic Pilot Science and Technology Special Project of the Chinese Academy of Sciences (XDA28020101) and the National Key Research and Development Programme of China (2021YFC1809205).

Data Availability Statement: The original contributions presented in the study are included in the article/supplementary material, further inquiries can be directed to the corresponding authors.

Conflicts of Interest: The authors declare that they have no known competing financial interests or personal relationships that could have appeared to influence the work reported in this paper.

References

1. Sow, A.Y.; Ismail, A.; Zulkifli, S.Z. Geofractionation of Heavy Metals and Application of Indices for Pollution Prediction in Paddy Field Soil of Tumpat, Malaysia. *Environ. Sci. Pollut. Res.* **2013**, *20*, 8964–8973. [CrossRef] [PubMed]
2. Vasudhevan, P.; Pu, S.; Ayyamperumal, R.; Manikandan, E.; Sujitha, S.B.; Singh, S.; Ponniah, J.M.; Dixit, S.; Thangavel, P. Pollution Assessment, Ecological Risk and Source Identification of Heavy Metals in Paddy Soils and Rice Grains from Salem, South India. *J. Hazard Mater. Adv.* **2024**, *17*, 100526. [CrossRef]
3. Saha, A.; Sen Gupta, B.; Patidar, S.; Hernández-Martínez, J.L.; Martín-Romero, F.; Meza-Figueroa, D.; Martínez-Villegas, N. A Comprehensive Study of Source Apportionment, Spatial Distribution, and Health Risks Assessment of Heavy Metal(Loid)s in the Surface Soils of a Semi-Arid Mining Region in Matehuala, Mexico. *Environ. Res.* **2024**, *260*, 119619. [CrossRef] [PubMed]
4. Oguntuase, M.A.; Oyebamiji, A.O.; Ayeni, O.M.; Adeyanju, O.T. A Comprehensive Environmental Analysis of Heavy Metal Contamination, Ecological Risks and Health Impacts in Ilokun Area, Ado-Ekiti, Nigeria. *Environ. Pollut. Manag.* **2024**, *1*, 118–127. [CrossRef]
5. Liu, H.; Zhang, Y.; Yang, J.; Wang, H.; Li, Y.; Shi, Y.; Li, D.; Holm, P.E.; Ou, Q.; Hu, W. Quantitative Source Apportionment, Risk Assessment and Distribution of Heavy Metals in Agricultural Soils from Southern Shandong Peninsula of China. *Sci. Total Environ.* **2021**, *767*, 144879. [CrossRef]
6. Zheng, S.; Wang, Q.; Yuan, Y.; Sun, W. Human Health Risk Assessment of Heavy Metals in Soil and Food Crops in the Pearl River Delta Urban Agglomeration of China. *Food Chem.* **2020**, *316*, 126213. [CrossRef]
7. Sun, L.; Guo, D.; Liu, K.; Meng, H.; Zheng, Y.; Yuan, F.; Zhu, G. Levels, Sources, and Spatial Distribution of Heavy Metals in Soils from a Typical Coal Industrial City of Tangshan, China. *CATENA* **2019**, *175*, 101–109. [CrossRef]
8. Zhang, L.; Xu, H.; Yang, H.; Du, S. Source Identification and Health Risk Assessment of Heavy Metals in the Soil from a Typical Agricultural Area in the Sanjiang Plain, China. *Chiang Mai J. Sci.* **2020**, *47*, 973–983.
9. Zhang, K.; Yang, J.; Ji, Y.; Xia, Y. Spatiotemporal Simulation and Predication of Heavy Metal(Loid) Concentrations in Coal Chemical Industrial Areas with a Soil Environmental Capacity Model. *Int. J. Coal Sci. Technol.* **2018**, *5*, 508–518. [CrossRef]
10. Fu, L.; Zhang, Z.; Zhang, Q.; Zhang, H. Spatial Distribution, Risk Assessment, and Source Identification of Pollutants around Gold Tailings Ponds: A Case Study in Pinggu District, Beijing, China. *Environ. Monit. Assess.* **2021**, *193*, 483. [CrossRef]
11. Su, C.; Meng, J.; Zhou, Y.; Bi, R.; Chen, Z.; Diao, J.; Huang, Z.; Kan, Z.; Wang, T. Heavy Metals in Soils from Intense Industrial Areas in South China: Spatial Distribution, Source Apportionment, and Risk Assessment. *Front. Environ. Sci.* **2022**, *10*, 820536. [CrossRef]
12. Dai, X.; Liang, J.; Shi, H.; Yan, T.; He, Z.; Li, L.; Hu, H. Health Risk Assessment of Heavy Metals Based on Source Analysis and Monte Carlo in the Downstream Basin of the Zishui. *Environ. Res.* **2024**, *245*, 117975. [CrossRef] [PubMed]
13. Li, D.; Lu, Q.; Cai, L.; Chen, L.; Wang, H. Characteristics of Soil Heavy Metal Pollution and Health Risk Assessment in Urban Parks at a Megacity of Central China. *Toxics* **2023**, *11*, 257. [CrossRef] [PubMed]
14. Mohammadi, A.A.; Zarei, A.; Esmaeilzadeh, M.; Taghavi, M.; Yousefi, M.; Yousefi, Z.; Sedighi, F.; Javan, S. Assessment of Heavy Metal Pollution and Human Health Risks Assessment in Soils Around an Industrial Zone in Neyshabur, Iran. *Biol. Trace Elem. Res.* **2020**, *195*, 343–352. [CrossRef]
15. Zhou, H.; Guo, X. Soil Heavy Metal Pollution Evaluation around Mine Area with Traditional and Ecological Assessment Methods. *J. Geosci. Environ. Prot.* **2015**, *03*, 28–33. [CrossRef]
16. Mishra, S.; Bharagava, R.N.; More, N.; Yadav, A.; Zainith, S.; Mani, S.; Chowdhary, P. Heavy metal contamination: An alarming threat to environment and human health. In *Environmental Biotechnology: For Sustainable Future*; Sobti, R.C., Arora, N.K., Kothari, R., Eds.; Springer: Singapore, 2019; pp. 103–125, ISBN 978-981-10-7283-3.
17. Saleem, M.; Sens, D.A.; Somji, S.; Pierce, D.; Wang, Y.; Leopold, A.; Haque, M.E.; Garrett, S.H. Contamination Assessment and Potential Human Health Risks of Heavy Metals in Urban Soils from Grand Forks, North Dakota, USA. *Toxics* **2023**, *11*, 132. [CrossRef]
18. Han, W.; Gao, G.; Geng, J.; Li, Y.; Wang, Y. Ecological and Health Risks Assessment and Spatial Distribution of Residual Heavy Metals in the Soil of an E-Waste Circular Economy Park in Tianjin, China. *Chemosphere* **2018**, *197*, 325–335. [CrossRef]
19. Pan, Y.; Han, W.; Shi, H.; Liu, X.; Xu, S.; Li, J.; Peng, H.; Zhao, X.; Gu, T.; Huang, C.; et al. Incorporating Environmental Capacity Considerations to Prioritize Control Factors for the Management of Heavy Metals in Soil. *J. Environ. Manag.* **2024**, *351*, 119820. [CrossRef]
20. Tian, K.; Li, M.; Hu, W.; Fan, Y.; Huang, B.; Zhao, Y. Environmental Capacity of Heavy Metals in Intensive Agricultural Soils: Insights from Geochemical Baselines and Source Apportionment. *Sci. Total Environ.* **2022**, *819*, 153078. [CrossRef]
21. Pan, Y.; Ding, L.; Xie, S.; Zeng, M.; Zhang, J.; Peng, H. Spatiotemporal Simulation, Early Warning, and Policy Recommendations of the Soil Heavy Metal Environmental Capacity of the Agricultural Land in a Typical Industrial City in China: Case of Zhongshan City. *J. Clean Prod.* **2021**, *285*, 124849. [CrossRef]
22. Singha, W.J.; Deka, H. Ecological and Human Health Risk Associated with Heavy Metals (HMs) Contaminant Sourced from Petroleum Refinery Oily Sludge. *J. Hazard. Mater.* **2024**, *476*, 135077. [CrossRef] [PubMed]
23. Wu, B.; Li, L.; Guo, S.; Li, Y. Source Apportionment of Heavy Metals in the Soil at the Regional Scale Based on Soil-Forming Processes. *J. Hazard. Mater.* **2023**, *448*, 130910. [CrossRef] [PubMed]

24. Liao, S.; Jin, G.; Khan, M.A.; Zhu, Y.; Duan, L.; Luo, W.; Jia, J.; Zhong, B.; Ma, J.; Ye, Z.; et al. The Quantitative Source Apportionment of Heavy Metals in Peri-Urban Agricultural Soils with UNMIX and Input Fluxes Analysis. *Environ. Technol. Innov.* **2021**, *21*, 101232. [CrossRef]
25. Adamiec, E.; Helios-Rybicka, E. Distribution of Pollutants in the Odra River System. Part IV. Heavy Metal Distribution in Water of the Upper and Middle Odra River, 1998–2000. *Pol. J. Environ. Stud.* **2002**, *11*, 669–673.
26. Li, X.; Ding, D.; Xie, W.; Zhang, Y.; Kong, L.; Li, M.; Li, M.; Deng, S. Risk Assessment and Source Analysis of Heavy Metals in Soil around an Asbestos Mine in an Arid Plateau Region, China. *Sci. Rep.* **2024**, *14*, 7552. [CrossRef]
27. Cheng, W.; Lei, S.; Bian, Z.; Zhao, Y.; Li, Y.; Gan, Y. Geographic Distribution of Heavy Metals and Identification of Their Sources in Soils near Large, Open-Pit Coal Mines Using Positive Matrix Factorization. *J. Hazard. Mater.* **2020**, *387*, 121666. [CrossRef]
28. Wu, J.; Li, J.; Teng, Y.; Chen, H.; Wang, Y. A Partition Computing-Based Positive Matrix Factorization (PC-PMF) Approach for the Source Apportionment of Agricultural Soil Heavy Metal Contents and Associated Health Risks. *J. Hazard. Mater.* **2020**, *388*, 121766. [CrossRef]
29. Wang, Z.; Zhang, J.; Watanabe, I. Source Apportionment and Risk Assessment of Soil Heavy Metals Due to Railroad Activity Using a Positive Matrix Factorization Approach. *Sustainability* **2022**, *15*, 75. [CrossRef]
30. Chandra, K.; Proshad, R.; Islam, M.; Idris, A.M. An Integrated Overview of Metals Contamination, Source-Specific Risks Investigation in Coal Mining Vicinity Soils. *Environ. Geochem. Health* **2023**, *45*, 7425–7458. [CrossRef]
31. Wang, H.; Yang, S.; Wang, Y.; Gu, Z.; Xiong, S.; Huang, X.; Sun, M.; Zhang, S.; Guo, L.; Cui, J.; et al. Rates and Causes of Black Soil Erosion in Northeast China. *CATENA* **2022**, *214*, 106250. [CrossRef]
32. Song, T.; An, Y.; Cui, G.; Tong, S.; He, J. Bioconcentrations and Health Risk Assessment of Heavy Metals in Crops in the Naoli River Basin Agricultural Area, Sanjiang Plain, China. *Environ. Earth Sci.* **2021**, *80*, 452. [CrossRef]
33. Song, T.; Su, X.; He, J.; Liang, Y.; Zhou, T. Source Apportionment and Health Risk Assessment of Heavy Metals in Agricultural Soils in Xinglonggang, Northeastern China. *Hum. Ecol. Risk Assess. Int. J.* **2018**, *24*, 509–521. [CrossRef]
34. Xie, X. GIS-Based Research on National-Scale and Regional-Scale Soil Organic Carbon Pools. Ph.D. Thesis, Nanjing Normal University, Nanjing, China, 2004.
35. *HJ/T 166-2004*; Ministry of Ecology and Environment the People's Republic of China. The Technical Specification for soil Environmental monitoring. China Quality Standards Publishing Media Co., Ltd.: Beijing, China, 2004.
36. Hakanson, L. An Ecological Risk Index for Aquatic Pollution Control.a Sedimentological Approach. *Water Res.* **1980**, *14*, 975–1001. [CrossRef]
37. Liang, F.; Pan, Y.; Peng, H.; Zeng, M.; Huang, C. Time-Space Simulation, Health Risk Warning and Policy Recommendations of Environmental Capacity for Heavy Metals in the Pearl River Basin, China. *Int. J. Environ. Res. Public. Health* **2022**, *19*, 4694. [CrossRef]
38. Dong, R.; Jia, Z.; Li, S. Risk Assessment and Sources Identification of Soil Heavy Metals in a Typical County of Chongqing Municipality, Southwest China. *Process Saf. Environ. Prot.* **2018**, *113*, 275–281. [CrossRef]
39. Xia, F.; Zhao, Z.; Niu, X.; Wang, Z. Integrated Pollution Analysis, Pollution Area Identification and Source Apportionment of Heavy Metal Contamination in Agricultural Soil. *J. Hazard. Mater.* **2024**, *465*, 133215. [CrossRef] [PubMed]
40. Chen, Z.; Xu, J.; Yang, F.; Hou, Z.; Ren, K.; Yu, L.; Yang, S.; Li, Z.; Zhang, X. Comprehensive Monitoring and Ecological Risk Assessment of Heavy Metals in Soil and Surface Water of Chishui River Basin in Upper Reaches of the Yangtze River. *Water.* **2023**, *15*, 2069. [CrossRef]
41. Xu, B.; Xu, H.; Zhao, H.; Gao, J.; Liang, D.; Li, Y.; Wang, W.; Feng, Y.; Shi, G. Source Apportionment of Fine Particulate Matter at a Megacity in China, Using an Improved Regularization Supervised PMF Model. *Sci. Total Environ.* **2023**, *879*, 163198. [CrossRef]
42. Shao, F.; Li, K.; Ouyang, D.; Zhou, J.; Luo, Y.; Zhang, H. Sources Apportionments of Heavy Metal(Loid)s in the Farmland Soils Close to Industrial Parks: Integrated Application of Positive Matrix Factorization (PMF) and Cadmium Isotopic Fractionation. *Sci. Total Environ.* **2024**, *924*, 171598. [CrossRef]
43. Singh, S.; Maiti, S.K.; Raj, D. An Approach to Quantify Heavy Metals and Their Source Apportionment in Coal Mine Soil: A Study through PMF Model. *Environ. Monit. Assess.* **2023**, *195*, 306. [CrossRef]
44. Taghavi, M.; Bakhshi, K.; Zarei, A.; Hoseinzadeh, E.; Gholizadeh, A. Soil Pollution Indices and Health Risk Assessment of Metal(Loid)s in the Agricultural Soil of Pistachio Orchards. *Sci. Rep.* **2024**, *14*, 8971. [CrossRef] [PubMed]
45. Yu, G.; Chen, F.; Zhang, H.; Wang, Z. Pollution and Health Risk Assessment of Heavy Metals in Soils of Guizhou, China. *Ecosyst. Health Sustain.* **2021**, *7*, 1859948. [CrossRef]
46. Liu, Z. Research on Heavy Metal Pollution and Risk of Agricultural Soils in Key Industrial and Agricultural Areas of Hexi Region. Ph.D. Thesis, Lanzhou University, Lanzhou, China, 2023.
47. Zhao, X.; Li, Z.; Wang, D.; Tao, Y.; Qiao, F.; Lei, L.; Huang, J.; Ting, Z. Characteristics, Source Apportionment and Health Risk Assessment of Heavy Metals Exposure via Household Dust from Six Cities in China. *Sci. Total Environ.* **2021**, *762*, 143126. [CrossRef] [PubMed]
48. Liu, X.; Chi, H.; Tan, Z.; Yang, X.; Sun, Y.; Li, Z.; Hu, K.; Hao, F.; Liu, Y.; Yang, S.; et al. Heavy Metals Distribution Characteristics, Source Analysis, and Risk Evaluation of Soils around Mines, Quarries, and Other Special Areas in a Region of Northwestern Yunnan, China. *J. Hazard. Mater.* **2023**, *458*, 132050. [CrossRef] [PubMed]
49. Liu, T.; Wang, Z. Contamination and Health Risk Assessment of Heavy Metals in Soil Surrounding an Electroplating Factory in JiaXing, China. *Sci. Rep.* **2024**, *14*, 4097. [CrossRef]

50. Jiang, H.-H.; Cai, L.-M.; Wen, H.-H.; Hu, G.-C.; Chen, L.-G.; Luo, J. An Integrated Approach to Quantifying Ecological and Human Health Risks from Different Sources of Soil Heavy Metals. *Sci. Total Environ.* **2020**, *701*, 134466. [CrossRef]
51. Luo, H.; Wang, Q.; Guan, Q.; Ma, Y.; Ni, F.; Yang, E.; Zhang, J. Heavy Metal Pollution Levels, Source Apportionment and Risk Assessment in Dust Storms in Key Cities in Northwest China. *J. Hazard. Mater.* **2022**, *422*, 126878. [CrossRef]
52. EMH. *China Soil Element Background Value. Environmental Monitoring Headquarters of the People's Republic of China*; China Environmental Science Press: Beijing, China, 1990.
53. Kang, Z.; Wang, S.; Qin, J.; Wu, R.; Li, H. Pollution Characteristics and Ecological Risk Assessment of Heavy Metals in Paddy Fields of Fujian Province, China. *Sci. Rep.* **2020**, *10*, 12244. [CrossRef]
54. Chen, X.; Zhang, H.; Wong, C.U.I. Spatial Distribution Characteristics and Pollution Evaluation of Soil Heavy Metals in Wulongdong National Forest Park. *Sci. Rep.* **2024**, *14*, 8880. [CrossRef]
55. Gong, C.; Lu, H.; Zhang, Z.; Wang, L.; Xia, X.; Wang, L.; Xiang, Z.; Shuai, L.; Ding, Y.; Chen, Y.; et al. Spatial Distribution Characteristics of Heavy Metal(Loid)s Health Risk in Soil at Scale on Town Level. *Sci. Rep.* **2022**, *12*, 19195. [CrossRef]
56. Cheng, B.; Wang, Z.; Yan, X.; Yu, Y.; Liu, L.; Gao, Y.; Zhang, H.; Yang, X. Characteristics and Pollution Risks of Cu, Ni, Cd, Pb, Hg and As in Farmland Soil near Coal Mines. *Soil Environ. Health* **2023**, *1*, 100035. [CrossRef]
57. Voltr, V.; Menšík, L.; Hlisnikovský, L.; Hruška, M.; Pokorný, E.; Pospíšilová, L. The Soil Organic Matter in Connection with Soil Properties and Soil Inputs. *Agronomy* **2021**, *11*, 779. [CrossRef]
58. Franzluebbers, A.J. Soil Organic Matter Stratification Ratio as an Indicator of Soil Quality. *Soil Tillage Res.* **2002**, *66*, 95–106. [CrossRef]
59. Zhao, J.; Cao, C.; Chen, X.; Zhang, W.; Ma, T.; Irfan, M.; Zheng, L. Source-Specific Ecological Risk Analysis and Critical Source Identification of Heavy Metal(Loid)s in the Soil of Typical Abandoned Coal Mining Area. *Sci. Total Environ.* **2024**, *947*, 174506. [CrossRef] [PubMed]
60. Wu, C.; Huang, F.; Wei, L.; Yi, S.; Wu, Y.; Huang, Z.; Yi, M.; Li, F. Do the Residual Metals in Multiple Environmental Media Surrounding Mines Pose Ecological and Health Risks? A Case of an Abandoned Mining Area in Central South China. *Environ. Res.* **2024**, *257*, 119279. [CrossRef]
61. Zerizghi, T.; Guo, Q.; Tian, L.; Wei, R.; Zhao, C. An Integrated Approach to Quantify Ecological and Human Health Risks of Soil Heavy Metal Contamination around Coal Mining Area. *Sci. Total Environ.* **2022**, *814*, 152653. [CrossRef]
62. Mai, X.; Tang, J.; Tang, J.; Zhu, X.; Yang, Z.; Liu, X.; Zhuang, X.; Feng, G.; Tang, L. Research Progress on the Environmental Risk Assessment and Remediation Technologies of Heavy Metal Pollution in Agricultural Soil. *J. Environ. Sci.* **2025**, *149*, 1–20. [CrossRef] [PubMed]
63. Wang, H.-Z.; Cai, L.-M.; Wang, Q.-S.; Hu, G.-C.; Chen, L.-G. A Comprehensive Exploration of Risk Assessment and Source Quantification of Potentially Toxic Elements in Road Dust: A Case Study from a Large Cu Smelter in Central China. *CATENA* **2021**, *196*, 104930. [CrossRef]
64. Zhao, G.; Ma, Y.; Liu, Y.; Cheng, J.; Wang, X. Source Analysis and Ecological Risk Assessment of Heavy Metals in Farmland Soils around Heavy Metal Industry in Anxin County. *Sci. Rep.* **2022**, *12*, 10562. [CrossRef]
65. Wang, P.; Han, G.; Hu, J.; Zhang, Q.; Tian, L.; Wang, L.; Liu, T.; Ma, W.; Li, J.; Zheng, H. Remarkable Contamination Characteristics, Potential Hazards and Source Apportionment of Heavy Metals in Surface Dust of Kindergartens in a Northern Megacity of China. *J. Hazard. Mater.* **2024**, *465*, 133295. [CrossRef]
66. Keshavarzi, A.; Kumar, V. Ecological Risk Assessment and Source Apportionment of Heavy Metal Contamination in Agricultural Soils of Northeastern Iran. *Int. J. Environ. Health Res.* **2019**, *29*, 544–560. [CrossRef] [PubMed]
67. Zhao, H.; Wu, Y.; Lan, X.; Yang, Y.; Wu, X.; Du, L. Comprehensive Assessment of Harmful Heavy Metals in Contaminated Soil in Order to Score Pollution Level. *Sci. Rep.* **2022**, *12*, 3552. [CrossRef]
68. Xiao, M.; Xu, S.; Yang, B.; Zeng, G.; Qian, L.; Huang, H.; Ren, S. Contamination, Source Apportionment, and Health Risk Assessment of Heavy Metals in Farmland Soils Surrounding a Typical Copper Tailings Pond. *Int. J. Environ. Res. Public Health* **2022**, *19*, 14264. [CrossRef]
69. Zhang, Y.; Song, B.; Zhou, Z. Pollution Assessment and Source Apportionment of Heavy Metals in Soil from Lead—Zinc Mining Areas of South China. *J. Environ. Chem. Eng.* **2023**, *11*, 109320. [CrossRef]
70. Chang, X.; Jia, Z.; Feng, J.; Duan, T.; Li, Y.-X. Refining the Diagnostics of Non-Point Source Metals Pollution to Urban Lakes Based on Interaction Normalized PMF Coupled with Bayesian Network. *Environ. Pollut.* **2022**, *304*, 119194. [CrossRef] [PubMed]
71. Strekopytov, S.; Brownscombe, W.; Lapinee, C.; Sykes, D.; Spratt, J.; Jeffries, T.E.; Jones, C.G. Arsenic and Mercury in Bird Feathers: Identification and Quantification of Inorganic Pesticide Residues in Natural History Collections Using Multiple Analytical and Imaging Techniques. *Microchem. J.* **2017**, *130*, 301–309. [CrossRef]
72. Yang, Y.; Christakos, G.; Guo, M.; Xiao, L.; Huang, W. Space-Time Quantitative Source Apportionment of Soil Heavy Metal Concentration Increments. *Environ. Pollut.* **2017**, *223*, 560–566. [CrossRef]
73. Hu, W.; Huang, B.; Shi, X.; Chen, W.; Zhao, Y.; Jiao, W. Accumulation and Health Risk of Heavy Metals in a Plot-Scale Vegetable Production System in a Peri-Urban Vegetable Farm near Nanjing, China. *Ecotoxicol. Environ. Saf.* **2013**, *98*, 303–309. [CrossRef]
74. Li, Z.; Niu, S. Modeling Pesticides in Global Surface Soils: Evaluating Spatiotemporal Patterns for USEtox-Based Steady-State Concentrations. *Sci. Total Environ.* **2021**, *791*, 148412. [CrossRef]
75. Huang, Y.; Zhang, X.; Li, Z. Analysis of Nationwide Soil Pesticide Pollution: Insights from China. *Environ. Res.* **2024**, *252*, 118988. [CrossRef]

76. Qi, H.; Zhao, B.; Li, L.; Chen, X.; An, J.; Liu, X. Heavy Metal Contamination and Ecological Risk Assessment of the Agricultural Soil in Shanxi Province, China. *R. Soc. Open Sci.* **2020**, *7*, 200538. [CrossRef] [PubMed]
77. Sun, W.; Ye, J.; Lin, H.; Yu, Q.; Wang, Q.; Chen, Z.; Ma, J.; Ma, J. Dynamic Characteristics of Heavy Metal Accumulation in Agricultural Soils after Continuous Organic Fertilizer Application: Field-Scale Monitoring. *Chemosphere* **2023**, *335*, 139051. [CrossRef]
78. Yang, S.; Zhou, Q.; Sun, L.; Qin, Q.; Sun, Y.; Wang, J.; Liu, X.; Xue, Y. Source to Risk Receptor Transport and Spatial Hotspots of Heavy Metals Pollution in Peri-Urban Agricultural Soils of the Largest Megacity in China. *J. Hazard. Mater.* **2024**, *480*, 135877. [CrossRef] [PubMed]
79. Wang, F.; Ouyang, W.; Hao, F.; Critto, A.; Zhao, X.; Lin, C. Multivariate Interactions of Natural and Anthropogenic Factors on Cd Behavior in Arable Soil. *RSC Adv.* **2015**, *5*, 41238–41247. [CrossRef]
80. Lian, Z.; Zhao, X.; Gu, X.; Li, X.; Luan, M.; Yu, M. Presence, Sources, and Risk Assessment of Heavy Metals in the Upland Soils of Northern China Using Monte Carlo Simulation. *Ecotoxicol. Environ. Saf.* **2022**, *230*, 113154. [CrossRef]
81. Wang, F.; Ouyang, W.; Hao, F.; Jiao, W.; Shan, Y.; Lin, C. Role of Freeze-Thaw Cycles and Chlorpyrifos Insecticide Use on Diffuse Cd Loss and Sediment Accumulation. *Sci. Rep.* **2016**, *6*, 27302. [CrossRef]
82. US EPA. *Soil Screening Guidance: Technical Background Document, Superfund US EPA*; United States Environmental Protection Agency: Washington, DC, USA, 1996.
83. Wang, D.; Zheng, L.; Ren, M.; Li, C.; Dong, X.; Wei, X.; Zhou, W.; Cui, J. Zinc in Soil Reflecting the Intensive Coal Mining Activities: Evidence from Stable Zinc Isotopes Analysis. *Ecotoxicol. Environ. Saf.* **2022**, *239*, 113669. [CrossRef]
84. Heilongjiang Provincial Department of Natural Resources. *Heilongjiang Provincial Mineral Resources Master Plan (2021–2025)*; Heilongjiang Provincial Department of Natural Resources of the People's Republic of China: Beijing, China, 2022.
85. Zhou, H.; Yue, X.; Chen, Y.; Liu, Y. Source-Specific Probabilistic Contamination Risk and Health Risk Assessment of Soil Heavy Metals in a Typical Ancient Mining Area. *Sci. Total Environ.* **2024**, *906*, 167772. [CrossRef]
86. Begum, B.A.; Biswas, S.K.; Hopke, P.K. Key Issues in Controlling Air Pollutants in Dhaka, Bangladesh. *Atmos. Environ.* **2011**, *45*, 7705–7713. [CrossRef]
87. Radziemska, M.; Fronczyk, J. Level and Contamination Assessment of Soil along an Expressway in an Ecologically Valuable Area in Central Poland. *Int. J. Environ. Res. Public Health* **2015**, *12*, 13372–13387. [CrossRef]
88. Shi, X.-M.; Liu, S.; Song, L.; Wu, C.-S.; Yang, B.; Lu, H.-Z.; Wang, X.; Zakari, S. Contamination and Source-Specific Risk Analysis of Soil Heavy Metals in a Typical Coal Industrial City, Central China. *Sci. Total Environ.* **2022**, *836*, 155694. [CrossRef] [PubMed]
89. Chen, H.; Teng, Y.; Lu, S.; Wang, Y.; Wang, J. Contamination Features and Health Risk of Soil Heavy Metals in China. *Sci. Total Environ.* **2015**, *512*, 143–153. [CrossRef] [PubMed]
90. Song, T.; Tong, S.; An, Y. Bioconcentrations and Health Risk Assessment of Heavy Metals in Paddy Fields—A Case Study in Naoli River Basin, Sanjiang Plain, China. *Arab. J. Geosci.* **2021**, *14*, 1988. [CrossRef]
91. Omar, N.A.; Praveena, S.M.; Aris, A.Z.; Hashim, Z. Health Risk Assessment Using in Vitro Digestion Model in Assessing Bioavailability of Heavy Metal in Rice: A Preliminary Study. *Food Chem.* **2015**, *188*, 46–50. [CrossRef] [PubMed]
92. Men, C.; Liu, R.; Xu, F.; Wang, Q.; Guo, L.; Shen, Z. Pollution Characteristics, Risk Assessment, and Source Apportionment of Heavy Metals in Road Dust in Beijing, China. *Sci. Total Environ.* **2018**, *612*, 138–147. [CrossRef]
93. Huang, J.; Wu, Y.; Sun, J.; Li, X.; Geng, X.; Zhao, M.; Sun, T.; Fan, Z. Health Risk Assessment of Heavy Metal(Loid)s in Park Soils of the Largest Megacity in China by Using Monte Carlo Simulation Coupled with Positive Matrix Factorization Model. *J. Hazard Mater* **2021**, *415*, 125629. [CrossRef]
94. An, L.; Liu, M.; Zhang, J.; Huang, L.; Chen, Z. Research progress on the source of arsenic in soil and the influencing factors of migration and release of arsenic. *Soil* **2020**, *52*, 234–246. [CrossRef]
95. Ma, G.; Guan, T.; Wang, Y.; Li, C. Effects of organic fertilizer application on Cd migration and accumulation in soil-radish system. *J. Hexi Univ.* **2024**, *40*, 55–61. [CrossRef]
96. MEPPRC. *Technical Guidelines for Risk Assessment of Contaminated Sites*; Ministry of Environmental Protection of the People's Republic of China: Beijing, China, 2014.
97. Zhao, R.; Guan, Q.; Luo, H.; Lin, J.; Yang, L.; Wang, F.; Pan, N.; Yang, Y. Fuzzy Synthetic Evaluation and Health Risk Assessment Quantification of Heavy Metals in Zhangye Agricultural Soil from the Perspective of Sources. *Sci. Total Environ.* **2019**, *697*, 134126. [CrossRef]
98. Chen, G.; Wang, X.; Wang, R.; Liu, G. Health Risk Assessment of Potentially Harmful Elements in Subsidence Water Bodies Using a Monte Carlo Approach: An Example from the Huainan Coal Mining Area, China. *Ecotoxicol. Environ. Saf.* **2019**, *171*, 737–745. [CrossRef]
99. Wei, X.; Gao, B.; Wang, P.; Zhou, H.; Lu, J. Pollution Characteristics and Health Risk Assessment of Heavy Metals in Street Dusts from Different Functional Areas in Beijing, China. *Ecotoxicol. Environ. Saf.* **2015**, *112*, 186–192. [CrossRef] [PubMed]
100. Kamunda, C.; Mathuthu, M.; Madhuku, M. Health Risk Assessment of Heavy Metals in Soils from Witwatersrand Gold Mining Basin, South Africa. *Int. J. Environ. Res. Public Health* **2016**, *13*, 663. [CrossRef] [PubMed]

processes

MDPI

Article

Assessment of Heavy Metal Contamination and Ecological Risk in Soil within the Zheng–Bian–Luo Urban Agglomeration

Xiaolong Chen [1,2], Hongfeng Zhang [1], Cora Un In Wong [1,*], Fanbo Li [3] and Sujun Xie [1]

1 Faculty of Humanities and Social Sciences, Macao Polytechnic University, Macao 999078, China
2 Department of Management, Henan Institute of Technology, Xinxiang 453000, China
3 School of Social Sciences, Tsinghua University, Beijing 100084, China
* Correspondence: corawong@mpu.edu.mo; Tel.: +86-00853-6534-1529

Abstract: As urbanization accelerates, the contamination of urban soil and the consequent health implications stemming from urban expansion are increasingly salient. In recent years, a plethora of cities and regions nationwide have embarked on rigorous soil geological surveys with a focus on environmental quality, yielding invaluable foundational data. This research aims to develop scientifically robust and rational land-use planning strategies while assessing the levels of heavy metal pollution and associated risks. The urban agglomeration encompassing Zhengzhou, Luoyang, and Kaifeng (referred to as Zheng–Bian–Luo Urban Agglomeration) in Henan Province was designated as the study area. Leveraging the Nemerow comprehensive index method alongside the Hakanson potential ecological risk assessment method, this study delved into the pollution levels and potential ecological ramifications of nine heavy metals, namely Cr, Mn, Ni, Cu, Zn, As, Cd, Pb, and Co. Research indicates that the hierarchy of individual potential ecological risks ranges from most to least significant as follows: Cd > Pb > Cr > Ni > Cu > Zn > As > Mn > Co. The concentrations of Cd in both Zhengzhou and Kaifeng surpassed the established background levels. Furthermore, the mean single-factor pollution index values for the heavy metals Cd and Zn exceeded 1, signifying a state of minor pollution. The Nemerow comprehensive index P of Cd and Zn is between $1 < P_{comp} \leq 2$, which is considered mild pollution. The comprehensive P values of the other seven metal elements are all less than 0.7, reaching a clean (alert) level. Predominantly, the primary potential risk factor in the superficial soil of the Zheng–Bian–Luo urban agglomeration is Cd, while the ecological risk implications associated with other heavy metal elements are comparatively minimal. The soil environmental quality within the designated study area remains secure, although certain localized areas pose potential risks of heavy metal pollution. A comprehensive assessment of the current state of soil heavy metal pollution is essential to establish a theoretical foundation and provide technical support for soil environmental protection, pollution mitigation, and sustainable utilization.

Keywords: heavy metals; pollution levels; risk assessment; Zheng–Bian–Luo urban agglomeration

1. Introduction

With the advancement of economic development and the escalation of population growth, the intensity of land use has significantly increased. Emissions from vehicular exhausts, industrial waste gases, agricultural practices, industrial mining operations, and the irrigation of crops with industrial wastewater have collectively contributed to the accumulation of heavy metals in the soil [1,2]. Urban soil, profoundly influenced by anthropogenic activities, constitutes an integral component of the ecosystem. It serves not only as a repository for pollutants but also as a potential source of contamination. Throughout the process of urbanization, relentless and recurrent human interventions have subjected urban soil to significant disturbances, culminating in pronounced environmental pollution challenges [3–5]. The ramifications of this trend are undeniable, surpassing the inherent resilience of soil ecosystems and posing significant socio-ecological challenges.

The progression has intensified the pressure on atmospheric and aquatic resources, exerting a significant impact on the health and quality of life of city dwellers [6–8]. Moreover, because of widespread and complex human interventions in urban soils, their ecological functions and biodiversity encounter significant challenges. These pressures could lead to irreversible ecological catastrophes over time. Therefore, proactive measures are essential to improve the quality of urban soil and reduce the pollution burdens it faces. This is essential not only to ensure the sustainable progression of urbanization but also to cultivate a more resilient and habitable urban environment for forthcoming generations.

The "National Soil Pollution Situation Survey Bulletin" unveiled in April 2023 reveals that the current exceedance rate of soil pollutant thresholds in Chinese urban areas stands at 16.1%. The predominant pollutants encompass a range of metal elements, notably Cd, Ni, As, Cu, Hg, Pb, among others [9,10]. As heavy metal elements infiltrate the soil, remediation efforts become more complex and challenging. These metals persist within the soil matrix, severely impairing both its physical and chemical properties. Furthermore, they inflict significant harm on soil enzyme and microbial activities. Such toxic contamination markedly undermines soil productivity. The most concerning aspect of heavy metal pollution lies in its latent and enduring nature [11–13]. The escalating severity of heavy metal pollution in soil poses threatens not only the ecological environment but also affects organisms across the food chain, thereby leading to potential risks to human health. Therefore, it is crucial to obtain a comprehensive understanding of the content and distribution of heavy metals in the soil [14,15]. Although China's exploration into heavy metals in urban soil commenced relatively recently, contemporary scholars have diligently investigated the content, distribution, and sources of soil heavy metal pollution in numerous large- and medium-sized cities, including Haikou, Beijing, Nanjing, and Shanghai [16–23]. The single factor index method was employed to assess the heavy metal pollution of the soil near the river, revealing severe pollution in the area, particularly from Cd [24]. An investigation into eight heavy metals within Beijing's urban green spaces revealed that Cd, Cu, Zn, and Ni are markedly influenced by anthropogenic activities. Pollution assessment outcomes indicate that, while Beijing's green spaces are predominantly clean, there exist minor potential ecological risks [25]. Utilizing Pb isotopes and multivariate statistical analyses to investigate the origins of heavy metals in the alluvial plain soil of the Pearl River Delta, the study determined that Pb primarily originates from gasoline combustion emissions. Furthermore, the sources of As, Cd, Hg, and Pb predominantly stem from coal combustion and the lead-zinc industry [26].

With global attention turning towards the soil environment, the presence of heavy metals in urban soil has become a central focus of research for both domestic and international scholars. Presently, numerous studies delve into the content, distribution, and sources of heavy metals in the soils of major urban centers such as Shanghai, Beijing, and Nanjing. However, research on smaller urban agglomerations is relatively limited. Positioned as the core of the rapidly growing regional economy within the Central Plains Economic Zone, the Zheng–Bian–Luo urban agglomeration has witnessed significant urban expansion and an increase in townships. Concurrently, this growth has intensified significant urban security challenges.

In recent years, Henan has undergone rapid urbanization, leading to the establishment of an industrial economic framework centered around sectors including medicine, textiles, chemicals, machinery, electronics, food, light industry, and building materials. Consequently, urban surface soils have been subjected to the discernible anthropogenic pollution. Notably, certain heavy metal elements, including Cd, Hg, Zn, and Cr, have manifested contamination [27,28]. The escalation of industrial development and urbanization within the Zheng–Bian–Luo urban agglomeration has amplified the influence of urban soil on both residents' living environments and human health [29,30]. Consequently, there is a pressing need to undertake pollution assessments and potential ecological hazard evaluations of heavy metals in the surface soil of the Zheng–Bian–Luo urban agglomeration. These efforts are crucial for improving the ecological environment of the soil in this urban conglomeration, carefully planning urban soil usage, establishing a resilient urban ecosystem, and safeguarding residents' health. This research provides essential foundational data and a scientific framework. Given these imperatives, investigating the sources, geochemical migration patterns, and the ecological impacts of heavy metals in urban soils is of paramount practical significance.

With the ongoing integration of the Zheng–Bian–Luo urban agglomeration, the regional urban ecological environment is facing escalating pressures, particularly from heavy metal pollution. Recent investigations indicate that specific zones within the Zheng–Bian–Luo urban agglomeration, encompassing soil, sediment, water, and near-surface dust, exhibit varying degrees of contamination by heavy metal elements such as Cd, Cu, Pb, and Zn. However, there is still a noticeable gap in comprehensive studies regarding the distribution characteristics of heavy metals across various functional zones—such as industrial, agricultural, and residential areas—in this region.

To address this research gap, the present study examines the soil of the Zheng–Bian–Luo urban agglomeration. We performed an in-depth analysis of the contents of Cd, Cr, Cu, Ni, Pb, and Zn, and further clarified the distribution characteristics of these heavy metals across various functional zones. Our objective is to establish a scientific foundation for the planning and development of the Central Plains urban agglomeration, with a particular emphasis on early warning systems for environmental pollution.

2. Materials and Methods

2.1. Study Area Overview

The Zheng–Bian–Luo urban agglomeration includes the three cities of Zhengzhou, Kaifeng, and Luoyang in Henan Province. Acting as the nucleus of Central Plains culture, it is also a key point within the "three points and one line" tourism corridor along the Yellow River in Henan Province (Figure 1). Geographically situated between 112°42′ and 114°14′ east longitude and between 34°16′ and 34°58′ north latitude, the region experiences a warm temperate continental-monsoon climate. It has an annual temperature of 14.3 °C and an annual rainfall of 640 mm [31,32].

The Zheng–Bian–Luo urban agglomeration strategically occupies a position in the Central Plains region, intersecting both the Beijing–Guangzhou Railway and the Longhai Railway. It acts as a crucial link connecting the Central Plains Economic Zone to the Yangtze River Delta Economic Zone. Situated in eastern China's heartland, this region is adjacent to the plains of the middle and lower reaches of the Yellow River, enjoying an advantageous location and superior transportation infrastructure. Since the advent of economic reforms and opening-up policies, the Zheng–Bian–Luo region has expedited its industrialization and urbanization processes, capitalizing on its advantageous geographical position. Currently, the "two cities and one prefecture" consisting of Zhengzhou, Kaifeng, and Luoyang have merged to form the central urban hub within the Central Plains region [33–35].

The Zheng–Bian–Luo urban agglomeration is located in the middle and lower reaches of the Yellow River Basin, characterized by intricate geological formations. Over the past two decades, the rapid industrialization and urbanization of the Zheng–Bian–Luo urban agglomeration have led to extensive industrial developments, exerting substantial environmental pressures on the region [36–39]. Situated in the "Central Plains", the Zheng–Bian–Luo conurbation hosts numerous metal mineral processing enterprises. Consequently, the sources of heavy metal contamination in its soil may differ from those observed in comparable urban areas domestically and internationally, that have been studied similarly.

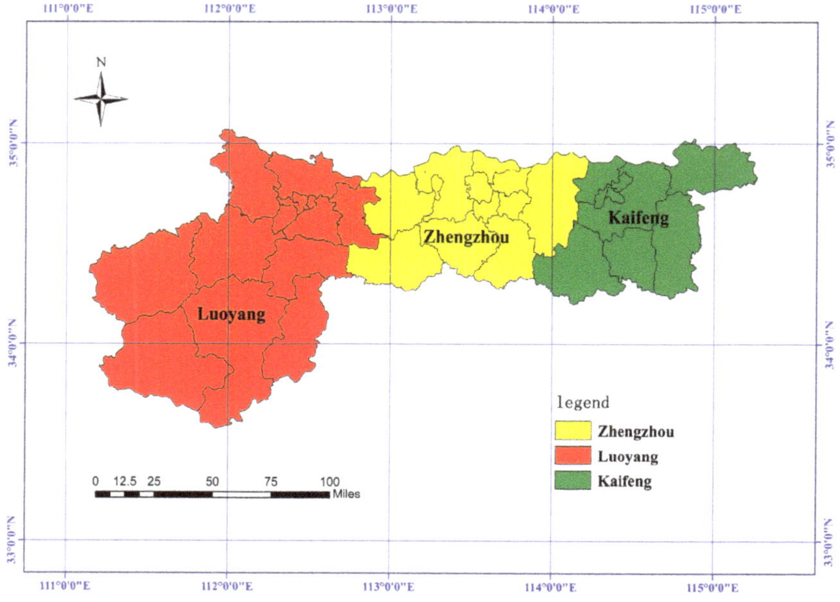

Figure 1. Geographic location map of Zheng–Bian–Luo urban agglomeration. (The illustration was crafted utilizing ArcGIS software, version 10.2. For further reference, the URL link is provided: https://www.arcgis.com/index.html, accessed on 1 March 2024).

2.2. Selection and Processing of Sampling Points

The research areas selected encompass Zhengzhou City, Luoyang City, and Kaifeng City, constituting the prominent "Three Points and One Line" tourist hub along the Yellow River in Henan Province. Based on land use classifications, these regions were demarcated into industrial zones, agricultural belts (adjacent to industrial zones), and residential precincts (encompassing urban green spaces).

The strategic placement of sampling points primarily adhered to two guiding principles: (1) variation in land use categories; and (2) sampling feasibility and convenience.

Taking into full consideration the relationship between rainfall and sampling depth, the annual rainfall in the area amounts to 640 mm. To avoid the direct impact of the physical and chemical properties of the surface soil on the surface runoff and pollutants caused by rainfall, the soil sampling points in the Zheng–Kai–Luo urban agglomeration were chosen to sample the 0–30 cm surface soil. In 2023, surface soil samples (0–30 cm depth) were meticulously procured from diverse functional zones across the aforementioned cities, culminating in a total of 300 samples. The distribution of these sampling points is as follows: Zhengzhou (100 sample points), Luoyang (100 sample points), and Kaifeng (100 sample points) (Figure 2). Each sampling point is uniquely numbered, ranging from 1 to 300.

2.3. Sample Processing

From March to July 2023, we meticulously collected soil samples using the plum blossom pile fixed-point technique. At each designated sampling point, three samples were procured from the surface soil layer spanning from 0 to 30 cm and subsequently blended to achieve homogeneity. Each collection yielded approximately 500 g of soil, meticulously stored in light-resistant plastic ziplock pouches to maintain the samples' integrity during preservation.

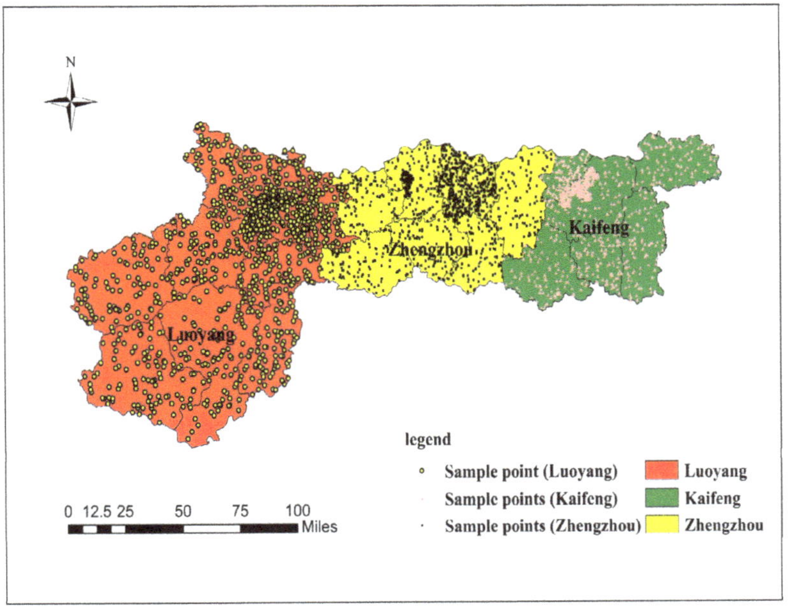

Figure 2. Spatial Distribution Map of Soil Sampling Sites in the Zheng–Bian–Luo Urban Agglomeration. (The illustration was crafted utilizing ArcGIS software, version 10.2. For further reference, the URL link is provided: https://www.arcgis.com/index.html, accessed on 1 March 2024).

After collection, the soil samples were transported to the laboratory and naturally dried in a shaded, cool environment. Subsequently, they were stored in a refrigerator maintained at 4 °C, protected from light exposure. Prior to analysis, the samples were ground using a mortar, filtered through a 0.15 mm mesh, placed in ziplock bags, and stored at −20 °C.

The concentrations of heavy metals in the soil samples were determined using the digestion combined with inductively coupled plasma-optical emission spectrometry (ICP-OES) methodology. In order to evaluate the effectiveness and accuracy of the sample digestion method, the author conducted a recovery rate test. By comparing the difference between the measured value and the certified value, the efficiency and reliability of the digestion method can be evaluated.

The digestion procedure is delineated as follows:

(1) Utilize a precision balance calibrated to one-ten-thousandth of a gram to weigh (0.1000 ± 0.0005) g of sediment sample, transferring it to a 25 mL digestion vessel made of polytetrafluoroethylene;
(2) In an acid-resistant fume hood, add 6 mL of nitric acid, 3 mL of hydrochloric acid, and 2 mL of hydrofluoric acid to the vessel. Allow the mixture to stand for 24 h to ensure the thorough mixing of the sample with these reagents;
(3) Execute the digestion process in accordance with the temperature ramping protocol detailed in Table 1, and subsequently allow the solution to cool upon the completion of the program;
(4) Once the temperature inside the digestion vessel has returned to ambient levels, pour all remaining substances into a 25 mL volumetric flask filled with distilled water. Rinse the digestion vessel and its lid with a small amount of distilled water, and then add these rinses to the volumetric flask. Fill the flask to the 25 mL mark with distilled water, allow it to stand for one hour, decant the supernatant, and proceed with measurements using ICP-OES.

ICP-OES analysis was used for the quantification of nine heavy metals in the samples, namely Cr, Mn, Ni, Cu, Zn, As, Cd, Pb, and Co.

Table 1. Digestion heating protocol.

Heating Duration/Minutes	Digestion Temperature/°C	Hold Time/Minutes
5	room temperature → 120	5
5	120 → 150	5
5	150 → 180	20

Note: The symbol "→" signifies the heating phase.

When performing ICP-OES (inductively coupled plasma–atomic emission spectrometry) analysis, the calculation of the limit of detection (LOD) and the limit of quantification (LOQ) is crucial to evaluate the sensitivity and suitability of the analytical method. And under the selected working conditions, the 2% nitric acid blank solution was measured 10 times in parallel according to the test method, and 3 times the standard deviation of the 10 measurement results was taken as the detection limit of each element. It can be seen that the detection limit of each element is 0.000214~5.578 μg/L, and the quantification limit is 0.000714~18.593 μg/L.

2.4. Research Methodology

2.4.1. Nemerow Comprehensive Index Methodology

The approach is based on the single-factor index methodology, but it has been refined to overcome its inherent limitation of only being applicable to areas contaminated by a single contaminant. Through the utilization of a single-factor pollution index, we can comprehensively assess the cumulative pollution magnitude of multiple contaminants in soil, enabling a detailed analysis of the multifaceted influence exerted by various pollutants on soil quality [40,41]. The mathematical expression for this integrated assessment is articulated as follows:

$$P_i = C_i/S_i \tag{1}$$

$$P_{\text{comp}} = \frac{\sqrt{(C_i/S_i)^2_{\max} + (C_i/S_i)^2_{ave}}}{2} \tag{2}$$

Within the formula: P_i represents the pollution value attributable to pollutant i; C_i represents the concentration of pollutant i; S_i represents the background level of pollutant i in the soil; $(C_i/S_i)_{max}$ stands for the peak single-factor pollution value; $(C_i/S_i)_{ave}$ corresponds to the average single-factor pollution value in the soil. The assessment criteria for are both shown in Tables 2 and 3.

Table 2. Criteria for single-factor index evaluation.

Rank	Single Factor Index	Pollution Level
I	$P_i \leq 1$	No pollution
II	$1 < P_i \leq 2$	slight pollution
III	$2 < P_i \leq 3$	light pollution
IV	$3 < P_i \leq 5$	Moderately polluted
V	$P_i > 5$	Heavy pollution

Table 3. Nemerow comprehensive index evaluation criteria.

Rank	Nemerow Comprehensive Index	Grade of Land Environmental Quality
I	$P_{comp} \leq 0.7$	Clean (Safe)
II	$0.7 < P_{comp} \leq 1$	Still clean (alert)
III	$1 < P_{comp} \leq 2$	light pollution
IV	$2 < P_{comp} \leq 3$	Moderately polluted
V	$P_{comp} > 3$	heavy pollution

2.4.2. Hakanson Potential Ecological Risk Assessment Methodology

The Hakanson methodology offers an efficient and effective approach for assessing the potential ecological consequences of heavy metal contamination in soil. Esteemed for its efficacy, this method finds extensive application in the realms of soil environmental quality assessment and research on soil pollution mitigation [42–44]. The formal expression for this methodology is articulated as follows:

$$RI = \sum_{i=1}^{n} E_f^i \tag{3}$$

$$E_f^i = T_f^i \times C_f^i \tag{4}$$

Within this framework, the subsequent evaluation metrics are encompassed:

$$C_f^i = C_d^i / C_n^i \tag{5}$$

$$C_H = \sum_{i=1}^{n} C_f^i \tag{6}$$

Let E_i^f denote the potential risk index associated with an individual metal; C_d^i represents the concentration of pollutant i in the soil; C_n^i signify the reference value for pollutant i in pollution assessment; C_H stand for the heavy metal pollution index; C_f^i indicates the index for pollution by a single metal; and T_f^i reflects the coefficient for the biotoxic response to various metals. The indices and corresponding degrees of harm are delineated in Table 4.

Table 4. Index of potential risk and classification of hazard degree.

Degree of Hazard	T_f^i	C_H	RI
Minor ecological risk	≤40	≤5	≤150
Moderate ecological risk	40–80	5–10	150–300
Significant ecological risk	80–160	10–20	300–600
High ecological risk	160–320	≥20	≥600
Severe ecological risk	≥320		

2.5. Data Processing and Analysis

The data were subjected to statistical analysis utilizing Excel 2019 and Origin 9.1, while spatial data and sampling points were visualized using ArcGIS 10.2.

3. Results and Analysis

3.1. Assessment of Soil Metal Content in the Zheng–Bian–Luo Urban Agglomeration

The findings regarding the heavy metal content across all samples are outlined in Table 5. Among soils sourced from diverse cities, only in Luoyang did the average arsenic (As) content surpass the background level [45,46], registering at 0.2 mg/kg. In Zhengzhou, especially in densely populated zones, concentrations of cadmium (Cd), copper (Cu), nickel (Ni), and zinc (Zn) surpassed the established background values. However, at other sampling sites, the heavy metal levels remained below the background thresholds [47,48]. In Kaifeng, the mean concentrations of Cd, chromium (Cr), Cu, and Zn all exceeded the background values [49]. Given the agricultural focus of this study, we employed risk screening values specific to agricultural soil pollution to evaluate the soil quality. The data revealed that the average Cd concentrations in Zhengzhou and Kaifeng exceeded the background levels by 1.77-fold and 10.37-fold, respectively. In conclusion, the Cd contamination in the soils across these cities is a matter of significant concern.

Table 5. Heavy metal concentrations in the soil of the Zheng–Bian–Luo urban agglomeration.

Area	Style	As	Cd	Co	Cr	Cu	Mn	Ni	Pb	Zn
Zhengzhou	Max	6.05	3.50	11.84	62.80	57.54	559.20	39.51	18.03	212.40
	Min	0.99	0.10	0.59	57.75	10.37	176.20	12.62	5.07	40.33
	Avg	2.79	0.11	3.65	60.27	25.34	373.40	24.63	14.64	106.80
Luoyang	Max	12.30	0.45	3.13	48.84	14.17	277	13.80	29.60	65.30
	Min	6.07	0.10	0.23	48.84	8.75	102	9.60	4.70	32.10
	Avg	9.20	0.24	1.68	48.84	10.07	166.70	11.08	14.90	43.70
Kaifeng	Max	10.90	5.79	4.22	96.30	118.3	731.60	26.02	124.50	750.00
	Min	2.60	1.40	2.15	38.40	25.90	310.80	17.11	15.20	150.30
	Avg	6.67	3.41	2.96	70.60	55.10	451.20	21.45	46.90	364.40
Agricultural land soil screening value		40	0.30	-	150	50	-	60	70	200
Soil background value		9	1.10	-	65	39	583	25	51	172

3.2. Correlation Assessment of Soil Heavy Metals in the Zheng–Bian–Luo Urban Agglomeration

Heavy metals in soil mainly originate from a variety of sources, including soil parent materials, agricultural practices, atmospheric deposition, and industrial emissions [50,51].

There is a potential correlation among heavy metals originating from the same sources. To ascertain the similarity in sources of these heavy metals, we conducted an analysis by calculating the correlation coefficients between them. The Spearman correlation analysis method was used to analyze the heavy metals in the soil of the Zheng–Kai-Luo urban agglomeration (Table 6). A high correlation coefficient suggests a common origin for these elements or the occurrence of composite pollution phenomena, whereas a low correlation coefficient implies a more intricate array of sources for these elements [52].

Table 6. Correlation assessment of heavy metals in the soil of the Zheng–Bian–Luo urban agglomeration.

Heavy Metal	As	Cd	Cr	Cu	Zn	Pb	Mn	Ni	Co
As	1								
Cd	0.03	1							
Cr	0.63 **	0.03	1						
Cu	0.45 **	0.2	0.62 **	1					
Zn	0.15	0.24	−0.12	0.15	0.28	1			
Pb	0.09	0.89 **	0.10 **	0.21	0.21	0.19 **	1		
Mn	0.25	0.06	0.13	0.35	0.26	0.06	0.13	1	
Ni	0.15	0.34	0.04	0.18	0.25	0.03	0.23	0.15	1
Co	0.03	0.21	0.16	0.05	0.17	0.04	0.28	0.33	0.27

Note: '**' denotes an extremely significant correlation at the 0.01 level, while '*' represents a significant correlation at the 0.05 level.

The assessment results of correlation coefficients for heavy metals in the soil of the Zheng–Bian–Luo urban agglomeration have been consolidated in Table 6. Within our study sample, lead (Pb) and cadmium (Cd) exhibited a pronounced correlation, registering a coefficient of 0.89. This suggests, at a significance level of 0.01, a shared source of contamination for these two elements. Furthermore, chromium (Cr) demonstrated a correlation of 0.63 with arsenic (As), copper (Cu) displayed a correlation of 0.45 with As, and the correlation between Cu and Cr stood at 0.62. These findings suggest a potential commonality in the pollution sources for these elements. Conversely, no significant correlations were observed between Cr and Cd, Cu and Cd, or As and Cd, indicating substantial disparities in their respective pollution sources.

3.3. Assessment and Analysis of Heavy Metal Pollution in the Soil of the Zheng–Bian–Lu Urban Agglomeration

The calculation results of both the single-factor index and the Nemerow comprehensive index for heavy metals in the soil of the Zheng–Bian–Luo urban agglomeration were recorded in Table 7. Upon scrutinizing the data, it is evident that the average single-factor pollution indices for Cd and Zn surpass a value of 1, suggesting a slight contamination by these two heavy metals. In contrast, the average index for Cu hovers near unity, specifically at 0.92. In terms of pollution severity, the average indices for Cd and Zn indicate mild pollution according to Level II criteria.

Table 7. Assessment index for heavy metal pollution in soil.

Area	Style	As	Cd	Co	Cr	Cu	Mn	Ni	Pb	Zn
Single-factor index	Max	0.46	0.87	0.78	0.69	0.69	0.41	0.26	0.16	0.39
	Min	1.35	1.84	1.51	1.23	1.85	1.74	1.33	0.76	0.75
	Avg	0.68	1.26	0.54	0.92	1.09	0.73	0.42	0.31	0.46
Nemerow composite index	P_{comp}	0.63	1.21	0.79	0.69	1.01	0.66	0.33	0.26	0.63
	Soil environmental quality classification	Clean (safe)	Light pollution	Still clean (alert)	Still clean (alert)	Light pollution	Clean (safe)	Clean (safe)	Clean (safe)	Clean (safe)

Additionally, upon computing the Nemerow comprehensive index, denoted by P comprehensive, it was observed that the P comprehensive values for Cd and Zn fell within the range of $1 < P_{comp} \leq 2$. This suggests a mild degree of pollution for these particular heavy metals. Conversely, the P values for the remaining seven metal elements all registered below 0.7, signifying a relatively clean or cautionary status.

3.4. Potential Ecological Risks of Heavy Metals in the Soil of Zheng–Bian–Luo Urban Agglomeration

According to the potential ecological risk index used for evaluating heavy metal contamination in surface soil (Table 8), the hierarchy of individual ecological risks, ranked in descending order, are as follows: Cd > Pb > Cr > Ni > Cu > Zn > As > Mn > Co. Across all sampling sites, the ecological risk index values for Pb, Cr, Ni, Cu, Zn, As, Mn, and Co are below 40, indicating minimal ecological risks. In particular, the ecological risk index for Cd ranges from 34.62 to 196.73, with 83.6% of the sampled sites exhibiting moderate ecological risks, while only a minority show severe risks.

Table 8. Potential ecological risk index for soil heavy metals.

Style	E_i^f									RI
	As	Cd	Cr	Cu	Zn	Pb	Mn	Ni	Co	
Max	0.81	34.62	4.747	2.01	1.09	5.38	0.43	3.67	0.34	50.79
Min	1.72	196.73	6.22	5.71	3.21	12.06	0.72	5.11	0.72	213.60
Avg	0.12	12.14	3.07	1.07	0.57	3.42	0.18	2.21	0.14	20.63
Mean risk level	Slight ecological risk	Slight ecological risk	Slight ecological risk	Slight ecological risk	Slight ecological risk	Slight ecological risk	Slight ecological risk	Slight ecological risk	Slight ecological risk	Slight ecological risk

The comprehensive potential ecological risk index (RI) spans from 20.63 to 213.60, mainly reflecting minor ecological risks that make up 98.4% of the total. These findings elucidate that cadmium (Cd) stands as the predominant risk factor in the surface soil of the Zheng–Bian–Luo urban agglomeration. In contrast, the ecological impacts of other heavy metal elements seem comparatively negligible. Collectively, the overall ecological risk within this region is assessed as moderate, albeit predominantly mild.

The bio-toxicity response coefficient of metals is a crucial metric used to evaluate the harmful effects of metals on various organisms, including humans, animals, and plants [53–55]. This coefficient primarily elucidates the correlation between the toxicological effects induced by specific metals and their concentrations. Leveraging this coefficient allows us to systematically quantify the potential health risks associated with exposure to these metals.

The biological toxicity response coefficients and associated potential hazard levels for various metals examined in this study are meticulously detailed in Table 9. Analysis of the data reveals that the toxicity response coefficients for the nine metallic elements listed therein all register below 40. Such findings signify that these metals can exert toxicological effects on organisms even at comparatively low concentrations, thereby suggesting a minimal ecological risk associated with their presence.

Table 9. Biotoxicity response coefficients of various metals.

Heavy Metal	As	Cd	Cr	Cu	Zn	Pb	Mn	Ni	Co	Degree
T_f^i	5	30	5	5	20	20	2	20	1	Slight ecological risk

This observation suggests that potential health hazards for humans or other organisms in everyday life or occupational settings may only become apparent when metal concentrations reach elevated levels [56–58]. Nonetheless, it remains imperative to acknowledge that even when the toxicity response coefficient falls below 40, metal exposure can still elicit detrimental effects on particularly vulnerable populations or ecosystems. Consequently, despite the reduced toxicity exhibited by these metals, a comprehensive risk assessment and management strategy is indispensable to safeguard both organisms and the environment.

4. Discussion

Upon examining the levels of heavy metals in the soil of the Zheng–Bian–Luo urban agglomeration, our findings indicate predominantly favorable soil quality within the study area. In isolated locations exhibiting moderate-to-severe pollution, the primary contaminants manifest as residues and oxidation states of iron and manganese, which exhibit limited mobility and absorption by plants. Currently, the potential adverse effects of these pollutants on both human health and environmental integrity appear to be relatively minimal.

The computed results for the single-factor index and the Nemerow comprehensive index pertaining to heavy metals in the soil of the Zheng–Bian–Luo urban agglomeration are presented below. The analysis reveals that the average single-factor pollution index values for the heavy metals Cd and Zn both exceed 1, signifying mild pollution levels, with an average value corresponding to Level II of mild pollution. Furthermore, the Nemerow comprehensive index (P) values for Cd and Zn range between $1 < P_{comp} \leq 2$, indicative of mild pollution. In contrast, the comprehensive P values for the remaining seven metal elements are all below 0.7, categorizing them at a relatively clean or alert level.

Elevated Cd concentrations in soil are often associated with anthropogenic activities, including sectors such as the chemical industry, battery manufacturing, mineral processing, and improper waste disposal practices. Specifically, the Luoyang Jianxi Industrial Zone, densely populated regions, and areas rich in historical human activity relics emerge as the most heavily contaminated zones. Noteworthy examples include the Luoyang Copper Processing Factory, Datang Power Plant, Luoyang Glass Factory, Baima Temple, and the Ru'ao areas. Contamination in the former locations may be attributed to mineral processing activities, whereas the latter encompasses remnants of ancient human settlements dating back to the Han and Wei dynasties, as well as activities from the Sui, Tang, and Song dynasties. Consequently, persistent or large-scale industrial emissions, illicit waste disposal, and improper application of pesticides and fertilizers are potential contributors to the heightened Cd levels observed in the soil.

Among the individual potential ecological risks assessed, Cd poses the most substantial risk, whereas Co presents a comparatively minor risk. Elevated Cd concentrations are often associated with specific anthropogenic activities, including the chemical industry, mineral processing, battery manufacturing, and improper waste disposal. Such activities can lead to the substantial release and accumulation of cadmium, thereby augmenting the risk of cadmium contamination within soils and ecosystems. In contrast, the primary sources of Co typically stem from natural geological processes, such as the inherent weathering and decomposition of cobalt ores. Given the relatively stable migration and transformation properties of Co within soil, its accumulation and enrichment in the environment are less pronounced.

The concentration of heavy metals in soil is influenced by a myriad of factors, resulting in a multifaceted array of sources. To enhance our comprehension and management of heavy metal pollution in soil, it is imperative to intensify relevant investigations and meticulously trace the origins of pollution. These efforts will provide a scientific foundation for the judicious utilization of soil resources and facilitate effective strategies for mitigating soil pollution amidst the backdrop of urbanization.

5. Conclusions

(1) In soil samples collected from Zhengzhou and Kaifeng, the average cadmium (Cd) concentrations surpassed the background values by 1.77-fold and 10.37-fold, respectively. Such pronounced cadmium contamination in these urban soil matrices warrants meticulous scrutiny and underscores the imperative for heightened attention to this environmental concern.

(2) Upon examining the correlation coefficients of heavy metals in the soil of the Zheng–Bian–Luo urban agglomeration, our findings unveil a significant association between Pb and Cd, yielding a coefficient of 0.89 at a significance level of 0.01. Such results suggest a common source of contamination for these two elements. Furthermore, the correlations between Cr and As, Cu and As, and Cu and Cr are determined to be 0.63, 0.45, and 0.62, respectively, implying potential shared sources of contamination for these elements. Conversely, the correlations between Cr and Cd, Cu and Cd, as well as As and Cd, are notably weak or even absent, indicating distinct sources of pollution for these metals.

(3) The average single factor pollution index for Cd and Zn surpasses 1, suggesting a state of slight pollution categorizing them under Level II mild pollution. Moreover, the Nemerow comprehensive index places the P comprehensive value for Cd and Zn within the $1 < P_{comp} \leq 2$ range, signifying a mild pollution level for these metals. In contrast, the comprehensive P values for the remaining seven metal elements fall below 0.7, suggesting pollution levels that either reach a clean state or warrant warning.

(4) In terms of individual potential ecological risks, the hierarchy, ranging from most to least significant, is as follows: Cd > Pb > Cr > Ni > Cu > Zn > As > Mn > Co. The ecological risk index for Cd oscillates between 34.62 and 196.73, suggesting a condition of mild ecological risk. The comprehensive potential ecological risk index, denoted by RI, spans from 20.63 to 213.60, predominantly signifying minimal potential ecological risks. Research findings elucidate that the predominant risk factor influencing the surface soil of the Zheng–Bian–Luo urban agglomeration is cadmium (Cd), with the remaining heavy metal elements exerting relatively negligible impacts on ecological risk levels.

6. Limitations and Future Directions

This study conducted a comprehensive exploration and assessment of heavy metal pollution levels in the soil of the Zheng–Bian–Luo urban agglomeration; however, several limitations warrant acknowledgment. Primarily, constrained by resources and scope, our investigation was confined to select heavy metal elements, thus failing to encompass all potential contributors to heavy metal pollution. Secondly, our study predominantly focused on a static pollution assessment, lacking a comprehensive exploration into the dynamic fluctuations of soil contamination over time.

In future research endeavors, it would be advisable to expand the range of heavy metal detection by incorporating additional trace elements or emerging pollutants. Moreover, the geographical scope of our study should include a broader array of counties surrounding urban agglomerations. By integrating soil types, topographical features, and other pertinent factors, our aspiration is to formulate a more sophisticated soil pollution assessment model. Such advancements aim to furnish robust foundational support for soil ecological sustainability research across various urban agglomerations nationwide.

Author Contributions: Conceptualization, X.C., H.Z., F.L. and C.U.I.W.; Data curation, X.C.; Formal analysis, X.C., H.Z. and F.L.; Methodology, X.C., H.Z. and C.U.I.W.; Software, X.C., F.L. and S.X.; Writing—original draft, X.C., H.Z. and C.U.I.W.; Writing—review and editing, X.C., H.Z., S.X. and C.U.I.W. All authors have read and agreed to the published version of the manuscript.

Funding: This research received no external funding.

Data Availability Statement: The original contributions presented in the study are included in the article, further inquiries can be directed to the corresponding author/s.

Conflicts of Interest: The authors declare no conflicts of interest.

References

1. Briseño-Bugarín, J.; Araujo-Padilla, X.; Escot-Espinoza, V.M.; Cardoso-Ortiz, J.; Flores de la Torre, J.A.; López-Luna, A. Lead (Pb) Pollution in Soil: A Systematic Review and Meta-Analysis of Contamination Grade and Health Risk in Mexico. *Environments* **2024**, *11*, 43. [CrossRef]
2. Man, Z.; Xie, C.; Qin, Y.; Che, S. Effects of Revetments on Nitrification and Denitrification Potentials in the Urban River–Riparian Interface. *Land* **2024**, *13*, 333. [CrossRef]
3. Phung, T.X.B.; Le, T.P.Q.; Da Le, N.; Hoang, T.T.H.; Nguyen, T.M.H.; Rochelle-Newall, E.; Nguyen, T.A.H.; Duong, T.T.; Pham, T.M.H.; Nguyen, T.D. Metal contamination, their ecological risk, and relationship with other variables in surface sediments of urban rivers in a big city in Asia: Case study of Hanoi, Vietnam. *Environ. Sci. Pollut. Res.* **2024**, *31*, 22362–22379. [CrossRef]
4. Verma, N.; Singh, G.; Ahsan, N. Assessment of spatiotemporal variations in water quality of the urban river reach, Yamuna, Delhi. *Water Air Soil Pollut.* **2023**, *234*, 571. [CrossRef]
5. Yang, X.; Leng, M.; Ge, X.; Wu, X.; Liu, H.; Lin, G.; Huang, Z.; Chen, Y. Characterization and Risk Assessment of Nutrient and Heavy Metal Pollution in Surface Sediments of Representative Lakes in Yangxin County, China. *Sustainability* **2024**, *16*, 2252. [CrossRef]
6. Liu, Y.; Shen, W.; Fan, K.; Pei, W.; Liu, S. Spatial Distribution, Source Analysis, and Health Risk Assessment of Heavy Metals in the Farmland of Tangwang Village, Huainan City, China. *Agronomy* **2024**, *14*, 394. [CrossRef]
7. Zhang, C.H.; Sears, C.G.; Guinn, B.; Zierold, K.M. Exploring environmental injustice in exposure to airborne lead released from industrial facilities in kentucky, united states. *Geogr. Rev.* **2024**, 1–18. [CrossRef]
8. Santos, R.S.S.; Marques, I.M.; de Oliveira, A.G.; de Oliveira, O.M.C.; Moreira, Í.T.A. Groundwater Contamination in a Rural Municipality of Northeastern Brazil: Application of Geostatistics, Geoprocessing, and Geochemistry Techniques. *Water Air Soil Pollut.* **2024**, *235*, 1–23. [CrossRef]
9. Zhang, J.; Peng, W.; Lin, M.; Liu, C.; Chen, S.; Wang, X.; Gui, H. Environmental geochemical baseline determination and pollution assessment of heavy metals in farmland soil of typical coal-based cities: A case study of Suzhou City in Anhui Province, China. *Heliyon* **2023**, *9*, e14841. [CrossRef]
10. Qiao, P.; Wang, S.; Lei, M.; Guo, G.; Yang, J.; Wei, Y.; Gou, Y.; Li, P.; Zhang, Z. Influencing factors identification and the nested structure analysis of heavy metals in soils in entire city and surrounding the multiple pollution sources. *J. Hazard. Mater.* **2023**, *449*, 130961. [CrossRef] [PubMed]
11. Zhang, Y.; Li, J.; Lu, F.; Wang, S.; Ren, Y.; Guo, S.; Wang, B.; Gao, W. The Enzyme Activity of Dual-Domain β-Propeller Alkaline Phytase as a Potential Factor in Improving Soil Phosphorus Fertility and Triticum aestivum Growth. *Agronomy* **2024**, *14*, 614. [CrossRef]
12. Xu, M.; Xiao, R.; Mei, C.; Chen, J.; Huang, Q.; Huang, F.; He, M. Rice husk biochar reduces Cd availability by affecting microbial community activity and structure in Cd-contaminated soils. *J. Soils Sediments* **2024**, *24*, 1764–1776. [CrossRef]
13. Chen, X.; Wong, C.U.I.; Zhang, H. Analysis and pollution evaluation of heavy metal content in soil of the Yellow River Wetland Reserve in Henan. *PeerJ* **2023**, *11*, e16454. [CrossRef]
14. Martínez-Morcillo, S.; Barrales, I.; Pérez-López, M.; Rodriguez, F.S.; Peinado, J.S.; Míguez-Santiyán, M.P. Mineral and potentially toxic element profiles in the soil-feed-animal continuum: Implications for public, environmental, and livestock health in three pasture-based sheep farming systems. *Sci. Total Environ.* **2024**, *919*, 170860. [CrossRef]
15. Quadri, S.S.A.; Ali, R.; Mesidor, A.; Shah, I.A.; Raqeeb, N.; Afzal, M.O.; Rizwan, M.; Rafique Sr, T.; Nosheen, S.; Shah, I.A. Arsenic in the Food Chain in Pakistan: Assessing Risks to Human Health and Ensuring Food Security Through Comprehensive Contamination Mitigation Strategies. *Cureus* **2024**, *16*, e54069. [CrossRef]
16. Guo, Y.; Yang, Y.; Li, R.; Liao, X.; Li, Y. Cadmium accumulation in tropical island paddy soils: From environment and health risk assessment to model prediction. *J. Hazard. Mater.* **2024**, *465*, 133212. [CrossRef]
17. Meng, D.; Shao, Y.; Luo, M.; Liu, Z.; Xu, D.; Ma, L. Risk Assessment and Source Analysis of Soil Heavy Metal (oid) s Pollution in Beijing, China. *Water Air Soil Pollut.* **2023**, *234*, 574. [CrossRef]

18. Wu, Z.; Xia, T.; Jia, X. Spatial variation and pollution risk assessment of heavy metals in industrial soil based on geochemical data and GIS-A case of an iron and steel plant in Beijing, China. *Int. J. Environ. Anal. Chem.* **2022**, 1–22. [CrossRef]
19. Shao, Y.; Yang, G.; Luo, M.; Xu, D.; Tazoe, H.; Yamada, M.; Ma, L. Multiple evaluation of typical heavy metals pollution in surface soil and road dust from Beijing and Hebei Province, China. *Bull. Environ. Contam. Toxicol.* **2022**, *109*, 317–322. [CrossRef] [PubMed]
20. Zhang, J.; Yu, F.; Pang, S.; Yu, Y. Spatial distribution and pollution assessment of potentially toxic elements in urban forest soil of Nanjing, China. *Pol. J. Environ. Stud.* **2019**, *28*, 3015. [CrossRef] [PubMed]
21. He, K.; Wang, J.; Geng, H.; Qin, Z.; Li, N.; Zhang, Y.; Yang, R.; Feng, S.; Wang, B. Will different land uses affect heavy metal pollution in soils of roadside trees? An empirical study from Shanghai. *Environ. Monit. Assess.* **2023**, *195*, 1388. [CrossRef]
22. Hu, X.-F.; Li, M.; He, Z.-C.; Cui, L.; Liu, R.; Wang, X.-D.; Wang, Z.-H. Magnetic responses to heavy metal pollution of the industrial soils in Shanghai: Implying the influences of anthropogenic magnetic dustfall on urban environment. *J. Appl. Geophys.* **2022**, *197*, 104544. [CrossRef]
23. Fei, X.; Christakos, G.; Xiao, R.; Ren, Z.; Liu, Y.; Lv, X. Improved heavy metal mapping and pollution source apportionment in Shanghai City soils using auxiliary information. *Sci. Total Environ.* **2019**, *661*, 168–177. [CrossRef]
24. Li, Y.; Zhu, Q.; Tang, X.; Wang, C.; Zhai, S. Ecological and health risk assessment of heavy metals in farmland in the South of Zhangbei County, Hebei Province, China. *Appl. Sci.* **2022**, *12*, 12425. [CrossRef]
25. Du, X.; Liang, H.; Fang, X.; Cui, S.; Li, J. Characteristics of colloids and their affinity for heavy metals in road runoff with different traffic in Beijing, China. *Environ. Sci. Pollut. Res.* **2021**, *28*, 20082–20092. [CrossRef]
26. Sheng, W.; Hou, Q.; Yang, Z.; Yu, T. Impacts of periodic saltwater inundation on heavy metals in soils from the Pearl River Delta, China. *Mar. Environ. Res.* **2023**, *187*, 105968. [CrossRef]
27. Tian, P.; Pan, Z.; Shen, Y.; Zhu, Y. Spatio-temporal pattern assessment of China's environmental performance and its spatial drivers: Evidence from city-level data over 2003–2019. *Environ. Sci. Pollut. Res.* **2024**, *31*, 15223–15256. [CrossRef]
28. Liu, X.; Yang, L. Spatiotemporal Distribution Characteristics of PM 2.5 Pollution and the Influential Meteorological Factors in the Henan Province, China, 2021. *Pol. J. Environ. Stud.* **2023**, *32*, 5211–5225. [CrossRef]
29. Zhang, B.; Liu, X.; Wei, W.; Li, X.; Zhu, H.; Chen, L. Environmental carrying capacity and ecological risk assessment of pesticides under different soil use types in the Central Plains Urban Agglomeration (CPUA), China. *Environ. Pollut.* **2024**, *341*, 122852. [CrossRef]
30. Wei, H.; Han, Q.; Yang, Y.; Li, L.; Liu, M. Spatial Heterogeneity of Watershed Ecosystem Health and Identification of Its Influencing Factors in a Mountain–Hill–Plain Region, Henan Province, China. *Remote Sens.* **2023**, *15*, 3751. [CrossRef]
31. Zhou, X.; Zhang, Y.; Wang, W.; Li, X.; Yang, H.; Sun, Y. Social vulnerability assessment under different extreme precipitation scenarios: A case study in Henan Province, China. *PLoS ONE* **2024**, *19*, e0299956. [CrossRef]
32. Qiao, Y.; Han, Z.; Zhang, B.; Wei, X.; Dong, C.; Liu, H. Gold Migration and Precipitation as Collaurum in Orogenic Gold Deposits: Constrains from Microscopic Gold Particles Observed in the Alteration Zone in Shanggong Gold Ore, Henan, China. *Minerals* **2024**, *14*, 327. [CrossRef]
33. Zhao, Y.; Zhou, L.; Luo, H.; Li, Y.; Li, X.; Pan, Y.; Huang, Y. Urban subsidence in rapid economic development: The case of Luoyang city, Henan Province. *All Earth* **2023**, *35*, 252–271. [CrossRef]
34. Wei, Z.; Wei, K.; Liu, J.; Zhou, Y. The relationship between agricultural and animal husbandry economic development and carbon emissions in Henan Province, the analysis of factors affecting carbon emissions, and carbon emissions prediction. *Mar. Pollut. Bull.* **2023**, *193*, 115134. [CrossRef]
35. Wei, Z.; Wei, K.; Liu, J. Decoupling relationship between carbon emissions and economic development and prediction of carbon emissions in Henan Province: Based on Tapio method and STIRPAT model. *Environ. Sci. Pollut. Res.* **2023**, *30*, 52679–52691. [CrossRef]
36. Sun, K.; Liu, C.; Yang, Z.; Han, J.; Xie, W.; Cao, Y. The impact of industrial structure on the environment pollution and its spatial-temporal differences in Henan, China. *J. Clean. Prod.* **2022**, *381*, 135127. [CrossRef]
37. Hei, W.; Li, X.; Gao, G.; Wang, S.; Zhang, R.; Wang, K. Air pollutants and CO_2 emissions in industrial parks and evaluation of their green upgrade on regional air quality improvement: A case study of seven cities in Henan Province. *Atmosphere* **2022**, *13*, 383. [CrossRef]
38. Zhang, H.; Song, Y.; Zhang, L. Pollution control in urban China: A multi-level analysis on household and industrial pollution. *Sci. Total Environ.* **2020**, *749*, 141478. [CrossRef]
39. Liu, Y.; Tian, Z.; He, X.; Wang, X.; Wei, H. Short-term effects of indoor and outdoor air pollution on the lung cancer morbidity in Henan Province, Central China. *Environ. Geochem. Health* **2022**, *44*, 2711–2731. [CrossRef]
40. Ogunkunle, C.O.; Fatoba, P.O. Pollution loads and the ecological risk assessment of soil heavy metals around a mega cement factory in southwest Nigeria. *Pol. J. Environ. Stud.* **2013**, *22*, 1011.
41. Brajer, V.; Mead, R.W.; Xiao, F. Searching for an Environmental Kuznets Curve in China's air pollution. *China Econ. Rev.* **2011**, *22*, 383–397. [CrossRef]
42. Li, Y.; Li, P.; Liu, L. Source identification and potential ecological risk assessment of heavy metals in the topsoil of the Weining Plain (northwest China). *Expo. Health* **2022**, *14*, 281–294. [CrossRef]
43. Akarsu, C.; Sönmez, V.Z.; Sivri, N. Potential ecological risk assessment of critical raw materials: Gallium, gadolinium, and germanium. *Arch. Environ. Contam. Toxicol.* **2023**, *84*, 368–376. [CrossRef]

44. Liu, B.; Xu, M.; Wang, J.; Wang, Z.; Zhao, L. Ecological risk assessment and heavy metal contamination in the surface sediments of Haizhou Bay, China. *Mar. Pollut. Bull.* **2021**, *163*, 111954. [CrossRef]
45. Chen, X.; Cui, F.; Wong, C.U.I.; Zhang, H.; Wang, F. An investigation into the response of the soil ecological environment to tourist disturbance in Baligou. *PeerJ* **2023**, *11*, e15780. [CrossRef]
46. Chen, Z.-Y.; Zhao, Y.-Y.; Chen, D.-L.; Huang, H.-T.; Zhao, Y.; Wu, Y.-J. Ecological risk assessment and early warning of heavy metal cumulation in the soils near the Luanchuan molybdenum polymetallic mine concentration area, Henan Province, central China. *China Geol.* **2023**, *6*, 15–26.
47. Yang, L.; Ge, S.; Liu, J.; Iqbal, Y.; Jiang, Y.; Sun, R.; Ruan, X.; Wang, Y. Spatial Distribution and Risk Assessment of Heavy Metal (oid) s Contamination in Topsoil around a Lead and Zinc Smelter in Henan Province, Central China. *Toxics* **2023**, *11*, 427. [CrossRef]
48. Chen, X.; Zhang, H.; Wong, C.U.I. Spatial distribution characteristics and pollution evaluation of soil heavy metals in Wulongdong National Forest Park. *Sci. Rep.* **2024**, *14*, 8880. [CrossRef]
49. Liu, D.; Li, Y.; Ma, J.; Li, C.; Chen, X. Heavy metal pollution in urban soil from 1994 to 2012 in Kaifeng City, China. *Water Air Soil Pollut.* **2016**, *227*, 154. [CrossRef]
50. Alloway, B.J. Sources of heavy metals and metalloids in soils. In *Heavy Metals in Soils: Trace Metals and Metalloids in Soils and Their Bioavailability*; Springer: Berlin, Germany, 2013; pp. 11–50. [CrossRef]
51. Shan, Y.; Tysklind, M.; Hao, F.; Ouyang, W.; Chen, S.; Lin, C. Identification of sources of heavy metals in agricultural soils using multivariate analysis and GIS. *J. Soils Sediments* **2013**, *13*, 720–729. [CrossRef]
52. Morabito, E.; Contini, D.; Belosi, F.; Stortini, A.M.; Manodori, L.; Gambaro, A. Atmospheric deposition of inorganic elements and organic compounds at the inlets of the Venice lagoon. *Adv. Meteorol.* **2014**, *2014*, 158902. [CrossRef]
53. Wang, W.; Wong, M.H.; Leharne, S.; Fisher, B. Fractionation and biotoxicity of heavy metals in urban dusts collected from Hong Kong and London. *Environ. Geochem. Health* **1998**, *20*, 185–198. [CrossRef]
54. Wang, W.-X.; Tan, Q.-G. Applications of dynamic models in predicting the bioaccumulation, transport and toxicity of trace metals in aquatic organisms. *Environ. Pollut.* **2019**, *252*, 1561–1573. [CrossRef]
55. Castillo, G.C.; Vila, I.C.; Neild, E. Ecotoxicity assessment of metals and wastewater using multitrophic assays. *Environ. Toxicol. Int. J.* **2000**, *15*, 370–375. [CrossRef]
56. Han, R.; Zhou, B.; Huang, Y.; Lu, X.; Li, S.; Li, N. Bibliometric overview of research trends on heavy metal health risks and impacts in 1989–2018. *J. Clean. Prod.* **2020**, *276*, 123249. [CrossRef]
57. Kuang, Y.; Chen, X.; Zhu, C. Characteristics of Soil Heavy Metal Pollution and Health Risks in Chenzhou City. *Processes* **2024**, *12*, 623. [CrossRef]
58. Mohammadi, A.A.; Zarei, A.; Esmaeilzadeh, M.; Taghavi, M.; Yousefi, M.; Yousefi, Z.; Sedighi, F.; Javan, S. Assessment of heavy metal pollution and human health risks assessment in soils around an industrial zone in Neyshabur, Iran. *Biol. Trace Element. Res.* **2020**, *195*, 343–352. [CrossRef]

processes

MDPI

Article

Characteristics of Soil Heavy Metal Pollution and Health Risks in Chenzhou City

Yingfeng Kuang [1,2,†], Xiaolong Chen [3,†] and Chun Zhu [1,*]

1 Faculty of Humanities and Arts, Macau University of Science and Technology, Macau 999078, China; 2109853mad30001@student.must.edu.mo
2 Faculty of Fine Arts and Design, Xiangnan University, Chenzhou 423000, China
3 Faculty of Humanities and Social Sciences, Macao Polytechnic University, Macao 999078, China; osakacool@hait.edu.cn
* Correspondence: czhu@must.edu.mo; Tel.: +853-6534-1529
† These authors have contributed equally to this work.

Abstract: The objective of this inquiry is to illuminate the attributes of heavy metal contamination and evaluate the potential ecological hazards inherent in the surface soil of Chenzhou City. A comprehensive analysis was conducted on 600 systematically collected soil samples within the study area, utilizing enrichment factors, geo-accumulation indices, comprehensive pollution indices, potential ecological hazard indices, and health risk assessment models to evaluate the degree of heavy metal contamination in the soil, potential ecological risks, and associated health hazards. The findings reveal that the average enrichment factor (EF) for each heavy metal is below 2, with the hierarchy from highest to lowest being Hg > Cd > Cu > Pb > Ni > Zn > Cr > As. Approximately 78.67% of soil samples exhibit no pollution to weak pollution levels based on heavy metal enrichment factors. Moreover, the comprehensive pollution index (IPI_N) indicates that 95.17% of soil samples are within safe and pollution-free levels, indicating an overall environmentally secure setting. However, 2.67% of samples display heightened potential ecological risk levels, primarily concentrated in the southwestern region of the study area, influenced by nearby industrial activities. Additionally, it is noteworthy that both the non-carcinogenic and carcinogenic health hazards emanating from soil heavy metals to adult individuals lie within tolerable thresholds. Among these, arsenic (As), chromium (Cr), and lead (Pb) have been discerned as the principal non-carcinogenic agents. It is of particular significance that only a solitary soil specimen, located in the southwestern quadrant of the investigative region, manifests detectable health perils for children.

Keywords: soil; heavy metals; potential ecological risks; health risks; Chenzhou

1. Introduction

Soil contamination has emerged as a ubiquitous global environmental concern, attracting considerable scrutiny, particularly in light of the pervasive infiltration of heavy metals [1]. Soil contamination by heavy metals can be delineated into natural and anthropogenic origins. The former emanates from inherent processes intrinsic to soil genesis, encompassing the dissolution of elements through mineral weathering and background sources [2]. Amidst the swift evolution of the Chinese economy and its hastened industrialization, diverse heavy metal constituents have permeated the soil via assorted conduits. This incursion not only disrupts the intrinsic ecological functionalities and equilibrium of the soil milieu, but also poses formidable hazards to human well-being, significantly constraining the sustainable progression of regional soil ecosystems.

The heavy metals entrenched within the soil manifest distinctive traits, encompassing concealment, temporal lag, accumulation, and facile enrichment. Perturbed by exogenous pollution, these heavy metals persist in accumulating within the soil, precipitating a diminution in the soil environment's carrying capacity and engendering substantial ecological and environmental perils [3,4]. Heavy metals residing within the soil possess the potential to engender risks to elevated trophic strata of organisms and human well-being via the transmission and enrichment within the food chain. The quandary of soil heavy metal pollution has garnered considerable focus from pertinent regulatory entities and presently stands as a paramount subject in the realm of environmental science research.

Anthropogenic sources serve as extrinsic agents influencing the heavy metal landscape within the soil. Human activities, encompassing industrial coal combustion, mining, metallurgy, the utilization of agricultural fertilizers and pesticides, emissions stemming from vehicular exhaust, and the disposal of electronic products, can induce varying degrees of heavy metal contamination in the soil [5,6]. Nevertheless, heavy metal pollution is distinguished by its formidable concealment, enduring presence, and bio-accumulative proclivity, rendering its remediation from the soil a formidable undertaking. Elevated concentrations of heavy metals not only impede the flourishing of plants within the soil, but also yield diminished crop output and compromised quality of agricultural products. Moreover, these metallic elements can infiltrate the human body through various pathways, precipitating a range of health complications, whether directly or indirectly.

Elevated consumption of lead (Pb) not only precipitates hypertension, but also imposes deleterious effects upon the skeletal, immune, and endocrine systems. It diminishes cognitive faculties in both juveniles and adults, concurrently compromising renal and cardiac functionality. Likewise, the excessive assimilation of copper (Cu), mercury (Hg), and chromium (Cr) engenders adverse health ramifications. Moreover, protracted exposure to cadmium (Cd) renders individuals increasingly susceptible to pulmonary carcinoma and skeletal fractures. Nickel (Ni), too, qualifies as a heavy metal carcinogen. Depending on the dosage and duration of exposure, it exhibits immunotoxic and carcinogenic properties. The accrual of nickel within the body can precipitate pulmonary fibrosis, and kidney and cardiovascular ailments, as well as respiratory malignancies. Furthermore, an estimated 20% of the global populace suffers from nickel allergies.

Henceforth, the principal objective of this investigation is to utilize Chenzhou City as a case study to examine the attributes of heavy metal pollution in urban surface soil and conduct an ecological health risk assessment. This endeavor holds immense importance for fostering the harmonious coexistence between urban inhabitants and their environment, and for promoting the sustainable progression of human society.

2. Literature Review

Presently, a multitude of scholars have employed a diverse array of research methodologies to undertake pertinent inquiries into soil heavy metal contamination across varied urban settings. With regards to research substance, their primary emphasis revolves around elucidating the attributes of heavy metal presence in urban soil, appraising pollution degrees, dissecting pollution origins, and appraising potential ecological hazards.

For instance, an investigation delved into the pollution status and spatial characteristics of heavy metals in soil across diverse verdant landscapes, including urban parklands, roadside greenery, and adjacent expanses of green. The findings unveiled the pollution levels of these heavy metals, all indicating a state of mild contamination [7]. The concentrations of zinc (Zn), chromium (Cr), nickel (Ni), manganese (Mn), copper (Cu), lead (Pb), and cadmium (Cd) in soils from various types and regions in Shanghai were assessed, revealing that their average concentrations exceeded the local background levels for soil elements [8]. Another investigation discerned that soil heavy metal contamination in industrial zones and bustling traffic corridors in Nanjing exhibits a relatively severe profile, in contrast with the comparatively lighter pollution levels observed in residential and parkland domains. The principal contributors to soil heavy metal pollution have been attributed to vehicular emissions and industrial discharges [9]. A study scrutinizing soil heavy metal pollution along thoroughfares, residential vicinities, parks, and other sectors in Guangzhou revealed that 61% of soil samples demonstrated moderate to severe levels of heavy metal pollution, with human activities discerned as the principal catalyst of this contamination [10].

The mean concentrations of heavy metals in the soil within Xi'an's industrial sectors, transportation hubs, commercial and transit amalgamated regions, residential quarters, cultural enclaves, and parklands surpass the baseline values of soil elements in Shaanxi Province [11].

Research on related underground heavy metal migration phenomena has also made progress. The migration behavior and patterns of persistent pollutants in soil–groundwater systems through heterogeneous distribution characteristics have been studied. The migration mode of heavy metals (HM) may be dominated by solute transport due to groundwater flow [12]. Studies have also been conducted in metallogenic zones along river basins. Self-organizing map (SOM) calculations and stable isotopes were employed to elucidate groundwater hydrochemistry and the migration of heavy metals, with particular emphasis on iron [13]. Additionally, there are pertinent studies in this domain. Through the establishment of an enhanced one-dimensional convection-diffusion model and air deposition model, alongside fitting experiments utilizing authentic urban soil data, the findings demonstrate that the refined model adeptly incorporates factors such as adsorption, decomposition, crop root absorption and discharge, topography, and air deposition. Consequently, it accurately delineates the transmission characteristics and diffusion process of heavy metal pollution [14]. The manner in which metals migrate within sewage sludge is of paramount importance, significantly influencing the assessment of their potential utilization. A study devised a metal mobility ecological risk index and an environmental risk index, with meticulous focus on the mobility of metals within the soil environment [15].

Within the domain of research methodologies, both domestic and international pursuits primarily deploy sophisticated frameworks, such as the geo-accumulation index, single-item pollution index, potential ecological risk index, Nemeiro comprehensive pollution index, and the enrichment factor method. These tools serve as instrumental means to evaluate the intricate landscape of heavy metal pollution within the soil [16,17]. Each of these methodologies possesses inherent advantages and drawbacks in its utilization. As a result, it is judicious to concurrently employ two or more methodologies during pollution assessment, thus augmenting the reliability of research findings. Furthermore, certain investigations employ mathematical statistical techniques such as correlation analysis and principal component analysis to scrutinize the origins of heavy metal concentrations.

In conclusion, considerable research endeavors have been dedicated to the exploration of heavy metal pollution in urban soil. Nevertheless, the focus of study locales primarily revolves around major metropolises and industrial hubs, with relatively limited attention directed towards small and medium-sized cities. The majority of research content delves into the characteristics of heavy metal pollution in soil, spatial variances, pollution degree assessments, and pollution source identification. However, studies concerning the ecological health risk assessment of heavy metal pollution in urban soil remain relatively scarce.

As Chenzhou's urbanization progresses, the issue of "urban maladies" has gained prominence, drawing increased attention from pertinent researchers towards urban heavy metal pollution. Currently, investigations into heavy metal pollution in Chenzhou City primarily center on assessing pollution levels in street dust, riverbeds, forest soil, and farmland soil flanking highways. Nonetheless, there is a dearth of research on the characteristics and evaluation of heavy metal pollution in urban surface soil, and a conspicuous lack of data regarding human exposure to soil heavy metals and investigations into ecological health risks.

This article aims to bridge this scholarly lacuna by selecting Chenzhou City as its focal point, delving into the complexities of heavy metal pollution within urban surface soils, and conducting a thorough ecological health risk assessment. The goal is to provide a theoretical framework supporting the amelioration and mitigation of urban soil pollution, the enhancement of ecological environment quality, and the safeguarding of resident health.

3. Materials and Methods

3.1. Study Area Overview

Chenzhou City resides in the southeastern domain of Hunan Province, cradled between the magnificent Nanling Mountains and the esteemed Luoxiao Mountains. Poised at the crossroads of the Yangtze River and Pearl River systems, its geographical coordinates span from 112°13′ to 114°14′ E longitude and from 24°53′ to 26°50′ N latitude (Figure 1). This locale, endowed with a subtropical monsoon climate, bestows upon its inhabitants a tableau of distinct seasons—a tapestry woven with sultry, rain-laden summers and crisp, arid winters. The annual mean temperature oscillates between 17 and 19 °C, while the average annual precipitation is within the range of 1300 to 1600 mm, adding to the region's atmospheric allure [18,19]. The topographical tapestry of Chenzhou unfolds in a rich panorama, where nearly three-quarters of the terrain is adorned with undulating mountains and hills. Embraced by majestic peaks to the southeast, the western expanse unfolds with more modest elevations, while the central precinct unveils a harmonious blend of hills, plains, and plateaus, painting a tableau of geographic diversity [20]. The undulating terrain gracefully descends from the southeast to the northwest, and this intricate interplay of topography and climate begets a myriad of soil types in the region. Notable among them are limestone red soil, granite red soil, and other variants. Chenzhou's distinctive landscape, characterized by predominant hills, plains with a plateau-like semblance, and an intricately contoured terrain, engenders a rich tapestry of diverse soil compositions [21].

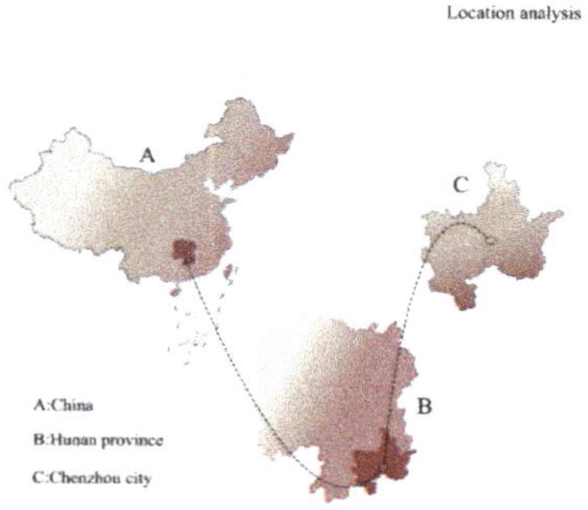

Figure 1. Geographic location analysis map of Chenzhou City.

3.2. Sample Acquisition and Analysis

3.2.1. Sample Acquisition

This investigation and exploration adopted a stratified, staggered, non-equilibrium methodology to strategically position sampling locations and meticulously collect surface soil specimens within the designated study area. The term "stratification" entails the division of the study area into distinct sampling units of varying magnitudes, delineated based on spatial dimensions. Notably, the sampling units of higher echelons harmoniously integrate with those of lower orders in this layered framework [22]. The term "staggered-unbalanced" denotes that the sampling units within higher strata are not evenly dispersed among lower tiers. Instead, a stochastic element is introduced by selectively choosing one or several units from each tier, thus perpetuating the hierarchical division into subsequent layers of sampling units [23].

After comprehensive consideration of the study area, grid density, the significance of statistical analysis data, efficiency, and economic feasibility, a grid with a sampling density of 8 km × 8 km was selected as the top-level cell, which contains multiple lower-level nested levels. Those of 4 km × 4 km, 2 km × 2 km, 1 km × 1 km, 500 m × 500 m, and 250 m × 250 m were randomly selected as target cells (Figure 2). According to the area of the study area, 30 top-level cells were deployed without affecting the layout of each level. In this case, each top-level sampling cell contains 20 survey samples. The specific locations of sample points in each sampling cell can be manually fine-tuned according to soil types, land types, etc., to ensure that when statistics are derived according to different classifications, the data satisfy the statistics requirements.

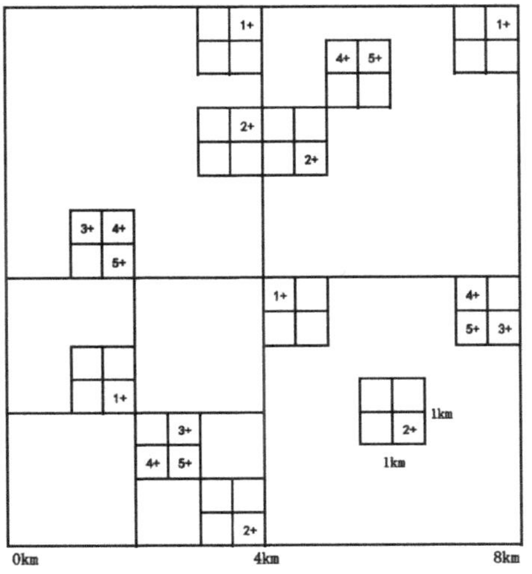

Figure 2. Layered, staggered, non-equilibrium mode sampling point deployment diagram (Note: The numbers represent the sample point numbers deployed within the 4 km × 4 km sampling unit).

At designated sampling nodes, with the GPS positioning point serving as the focal point, 3–5 sampling nodes are delineated within a radius of 20–50 m in all directions. These nodes were amalgamated into composite samples, constituting a total of 600 soil specimens extracted from depths ranging within 0–20 cm beneath the surface (see Figure 3). Positioned across diverse land use categories, the sampled areas encompass 460 parcels of arable land, 13 parcels of orchards, 5 parcels of pastures, 72 parcels of woodlands, 42 parcels of urbanized land, and 8 parcels of miscellaneous land (refer to Figure 4).

When sampling, unrepresentative plots such as ditches, field ridges, and roadbeds were avoided. Litter on the surface was manually removed. Where there were plant growth points, the plants and their roots were removed first. Foreign matter such as gravel was removed at the sampling site. Attention was paid to cleaning sampling tools in time to avoid cross-contamination.

In this study, after digging with shovels and soil augers, bamboo chips were used to scrape off the parts in contact with the metal samplers, and then bamboo chips were used to collect samples. For samples of contaminated soil, appropriate protective measures were taken according to the nature of the pollutants to avoid direct contact with the human body. Following meticulous blending, 1.0–1.5 kg was retained and transferred into a specimen pouch. The soil specimen underwent a process of natural desiccation, pulverization, and filtration through a 10-mesh nylon sieve before being dispatched to the laboratory for meticulous analysis and evaluation. The procedures for soil sampling and processing adhered rigorously to the specifications outlined in DZ/T 0295-2016 [24,25].

3.2.2. Sample Testing and Quality Control

Soil sample analysis test indicators include As, Cd, Cr, Hg, Pb, Ni, Zn, Cu, soil acidity and alkalinity (pH value) indices, Sc, and other element indices.

The methodologies employed for the analysis and testing of soil heavy metals were as follows: As and Hg were scrutinized utilizing atomic fluorescence spectrometry (AFS) [26,27];

Cd, Ni, Cu, Pb, Zn, and Se were scrutinized through the sophisticated technique of inductively coupled plasma mass spectrometry (ICP-MS) [28,29]. Cr was analyzed by X-ray fluorescence spectroscopy (XRF) [30], and the pH value was analyzed using the potentiometric method (POT) [31]. In the course of analysis and testing, soil samples of national premier quality were enlisted to oversee the precision of the procedures. The integrity of the analysis and testing was ensured through the utilization of repetitive samples for precision monitoring. Remarkably, the first-tier reference materials for all analytically tested elements achieved an unblemished 100% success rate, attesting to the rigor of the process. Furthermore, the overall success rate for repetitive sample examination stood at an impressive 99.7%. In parallel, the overall success rate for scrutinizing abnormal point samples reached a commendable 98.9%. The quality of sample analysis and testing, alongside the detection thresholds for each elemental index, aligned impeccably with the technical specifications for ecological geochemical evaluation sample analysis, thereby affirming the reliability of the analytical data.

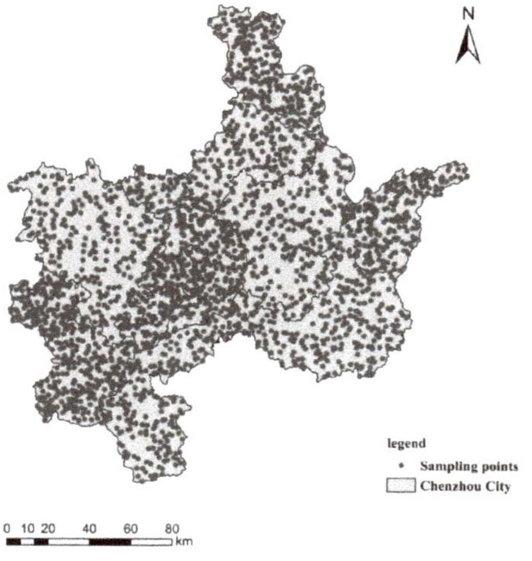

Figure 3. Distribution of soil sampling points in Chenzhou City.

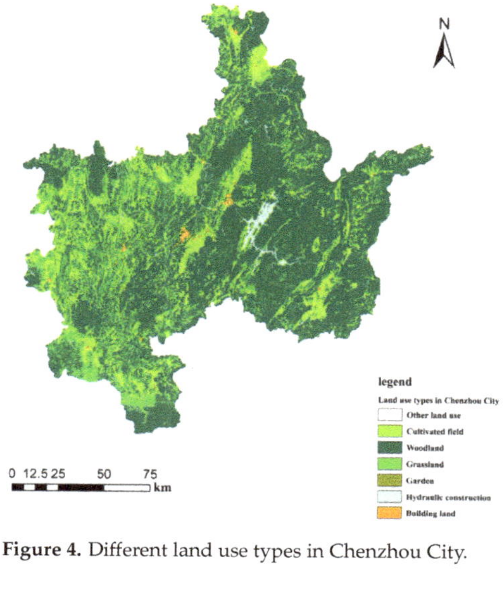

Figure 4. Different land use types in Chenzhou City.

3.3. Evaluation Methods

3.3.1. Enrichment Coefficient

The enrichment factor (EF) represents the enrichment degree of a certain element or substance in a certain area, and is usually used to describe pollution levels or the accumulation of certain chemical elements in the ecosystem [32,33]. The extent of the contribution from anthropogenic and natural origins to the elemental composition of particulate matter was examined and assessed, and pivotal metrics indicative of pollution severity and its sources were quantitatively appraised. This entailed the selection of elements meeting specific criteria as reference or standard elements. The enrichment factor is defined as the quotient between the concentration of contaminating elements within the sample and the concentration of reference elements, juxtaposed with the ratio of these concentrations in the ambient background region [34]. The calculation formula is:

$$EF = \frac{(C_i/C_r)_{sample}}{(C_i/C_r)_{baseline}} \quad (1)$$

Within the equation EF denotes the enrichment factor; Ci signifies the concentration of element i; Cr denotes the concentration of the designated reference element; and "sample" and "baseline" delineate the sample and background, respectively. In this investigation, Sc was chosen as the reference element due to the discernible influence of human activities on its source. The Sc content within the Earth's crust is relatively stable, thereby minimizing its susceptibility to external perturbations [35,36]. The relationship between soil heavy metals and Sc content can serve as a criterion for discerning the extent to which an element is influenced by human activities. Southerland classifies the enrichment factor (EF) into five distinct levels [37,38]. An enrichment factor (EF) less than 2 signifies marginal contamination, while EF values between 2 and 5 denote moderate pollution. EF values falling between 5 and 20 indicate substantial pollution, whereas values ranging from 20 to 40 suggest severe contamination. EF values surpassing 40 denote egregious pollution levels. The categorization of EF is elaborated upon in Table 1.

Table 1. Enrichment factor (EF) classification.

EF Classification	EF	The Degree of Enrichment	Source
1	EF < 2	weak pollution	crustal or soil source
2	2 < EF < 5	moderate pollution	Natural and man-made sources work together
3	5 < EF < 20	significant pollution	man-made pollution sources
4	20 <EF < 40	high pollution	man-made pollution sources
5	EF > 40	extreme pollution	man-made pollution sources

3.3.2. Geo-Accumulation Index

The geo-accumulation index, conceived by the esteemed German scientist G. Müller, serves as a metric to quantify the degree of accumulation of heavy metals or other pollutants within soil matrices [39,40]. In contrast to alternative methodologies for assessing pollution, the geo-accumulation index method stands as distinct because of its inclusion of factors influencing shifts in baseline values due to natural diagenesis. This conceptual framework finds primary application within the domains of environmental science and soil science [41]. Geo-accumulation indices are frequently employed to evaluate the magnitude of soil contamination and aid in the formulation of suitable soil management and remediation strategies. The computational formula is as follows:

$$I_{geo} = \log_2\left[\frac{C_i}{k \times S_i}\right] PI_i = C_i/T_i \quad (2)$$

In the equation, Igeo represents the geo-accumulation index of heavy metal i, where Ci denotes the measured concentration of heavy metal i in soil, and Si stands for the background concentration of element i. The correction coefficient, denoted by k, typically assumes a value of 1.5. Taking into account the genetic type and geomorphology of the surface soil within the study region, alongside the evolutionary dynamics of the environment, this investigation employed the soil environmental background values derived from a comprehensive survey conducted in the northern Haihe Plain area as reference points. These background values serve as benchmarks for comparison. The geo-accumulation index is subsequently categorized into seven tiers based on the Igeo values, as delineated in Table 2.

Table 2. Classification of the geo-accumulation index (Igeo).

Classification	Igeo	Meaning
1	$I_{geo} \leq 0$	No pollution
2	$0 < I_{geo} \leq 1$	No pollution to moderate pollution
3	$1 < I_{geo} \leq 2$	Moderately polluted
4	$2 < I_{geo} \leq 3$	Moderate to strong pollution
5	$3 < I_{geo} \leq 4$	strong pollution
6	$4 < I_{geo} \leq 5$	Strong pollution to extremely strong pollution
7	$I_{geo} > 5$	Extremely polluted

3.3.3. Comprehensive Pollution Index

The pollution index (PIi) and comprehensive pollution index (IPI_N) of each soil heavy metal are employed to assess the degree of soil contamination [42,43]. The calculation formula is delineated as follows:

$$PI_i = C_i / T_i \quad (3)$$

$$IPI_N = \left[\left(IPI_{avg}{}^2 + IPI_{max}{}^2 \right) / 2 \right]^{1/2} \quad (4)$$

In the equation, Ci represents the measured concentration of element i in the surface soil, while Ti signifies the target management value for trace element i in soil. This assessment relies upon the screening threshold outlined in the Soil Environmental Quality Standard for Agricultural Land and Soil Pollution Risk Management and Control (GB15618-2018) [44]. IPI_{avg} represents the mean value of all pollution indices (PIi) of heavy metals within the surveyed soil, while IPI_{max} denotes the highest recorded value. Based on the magnitude of the IPI_N value, which reflects the extent of soil contamination, it is categorized into five levels ranging from absence of pollution to severe pollution. The corresponding relationship between IPI_N and pollution severity is outlined as follows: IPI_N values less than or equal to 0.7 indicate safety, those exceeding 0.7 but not surpassing 1.0 signify an early warning, while values between 1.0 and 2.0 indicate mild pollution. Similarly, IPI_N values ranging from 2.0 to 3.0 suggest moderate pollution, and values exceeding 3.0 indicate severe pollution. The parameters of IPI_N are detailed in Table 3.

Table 3. Classification of the integrated pollution index (IPI_N).

Classification	IPI_N	Meaning
1	$IPI_N \leq 0.7$	Safety
2	$0.7 < IPI_N \leq 1.0$	Early warning
3	$1.0 < IPI_N \leq 2.0$	Light pollution
4	$2.0 < IPI_N \leq 3.0$	Moderately polluted
5	$IPI_N \geq 3.0$	Heavy pollution

3.3.4. Potential Ecological Risk Index

The potential ecological risk index (RI) method stands as the prevailing approach for evaluating the magnitude of soil heavy metal contamination and its concomitant ecological ramifications. This method, pioneered by the esteemed Swedish geochemist, Lars Hakanson, embodies a cornerstone in environmental assessment [45]. This method not only considers the concentration of heavy metals, but also integrates ecological, environmental, and toxicological effects, rendering it a widely adopted approach for ecological risk assessment. Its computational formula is delineated as follows:

$$RI = \sum_{i=1}^{n} E_r^i = \sum_{i=1}^{n} (T_i^T \times C_f^i) = \sum_{i=1}^{n} (T_r^i \times \frac{C_i}{C_n^i}) \tag{5}$$

In the formula, C_f^i represents the pollution index of a specific metal; Ci denotes the measured concentration of a specific heavy metal within the soil; and C_n^i represents the reference concentration of a specific heavy metal. Given the geographical context of the study area nestled within the middle and lower reaches of the Yangtze River Plain, the fluctuation in heavy metal concentrations within its soil is predominantly influenced by the underlying soil parent material and geological backdrop. Consequently, this investigation adopts the eastern Gansu Plain as a comparative model due to its similar genetic characteristics and analogous material origins to those found in the study area. The soil environmental baseline in the eastern territory serves as the benchmark value; E_r^i represents a singular potential ecological risk index for heavy metals; T_r^i signifies the toxicity response parameter of a particular heavy metal.

The toxicity factor increases with the increase in metal toxicity. The toxicity response parameters of each heavy metal are 1 for Zn; 2 for Cr and Mn; 5 for Cu, Ni, and Pb; 10 for As; 30 for Cd; and 40 for Hg.

RI represents the cumulative potential ecological risk index. Based on E_r^i and RI, the classification is conducted for both the potential ecological hazards attributed to individual factors and the comprehensive total potential ecological hazards [46,47] (Table 4).

Table 4. Hakanson potential ecological hazard evaluation indicators.

Index	Ecological Hazard					
	Slight	Medium	Powerful	Very Strong	Extremely Strong	Slight
E_r^i	<40	40–80	80–160	160–320	≥320	<40
RI	<150	150–300	300–600	600–1200	≥1200	<150

3.3.5. Health Risk Assessment

Health risk assessment entails establishing a nexus between human well-being and environmental contamination, while quantitatively scrutinizing and elucidating the perils posed by environmental pollution to human health [48]. The assessment of health risks posed by soil heavy metals entails a meticulous inquiry into the potential detrimental effects these elements may inflict upon human well-being. The accumulation of excessive quantities of heavy metals such as arsenic (As), cadmium (Cd), chromium (Cr), mercury (Hg), lead (Pb), nickel (Ni), zinc (Zn), copper (Cu), and scandium (Sc) within soil matrices may precipitate adverse consequences for both the local ecosystem and human health [49].

Soil heavy metals permeate the human organism through three primary avenues: direct oral ingestion, respiratory inhalation, and dermal contact. They pose both non-carcinogenic and carcinogenic hazards to human health. Presently, the predominant international framework for assessing health risks comprises carcinogenic risk assessment models and non-carcinogenic risk assessment models [50,51]. The formulas for their calculation are as follows:

$$HQ = \sum HQ_i = \sum \frac{ADD_{iing} + ADD_{iinh} + ADD_{iderm}}{RfD_i} \tag{6}$$

$$CR = \sum CR_i = \sum (ADD_{iing} + ADD_{iinh} + ADD_{iderm}) \times SF \tag{7}$$

In the formulas, HQ signifies the composite index of non-carcinogenic health risks associated with all heavy metals, where HQ < 1 denotes that the non-carcinogenic risk posed by heavy metals is negligible, whereas any deviation from this indicates a non-carcinogenic risk; CR denotes the carcinogenic health risk index associated with all heavy metals, and CR values below 1×10^{-6} indicate it will not have a significant impact on health. A CR value between 1×10^{-6} and 1×10^{-4} is with the acceptable range. A CR value exceeding 1×10^{-4} indicates that exposure to a specific environment will cause an individual to have a certain risk of cancer.

ADD_{iing}, ADD_{iinh}, and ADD_{iderm} symbolize the mean daily exposure of a heavy metal via oral ingestion, respiratory inhalation, and dermal contact, respectively. RfDi represents the non-carcinogenic average daily intake of heavy metal "i", while SF symbolizes the carcinogenic slope factor.

Within the framework of the soil heavy metal health risk assessment model, the average daily exposure to heavy metals varies between children and adults. It becomes imperative to factor in the average exposure for each individual child and adult, subsequently distributing this exposure uniformly over their entire lifespan. The calculation formulas are as follows:

$$HQ = \sum HQ_i = \sum \frac{ADD_{iing} + ADD_{iinh} + ADD_{iderm}}{RfD_i} \quad (8)$$

$$CR = \sum CR_i = \sum (ADD_{iing} + ADD_{iinh} + ADD_{iderm}) \times SF \quad (9)$$

$$HQ = \sum HQ_i = \sum \frac{ADD_{iing} + ADD_{iinh} + ADD_{iderm}}{RfD_i} \quad (10)$$

In the equations, IngR and InhR denote the daily rates of soil intake and soil inhalation, respectively; Ci signifies the concentration of a particular contaminant heavy metal within the soil; EF denotes the frequency of exposure; ED represents the duration of exposure in years; BW denotes the mean body weight; AT represents the mean duration of exposure; PEF signifies the emission factor for surface dust; SA denotes the surface area of exposed skin; SL represents the coefficient of skin adhesion; and ABS signifies the factor of skin absorption.

The exposure parameters, along with the reference dose (RfD) and slope factor (SF) values pertaining to various exposure pathways within the soil heavy metal health risk assessment model, are drawn from the prescribed values delineated in HJ 25.3-2019, issued by the Ministry of Ecology and Environment of China, and the "Outline of the Chinese Population Exposure Parameters Manual (Children's Volume)" [52–54]. The reference measurements and exposure parameters for various exposure pathways in this investigation are elaborated upon in Tables 5 and 6.

Table 5. The reference quantities and carcinogenic slope factors for diverse heavy metal exposure pathways.

Heavy Metal	Reference Metrology (RfD) [mg/(kg d)]				Carcinogenic Slope Factor (SF) [(kg d)/mg]		
	Oral Intake	Skin Contact	Breathing Inhalation (Aldult)	Breathing Inhalation (Children)	Oral Intake	Skin Contact	Breathing Inhalation
As	3.0×10^{-4}	3.0×10^{-4}	3.52×10^{-6}	5.86×10^{-6}	1.5	1.5	4.3×10^{-3}
Cd	1.0×10^{-3}	2.5×10^{-5}	2.35×10^{-6}	3.91×10^{-6}	6.1	6.1	6.3
Cr	3.0×10^{-3}	7.5×10^{-5}	2.35×10^{-5}	3.91×10^{-5}	-	-	-
Cu	4.0×10^{-2}	4.0×10^{-2}	-	-	-	-	-
Hg	3.0×10^{-4}	2.1×10^{-5}	7.04×10^{-5}	1.17×10^{-5}	-	-	-
Ni	2.0×10^{-2}	8.0×10^{-4}	2.11×10^{-5}	3.52×10^{-5}	-	-	-
Pb	3.5×10^{-3}	5.3×10^{-4}	8.21×10^{-5}	1.37×10^{-4}	-	-	-
Zn	3.0×10^{-1}	3.0×10^{-1}	-	-	-	-	-

Table 6. Parameters for exposure risks associated with heavy metals.

Symbol	Reference Name	Unit	Adult (Reference Value)	Children (Reference Value)
ED	years of exposure	A	24	6
BW	average weight	Kg	61.80	19.20
EF	exposure frequency	d/a	350	350
AT	average exposure time	d	Carcinogenic 27,740 Non-carcinogenic 9125	Carcinogen 27,740 Non-carcinogenic 9125
IngR	daily soil intake	mg/d	100	200
InhR	daily respiratory volume	m^3/d	14.50	7.50
SA	exposed skin surface area	cm^2	5373.99	2848.01
SL	skin adhesion coefficient	$mg/(cm^2\ d)$	0.07	0.20
PEF	surface emission factor	m^3/kg	13.6×10^9	13.6×10^9
ABS	skin absorption factor	dimensionless	As: 0.03 Cd: 0.001 Hg: 0.05 Ni: 0.001	Cr:0.001 Cu: 0.06 Pb:0.006 Zn: 0.02

4. Results and Discussion

4.1. Examination of Soil Heavy Metal Levels and Contamination Severity

The average contents of heavy metals As, Cd, Cr, Cu, Hg, Ni, Pb, and Zn in the surface soil of Chenzhou in the study area were 11.2, 0.23, 69.7, 31.1, 0.049, 32.9, 27.81, and 82.9 mg/kg, respectively (Table 7). The values ranging from 7.6 to 9.3 were mainly alkaline and strongly alkaline. The coefficient of CV serves as an indicator of the uniformity and extent of variation among elements present in the soil. In the study area, the soil heavy metals As, Cd, Cu, Hg, and Pb exhibit notable variability. Notably, cadmium displays the most pronounced coefficient of variation and highest level of heterogeneity, potentially influenced by anthropogenic activities. Conversely, Cr, Ni, and Zn demonstrate moderate variability.

Table 7. Characteristics of heavy metal content in surface soil in Chenzhou.

Item	Element (mg/kg)								pH
	As	Cd	Cr	Cu	Hg	Ni	Pb	Zn	
Maximum value	44.80	2.73	107.00	156.0	0.274	58.90	198.0	329.0	9.30
Minimum value	3.90	0.006	47.90	11.60	0.015	13.50	8.90	33.80	7.60
Average value	11.20	0.230	69.70	31.10	0.049	32.90	27.80	82.90	8.30
Coefficient of variation (%)	36.30	83.50	15.80	52.30	53.70	22.50	47.30	32.20	2.80
Agricultural land filter value	25.00	0.60	250.00	100.00	3.40	190	170.00	300.00	/
Agricultural land control value	100.00	4.00	1300.00	/	6.00	/	1000	/	/
Background values	10.80	0.16	66.30	22.70	0.028	28.90	21.20	73.80	8.22

As per the guidelines stipulated in GB15618-2018 [55] Soil Environmental Quality Agricultural Land Soil Pollution Risk Management and Control Standards, the concentrations of Cr, Hg, Ni, and Zn in all 600 soil samples within the study area fell below the threshold designated for agricultural land soil pollution risk screening. This observation suggests a safe and uncontaminated level of risk. While the levels of As, Cd, Cu, and Pb in certain soil samples surpassed the specified threshold for agricultural land soil pollution risk screening, they remained below the threshold for agricultural land soil pollution risk control. Notably, cadmium concentrations exceeded the agricultural land soil pollution risk screening threshold at 22 sampling points, predominantly situated in the southwest region of Chenzhou.

The heavy metals arsenic (As), cadmium (Cd), chromium (Cr), copper (Cu), lead (Pb), nickel (Ni), and zinc (Zn) in the study region exhibit a pronounced positive correlation with the content of Sc ($p < 0.01$), and there exists a positive correlation between mercury (Hg) and the content of Sc ($p < 0.05$) (Figure 5). Hence, in the evaluation of soil heavy metal enrichment factors within the study area, it is judicious to designate Sc as the reference element. Sc stands out due to its solitary natural geological provenance, its intimate association with pollutants, and its consistent spatial dispersion [56,57].

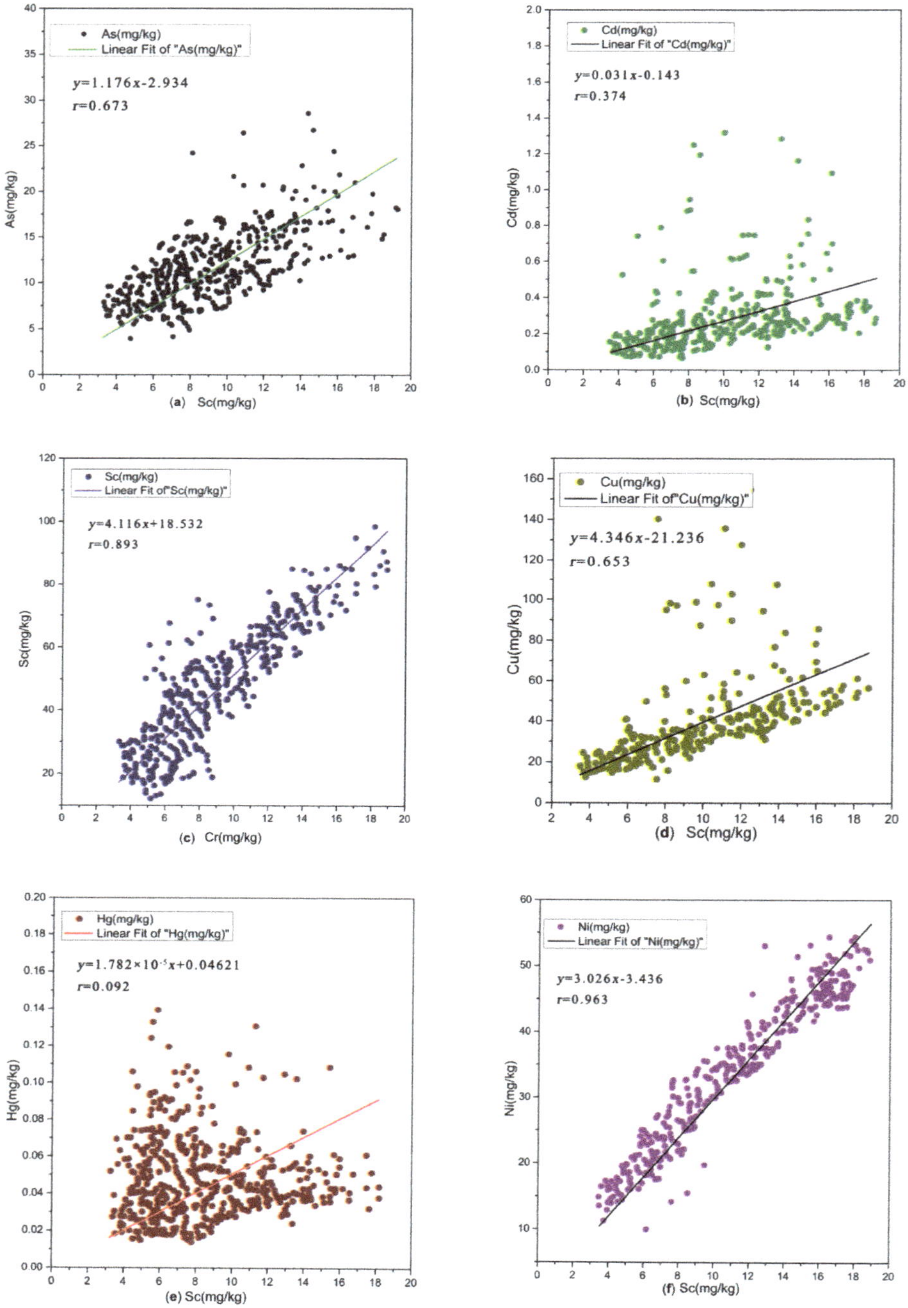

Figure 5. *Cont.*

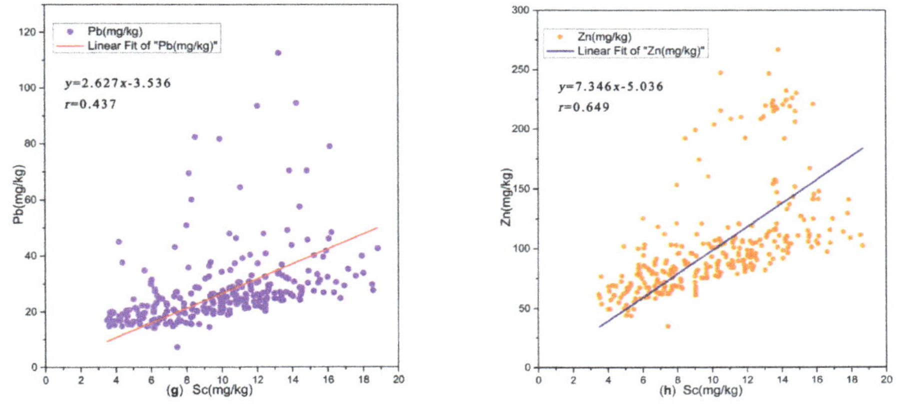

Figure 5. Correlation between heavy metals and Sc content in soil in the study area. ((**a**): Correlation between As and Sc content; (**b**): Correlation between Cd and Sc content; (**c**): Correlation between Cr and Sc content; (**d**): Correlation between Cu and Sc content; (**e**): Correlation between Hg and Sc content; (**f**): Correlation between Ni and Sc content; (**g**): Correlation between Pb and Sc content; (**h**): Correlation between Zn and Sc content).

The assessment of soil heavy metal enrichment factors within the study locale, as delineated in Table 8, indicates that the enrichment factors for chromium (Cr) and nickel (Ni) across all soil samples were below 2, suggesting a general state of minimal to negligible pollution. The manifestations of pollution wrought by mercury (Hg) and cadmium (Cd) were large, with their impact weightier than that of other heavy metals, and a multitude of samples exhibiting contamination. Specifically, there are 94 sample points demonstrating moderate contamination by Hg, and 26 by Cd, constituting 15.67% and 4.33% of the total samples, respectively. Moreover, seven samples evinced significant pollution levels, accounting for 1.17% of the total, while no samples registered highly or extremely contaminated levels.

Table 8. Classification statistics of heavy metal enrichment factors in surface soil within Chenzhou City.

Heavy Metal	EF Average	Mean Value of EF for Different Land Use Types							Number of Samples at Each Level				
									EF < 2	2 < EF < 5	5 < EF < 20	20 < EF < 40	EF > 40
		Cultivated Field	Garden	Woodland	Grassland	Hydraulic Construction	Building Land	Other Land Use	Weak Pollution	Medium Pollution	Significant Contamination	Highly Polluted	Extremely Polluted
As	0.83	0.83	0.77	0.85	0.84	0.970	0.85	0.68	596	4	0	0	0
Cd	1.17	1.20	0.99	0.96	1.00	0.731	1.25	1.03	567	26	7	0	0
Cr	0.86	0.86	0.86	0.84	0.86	0.749	0.86	0.88	600	0	0	0	0
Cu	1.08	1.09	0.97	0.98	0.89	0.963	1.20	0.93	583	16	1	0	0
Hg	1.46	1.49	1.67	1.11	1.07	2.304	1.60	1.21	499	94	7	0	0
Ni	0.92	0.92	0.93	0.91	0.91	0.908	0.91	0.90	600	0	0	0	0
Pb	1.06	1.08	0.94	0.95	0.97	0.973	1.05	1.00	584	15	1	0	0
Zn	0.91	0.91	0.90	0.85	0.82	0.764	0.99	0.79	595	5	0	0	0

The enrichment factors of Pb and Cu each reached a significant level of pollution in one sample. The average soil heavy metal enrichment factors in descending order were Hg > Cd > Cu > Pb > Ni > Zn > Cr > As. The factor mean values were generally at no pollution to weak pollution levels (Figure 6).

The statistical analysis of soil heavy metal enrichment factors across various land use categories in the study area reveals that the mean disparities in chromium (Cr) and nickel (Ni) enrichment factors remain relatively stable across different land use types. This suggests that the spatial dispersion of these heavy metals is predominantly influenced by geological background variables. Conversely, in comparison to other land use types, construction areas exhibit the highest average enrichment factors for cadmium (Cd), copper (Cu), and zinc (Zn), indicative of the significant impact of human activities on the enrichment levels of these metals.

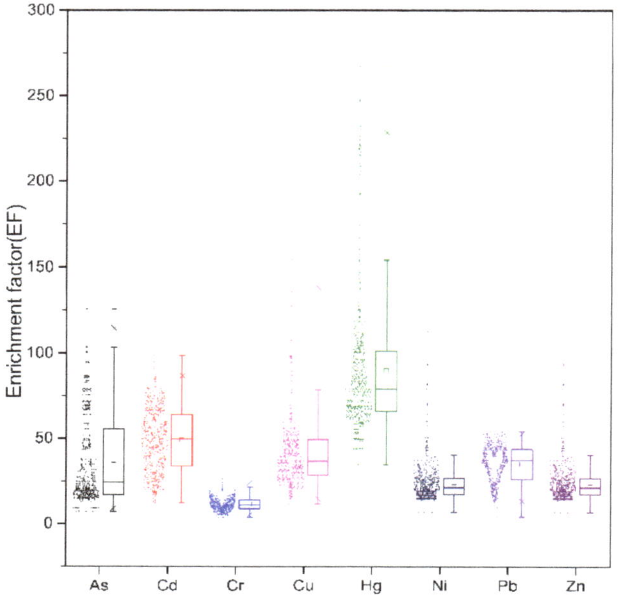

Figure 6. Box plot of soil heavy metal enrichment factors.

Table 9 delineates the assessment outcomes regarding soil heavy metal accumulation indices in the designated region. The descending order of average accumulation index values for heavy metals is as follows: mercury (Hg), cadmium (Cd), copper (Cu), lead (Pb), nickel (Ni), zinc (Zn), chromium (Cr), and arsenic (As), aligning with the enrichment factor evaluation results. Notably, the investigation reveals significant soil pollution attributable to mercury (Hg) and cadmium (Cd) within the area. Mercury-contaminated samples constitute 44.50%, 6.33%, and 1.00% of the total samples, respectively, while Cd-contaminated samples comprise 22.83% of the total, including one sample reaching a level of severe pollution. However, no samples exhibit heavy to extremely severe pollution. On the other hand, the geo-accumulation index (Igeo) of arsenic (As), chromium (Cr), nickel (Ni), lead (Pb), and zinc (Zn) generally indicates a non-polluted state, with a minor presence of copper (Cu) and lead (Pb) samples exhibiting light to moderate pollution levels. Samples displaying severe contamination or above are absent.

The comprehensive soil heavy metal pollution index (IPI_N) within the study region spans from 0.2 to 3.32, with an average of 0.39, underscoring a relatively low level of soil heavy metal pollution and an overall environment that is deemed relatively safe.

The assessment outcomes of the comprehensive pollution index (PIN) reveal that the IPI_N of 571 samples within the study area indicates a level of safety and absence of pollution, constituting 95.17% of the total sample pool. The IPI_N of 14 samples reached an early warning status, representing 2.33% of the total sample set. Among these, twelve samples exhibited a light contamination level, while two samples showed moderate contamination, and only one sample reached a severe contamination level. Thus, the proportion of samples in a contaminated state amounted to 2.50% of the total samples.

Of the aforementioned fifteen samples exhibiting contamination, thirteen samples were dominated by cadmium (Cd) in terms of the IPI_{max} of the comprehensive soil heavy metal pollution index, while the remaining two contaminated samples were predominantly influenced by copper (Cu) in this regard. This suggests that cadmium (Cd) primarily contributes to the contamination level surpassing the threshold in these samples.

Table 9. Classification statistics of heavy metal accumulation indices in surface soil within Chenzhou City.

Heavy metal	Exponential Average	Number of Samples at Each Level						
		$I_{geo} < 0$ (No Pollution)	$0 < I_{geo} \leq 1$ (Light Pollution)	$1 < I_{geo} \leq 2$ (Medium Pollution)	$2 < I_{geo} \leq 3$ (Medium–Heavy Pollution)	$3 < I_{geo} \leq 4$ (Heavy Pollution)	$4 < I_{geo} \leq 5$ (Heavy–Very Heavy Pollution)	$5 \leq I_{geo}$ (Extremely Polluted)
As	−0.62	545	52	3	0	0	0	0
Cd	−0.24	463	106	22	8	1	0	0
Cr	−0.53	594	6	0	0	0	0	0
Cu	−0.26	441	139	18	2	0	0	0
Hg	−0.06	289	267	38	6	0	0	0
Ni	−0.44	538	62	0	0	0	0	0
Pb	−0.28	507	80	11	2	0	0	0
Zn	−0.47	547	47	6	0	0	0	0

The soil heavy metal content is affected by different land use patterns. The ranking of the average soil heavy metal comprehensive pollution index IPI_N of each land use pattern in the region is construction land (0.425) > cultivated land (0.397) > water and hydraulic construction land (0.391) > forested land (0.374) > artificial grass (0.344) > garden (0.308). The statistical results further reflect that the intensity of human production activities directly affects the degree of soil heavy metal pollution.

4.2. Evaluation of Prospective Ecological Hazards Associated with Soil Heavy Metals

The assessment outcomes regarding the potential ecological risk extent of soil within the study domain, as elucidated in Table 10, indicate that the single-factor potential ecological risk indices of chromium (Cr), copper (Cu), nickel (Ni), and zinc (Zn) across all soil samples in the region fall below 40, signifying a minor ecological risk. The potential ecological risk index for arsenic (As) ranges from 3.64 to 41.52, while, for lead (Pb), it varies from 2.06 to 45.93, indicating slight to moderate ecological risks, but primarily slight. Conversely, the potential ecological risk index for cadmium (Cd) ranges from 11.98 to 511.26, encompassing slight to very severe ecological risks, with predominantly minor risks constituting 67.00% of the total samples, followed by samples exhibiting moderate and severe ecological risks accounting for 27.67% and 3.50% of the total samples, respectively. Regarding mercury (Hg), the risk index spans from 21.18 to 391.12, with a distribution ranging from minor to very severe ecological risks, with mainly medium risks, constituting 61.17% of the total samples, while samples with severe ecological risks make up 22.83% of the total samples. Consequently, the primary potential ecologically hazardous heavy metal elements present in the soil of the study area are mercury (Hg) and cadmium (Cd).

The range of distribution for the overall potential ecological index (RI) of soil heavy metals spans from 58.21 to 837.53. Within this range, ecological risks vary from slight to severe, predominantly comprising slight and moderate risks, which constitute 69.67% and 27.33% of the total, respectively. Additionally, a minority of sample points exhibit strong potential ecological risks, comprising 2.67% of the total sample set.

The spatial distribution map of RI (Figure 7) illustrates that Linwu County, situated in the southwestern region of Chenzhou, exhibits the most elevated ecological risk. In the local soils there exist formidable potential ecological risks alongside a minor presence of exceedingly potent ecological risks. The inquiry has revealed the presence of localized areas with significant potential ecological risks within Linwu County. Enterprises engaged in the reclamation of non-ferrous metal waste and mining activities are prevalent. Linwu

County boasts a mining legacy spanning over four centuries. However, the operations of mining enterprises have inflicted significant environmental pollution upon the mining locale. Following enterprise production, myriad tailings and waste ores remain, extensively scattered throughout the rainfall catchment area spanning hundreds of square kilometers in the upper reaches of the Wushui River. This accumulation has precipitated severe heavy metal pollution and latent safety perils, including landslides and debris flows, within the soil environment of the mining zone and downstream areas. Prolonged industrial activities have escalated ecological risks associated with heavy metals such as cadmium (Cd) and mercury (Hg) in the surrounding soil.

Table 10. Classification statistics of the potential ecological risk index of soil heavy metals in Chenzhou City.

Hazard Index	Metal Element	Distribution Area	Number of Samples at Each Level				
			Light	Middle	Powerful	Very Strong	Extremely Strong
E_i	As	3.64–41.52	599	1	0	0	0
	Cd	11.98–511.26	402	166	21	9	2
	Cr	1.45–3.21	600	0	0	0	0
	Cu	2.56–34.35	600	0	0	0	0
	Hg	21.18–391.12	81	367	137	14	1
	Ni	2.33–10.18	600	0	0	0	0
	Pb	2.09–46.58	599	1	0	0	0
	Zn	0.46–4.46	600	0	0	0	0
RI		58.21–837.53	418	164	16	2	0

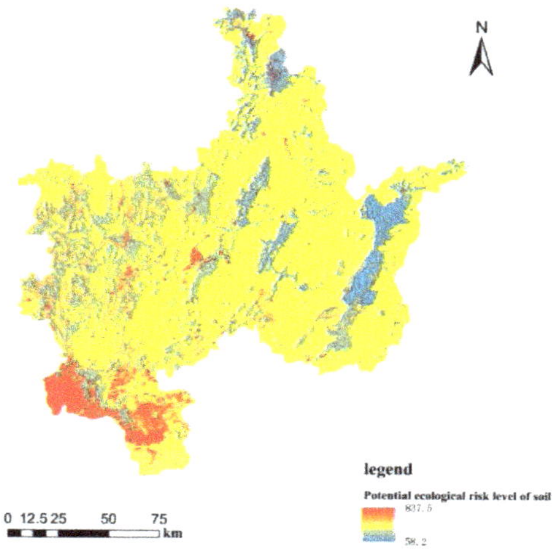

Figure 7. Distribution of soil potential ecological risk levels in the study area.

4.3. Health Risk Appraisal

4.3.1. Non-Carcinogenic Health Hazard Evaluation

The outcomes of the daily exposure risk assessment within the non-carcinogenic health risk model, as delineated in Table 11, reveal a notable disparity. Specifically, the mean daily exposure to soil heavy metals via direct oral ingestion, primarily through crops, far surpasses the average daily exposure via skin contact and inhalation.

The hierarchy of non-carcinogenic average daily exposure, ranging from most to least significant, through the three distinct pathways, is as follows: $ADD_{ing} > ADD_{derm} > ADD_{inh}$. In sequence from greatest to least, the average daily intake of various heavy metals through the three pathways unfolds as follows: Zn > Cr > Ni > Cu > Pb > As> Cd > Hg. The solitary-path average daily exposure and cumulative daily exposure of children to all soil heavy metals surpassed those of adults.

Table 11. The mean daily exposure to non-carcinogenic heavy metals in soil [mg/(kg·d)].

Metal Element	Adult				Children			
	ADD_{ing}	ADD_{inh}	ADD_{derm}	ADD_{adult}	ADD_{ing}	ADD_{inh}	ADD_{derm}	ADD_{child}
As	1.66×10^{-5}	1.77×10^{-9}	1.87×10^{-6}	1.85×10^{-5}	4.33×10^{-5}	2.51×10^{-9}	4.16×10^{-6}	4.75×10^{-5}
Cd	3.48×10^{-7}	3.71×10^{-11}	1.31×10^{-9}	3.50×10^{-7}	9.09×10^{-7}	5.26×10^{-11}	2.91×10^{-9}	9.12×10^{-7}
Cr	1.04×10^{-4}	1.10×10^{-8}	3.90×10^{-7}	1.04×10^{-4}	2.70×10^{-4}	1.56×10^{-8}	8.64×10^{-7}	2.71×10^{-4}
Cu	4.61×10^{-5}	4.92×10^{-9}	1.04×10^{-5}	5.65×10^{-5}	1.20×10^{-4}	6.96×10^{-9}	2.31×10^{-5}	1.43×10^{-4}
Hg	7.23×10^{-8}	7.71×10^{-12}	1.36×10^{-8}	8.60×10^{-8}	1.89×10^{-7}	1.09×10^{-11}	3.02×10^{-8}	2.19×10^{-7}
Ni	4.89×10^{-5}	5.21×10^{-9}	1.84×10^{-7}	4.90×10^{-5}	1.27×10^{-4}	7.38×10^{-9}	4.08×10^{-7}	1.28×10^{-4}
Pb	4.14×10^{-5}	4.41×10^{-9}	9.35×10^{-7}	4.23×10^{-5}	1.08×10^{-4}	6.25×10^{-9}	2.07×10^{-6}	1.10×10^{-4}
Zn	1.23×10^{-4}	1.31×10^{-8}	9.27×10^{-6}	1.32×10^{-4}	3.21×10^{-4}	1.86×10^{-8}	2.06×10^{-5}	3.42×10^{-4}
ADD	3.80×10^{-4}	4.05×10^{-8}	2.31×10^{-5}	4.03×10^{-4}	9.92×10^{-4}	5.74×10^{-8}	5.12×10^{-5}	1.04×10^{-3}

The outcomes of the non-carcinogenic health risk assessment reveal a distinctive order in the non-carcinogenic health risk index across diverse exposure routes: $HQ_{ing} > HQ_{derm} > HQ_{inh}$. This hierarchy underscores that oral ingestion serves as the predominant avenue for non-carcinogenic risks associated with soil heavy metals. Indeed, non-carcinogenic health risks exhibit a correlation with exposure pathways. Notably, whether pertaining to children or adults, the mean value of each soil heavy metal's non-carcinogenic risk index registers below 1. The ranking of non-carcinogenic risk for different heavy metals, from highest to lowest, delineates as follows: As > Cr > Pb > Ni > Cu > Hg > Zn > Cd. This sequence reflects the varied conditions prevailing within the study area. Importantly, individual soil heavy metals do not present non-carcinogenic risks to human health.

The non-carcinogenic health risk index, denoted as HQ_{adult}, spans from 0.078 to 0.371 among adults, with an average of 0.122. All HQ_{adult} values fall below 1, signifying that the non-carcinogenic health risks posed by the eight heavy metals in the soil to adults are negligible and may be disregarded.

The pediatric heavy metal non-carcinogenic health risk index, denoted as HQ_{child}, spans from 0.198 to 1.007, with an average value of 0.310. Notably, one sample exhibits an HQ_{child} greater than 1, suggesting that heavy metals present in 99.83% of the study area's samples pose minimal non-carcinogenic health risks to children and may be deemed inconsequential. Nonetheless, a mere 0.17% of isolated soil samples harbor heavy metals that entail non-carcinogenic health risks to children.

The non-carcinogenic risk index for children surpasses that of adults, owing to their distinctive behavioral and physiological traits, rendering them more susceptible to environmental contaminants per unit of body weight than adults. Upon examining the average composition ratio of the non-carcinogenic risk index HQ, as delineated in Table 12, it becomes evident that arsenic (As), chromium (Cr), and lead (Pb) stand out as the primary non-carcinogenic factors among heavy metals in the region. Indeed, the collective HQ_i values of these three elements collectively represent over 95% of the non-carcinogenic risk index HQ.

Table 12. Contribution rate of soil heavy metals HQ for adults and children.

	As	Cd	Cr	Cu	Hg	Ni	Pb	Zn
Adult	50.88	0.34	32.95	1.16	0.73	2.39	11.19	0.36
Children	51.11	0.33	32.87	1.16	0.67	2.28	11.22	0.37

4.3.2. Cancer Health Risk Assessment

In light of the absence of established carcinogenic slope factors pertaining to the remaining six heavy metals, save for arsenic (As) and cadmium (Cd), this inquiry exclusively delved into the carcinogenic hazards emanating from soil contact with As and Cd. Correspondingly, the findings from the daily exposure risk assessment employing the non-carcinogenic health risk model remain unaltered. The hierarchy of average daily carcinogenic exposure, in descending order, delineates ingestion as surpassing dermal contact, followed by inhalation. Notably, the primary contributor to the cancer risk index (CR) emerges as the direct oral ingestion of soil-borne heavy metals through the consumption of crops.

The assessment outcomes derived from the cancer health risk model within the study locale reveal that the adult cancer risk index value (CR_{adult}) is dispersed within the range of 4.40×10^{-6} to 4.22×10^{-5}, with an average value of 9.82×10^{-6}. Based on the CR_{adult} value of 600 samples in the study area within the acceptable range of 10^{-6}~10^{-4}, soil exposure will not cause significant carcinogenic health risks to adults.

The cancer risk index values for children (CR_{child}) are distributed across a spectrum between 1.13×10^{-5}~1.09×10^{-4}, with an average value of 2.53×10^{-5}. Among them, the CR_{child} values of 599 samples in the study area are between 10^{-6}~10^{-4}, and the soil exposure is not significant. It will not engender considerable carcinogenic health hazards for children. The cancer risk index (CR_{child}) value of a singular soil sample fell beyond the anticipated range, exhibiting an excess rate of 0.17%.

Within this set, the individual child cancer risk indices pertaining to each heavy metal, arsenic (As) and cadmium (Cd), fall within the acceptable spectrum. Thus, on the whole, the cumulative cancer risk attributed to soil heavy metals within the study vicinity remains within acceptable bounds. Only isolated soil samples containing heavy metals pose carcinogenic health risks to children, warranting due attention.

5. Conclusions

(1) The variation coefficients of soil elements such as arsenic (As), cadmium (Cd), copper (Cu), mercury (Hg), and lead (Pb) in the study area exceed 36%, rendering them highly variable constituents. Notably, cadmium exhibits the highest variation coefficient, displaying localized enrichment characteristics attributed to external influences.

(2) The mean heavy metal enrichment factor (EF) value for each soil sample was below 2, indicating an overall absence to minimal pollution level in the soil. Evaluation using the geo-accumulation index (Igeo) suggests that soil samples exhibit generally pollution-free levels of arsenic (As), chromium (Cr), nickel (Ni), lead (Pb), and zinc (Zn). Moreover, the comprehensive soil pollution index (IPI_N) indicates that 95.17% of soil samples in the area are within safe and pollution-free levels, underscoring the overall environmental safety of the region.

(3) The soil within the region predominantly exhibits mild and moderate levels of potential ecological risk, comprising 69.67% and 27.33% respectively. A minority of samples, constituting 2.67% of the total, demonstrate pronounced potential ecological risk levels, primarily concentrated within the research area. In the southwestern sector of the district, influenced by the waste nonferrous metal recycling, mining, and other associated industrial operations of neighboring enterprises, the ecological hazards posed by heavy metals such as cadmium (Cd) and mercury (Hg) in the surrounding soil are relatively elevated.

(4) The human health risk assessment outcomes pertaining to heavy metals indicate that oral ingestion constitutes the primary pathway of exposure to heavy metals present in soil. In the study area, arsenic (As), chromium (Cr), and lead (Pb) emerge as the principal non-carcinogenic factors within the soil matrix. The non-carcinogenic health risk index (HQ_{adult}) for adults remains consistently below 1, rendering the associated hazards negligible. However, the non-carcinogenic health risk index (HQ_{child}) for children exceeds 1 in one sample, signifying a certain level of non-carcinogenic risk to children. This underscores the heightened vulnerability of children to the threats posed by heavy metals compared to adults.

6. Future Directions and Research Opportunities

In contemplating future trajectories and avenues for research concerning soil heavy metal risk assessment in Chenzhou, it is imperative to accentuate the significance of bioavailability scores in evaluating heavy metal risks. These scores offer critical insights into the accessibility and potential harm posed by heavy metals present in soil to biological organisms. Moving forward, further exploration into refining and standardizing methodologies for determining bioavailability scores across diverse environmental contexts holds great promise. Additionally, the examination of the effect of various factors such as soil properties, land use patterns, and climatic conditions on heavy metal bioavailability warrants diligent attention. Environmental purification can be conducted using organisms and using plants. It is hoped that, in the future, progress can be made in the decomposition of heavy metals in soil. Moreover, the integration of advanced techniques such as molecular biology and spectroscopy with traditional approaches can enrich our comprehension of heavy metal dynamics in soil ecosystems. Embracing interdisciplinary collaboration and harnessing cutting-edge technologies will undoubtedly pave the way for more comprehensive and precise assessments of soil heavy metal risks, thereby contributing to the formulation of effective mitigation strategies and the preservation of environmental and human health in Chenzhou and beyond.

Author Contributions: Conceptualization, C.Z.; Data curation, X.C.; Formal analysis, Y.K., X.C. and C.Z.; Methodology, Y.K. and C.Z.; Software, Y.K. and X.C.; Writing—original draft, Y.K. and X.C.; Writing—review and editing, Y.K., X.C. and C.Z. All authors have read and agreed to the published version of the manuscript.

Funding: This research received no external funding.

Data Availability Statement: The data presented in this study are available on request from the corresponding author.

Conflicts of Interest: The authors declare no conflicts of interest.

References

1. Kutralam-Muniasamy, G.; Pérez-Guevara, F.; Martínez, I.E.; Shruti, V. Overview of microplastics pollution with heavy metals: Analytical methods, occurrence, transfer risks and call for standardization. *J. Hazard. Mater.* **2021**, *415*, 125755. [CrossRef]
2. Tapia-Gatica, J.; Selles, I.; Bravo, M.A.; Tessini, C.; Barros-Parada, W.; Novoselov, A.; Neaman, A. Global issues in setting legal limits on soil metal contamination: A case study of Chile. *Chemosphere* **2022**, *290*, 133404. [CrossRef]
3. Peng, J.-Y.; Zhang, S.; Han, Y.; Bate, B.; Ke, H.; Chen, Y. Soil heavy metal pollution of industrial legacies in China and health risk assessment. *Sci. Total Environ.* **2022**, *816*, 151632. [CrossRef]
4. Peng, C.; Zhang, K.; Wang, M.; Wan, X.; Chen, W. Estimation of the accumulation rates and health risks of heavy metals in residential soils of three metropolitan cities in China. *J. Environ. Sci.* **2022**, *115*, 149–161. [CrossRef] [PubMed]
5. Liu, P.; Wu, Q.; Wang, X.; Hu, W.; Liu, X.; Tian, K.; Xie, E.; Zhao, Y.; Huang, B.; Yoon, S.J. Spatiotemporal variation and sources of soil heavy metals along the lower reaches of Yangtze River, China. *Chemosphere* **2022**, *291*, 132768. [CrossRef] [PubMed]
6. Anaman, R.; Peng, C.; Jiang, Z.; Liu, X.; Zhou, Z.; Guo, Z.; Xiao, X. Identifying sources and transport routes of heavy metals in soil with different land uses around a smelting site by GIS based PCA and PMF. *Sci. Total Environ.* **2022**, *823*, 153759. [CrossRef] [PubMed]
7. Chen, T.-B.; Zheng, Y.-M.; Lei, M.; Huang, Z.-C.; Wu, H.-T.; Chen, H.; Fan, K.-K.; Yu, K.; Wu, X.; Tian, Q.-Z. Assessment of heavy metal pollution in surface soils of urban parks in Beijing, China. *Chemosphere* **2005**, *60*, 542–551. [CrossRef] [PubMed]
8. Wu, S.; Liu, X.; Liu, M.; Chen, X.; Liu, S.; Cheng, L.; Lin, X.; Li, Y. Sources, influencing factors and environmental indications of PAH pollution in urban soil columns of Shanghai, China. *Ecol. Indic.* **2018**, *85*, 1170–1180. [CrossRef]
9. Zhou, W.; Li, R.; Yue, H.; Shen, S.; Zhu, H.; Xie, X. Characteristics and assessment of heavy metal pollution in vegetable soils in Jiangbei area of Nanjing City. *Trans. Atmos. Sci.* **2009**, *32*, 574–581.
10. Zhou, H.; Ouyang, T.; Guo, Y.; Peng, S.; He, C.; Zhu, Z. Assessment of soil heavy metal pollution and its ecological risk for city parks, vicinity of a landfill, and an industrial area within Guangzhou, South China. *Appl. Sci.* **2022**, *12*, 9345. [CrossRef]
11. Chen, X.; Lu, X. Contamination characteristics and source apportionment of potentially toxic elements in the topsoil of Huyi District, Xi'an City, China. *Environ. Earth Sci.* **2021**, *80*, 153–160. [CrossRef]
12. Shi, W.; Li, T.; Feng, Y.; Su, H.; Yang, Q. Source apportionment and risk assessment for available occurrence forms of heavy metals in Dongdahe Wetland sediments, southwest of China. *Sci. Total Environ.* **2022**, *815*, 152837. [CrossRef]

13. Qiao, P.; Wang, S.; Li, J.; Zhao, Q.; Wei, Y.; Lei, M.; Yang, J.; Zhang, Z. Process, influencing factors, and simulation of the lateral transport of heavy metals in surface runoff in a mining area driven by rainfall: A review. *Sci. Total Environ.* **2023**, *857*, 159119. [CrossRef]
14. Li, Z.; Lu, W.; Huang, J. Monitoring, Diffusion and Source Speculation Model of Urban Soil Pollution. *Processes* **2020**, *8*, 1339. [CrossRef]
15. Janaszek, A.; Silva, A.F.d.; Jurišević, N.; Kanuchova, M.; Kozáková, Ľ.; Kowalik, R. The Assessment of Sewage Sludge Utilization in Closed-Loop Economy from an Environmental Perspective. *Water* **2024**, *16*, 383. [CrossRef]
16. Williams, J.A.; Antoine, J. Evaluation of the elemental pollution status of Jamaican surface sediments using enrichment factor, geoaccumulation index, ecological risk and potential ecological risk index. *Mar. Pollut. Bull.* **2020**, *157*, 111288. [CrossRef]
17. Günal, E.; Budak, M.; Kılıç, M.; Cemek, B.; Sırrı, M. Combining spatial autocorrelation with artificial intelligence models to estimate spatial distribution and risks of heavy metal pollution in agricultural soils. *Environ. Monit. Assess.* **2023**, *195*, 317. [CrossRef] [PubMed]
18. Xiaofeng, W.; Wenbo, L.; Qipeng, X.; Yanbin, M.; Bin, Q.; Zeyan, L.; Liping, H. Evaluation of soil environmental quality of cultivated land in the Donghe River Basin of Chenzhou City, Hunan Province by three methods. *J. Environ. Occup. Med.* **2022**, *39*, 684–689.
19. Zheng, B.; Li, J.; Chen, X.; Luo, X. Evaluating the Effects of Roof Greening on the Indoor Thermal Environment throughout the Year in a Chinese City (Chenzhou). *Forests* **2022**, *13*, 304. [CrossRef]
20. Xue, B.; Liu, B.; Yang, Q.; Sun, X.; Wang, W.; Li, L. Formalizing an evaluation-prediction based roadmap towards urban sustainability: A case study of Chenzhou, China. *Habitat. Int.* **2021**, *112*, 102376. [CrossRef]
21. Luo, X.; Wang, J.; Li, J. Study on the Effect of Vegetation Coverage on Urban Cooling and Energy Conservation: A Case Study of a Typical Hilly City, Chenzhou, China. *Buildings* **2022**, *12*, 640. [CrossRef]
22. Zhu, L.; Carlin, B.P. Comparing hierarchical models for spatio-temporally misaligned data using the deviance information criterion. *Stat. Med.* **2000**, *19*, 2265–2278. [CrossRef]
23. Cui, X.; Mei, C.; Qin, Y.; Yan, G.; Fu, Q. In-motion alignment for low-cost SINS/GPS under random misalignment angles. *J. Navig.* **2017**, *70*, 1224–1240. [CrossRef]
24. Fan, C.-Z.; Liu, Y.-B.; Yuan, J.-H.; Guo, W.; Sun, D.-Y.; Liu, C.-H. Characteristics of Heavy Metals and Other Elements in Atmospheric Dry and Wet Deposition and Health Risk Assessment of a Typical Industrial and Mining City in Central Yunnan. *Huan Jing Ke Xue Huanjing Kexue* **2022**, *43*, 3923–3933. [PubMed]
25. Li, M.; Yang, B.; Ju, Z.; Qiu, L.; Xu, K.; Wang, M.; Chen, C.; Zhang, K.; Zhang, Z.; Xiang, S. Do high soil geochemical backgrounds of selenium and associated heavy metals affect human hepatic and renal health? Evidence from Enshi County, China. *Sci. Total Environ.* **2023**, *883*, 163717. [CrossRef]
26. Parker, C.; Rees, W. Fluorescence spectrometry. A review. *Analyst* **1962**, *87*, 83–111. [CrossRef]
27. Chen, Y.; Zhang, Q.; Zhang, L.; Wang, Y.; Song, Y.; Li, Y.; Yin, Y.; Cai, Y. An improved method for rapid and safe preparation and measurement of dimethylmercury using gas chromatography-atomic fluorescence spectrometry. *J. Chromatogr. A* **2023**, *1712*, 464472. [CrossRef]
28. Roberts, N.M.; Drost, K.; Horstwood, M.S.; Condon, D.J.; Chew, D.; Drake, H.; Milodowski, A.E.; McLean, N.M.; Smye, A.J.; Walker, R.J. Laser ablation inductively coupled plasma mass spectrometry (LA-ICP-MS) U–Pb carbonate geochronology: Strategies, progress, and limitations. *Geochronology* **2020**, *2*, 33–61. [CrossRef]
29. Bolea-Fernandez, E.; Rua-Ibarz, A.; Velimirovic, M.; Tirez, K.; Vanhaecke, F. Detection of microplastics using inductively coupled plasma-mass spectrometry (ICP-MS) operated in single-event mode. *J. Anal. At. Spectrom.* **2020**, *35*, 455–460. [CrossRef]
30. Weindorf, D.C.; Chakraborty, S. Portable X-ray fluorescence spectrometry analysis of soils. *Soil Sci. Soc. Am. J.* **2020**, *84*, 1384–1392. [CrossRef]
31. Cui, M.-J.; Lai, H.-J.; Hoang, T.; Chu, J. One-phase-low-pH enzyme induced carbonate precipitation (EICP) method for soil improvement. *Acta Geotech.* **2021**, *16*, 481–489. [CrossRef]
32. Chung, M.-T.; Peng, Y.-Y.; Nazir, A.; Wang, Y.-C.; Wang, P.-L.; Shiao, J.-C. Species-specific trophic enrichment factor of stable nitrogen and carbon isotopes in fish otolith organic matter. *Mar. Freshw. Res.* **2023**, *74*, 956–968. [CrossRef]
33. Lauderdale, L.K.; Shorter, K.A.; Zhang, D.; Gabaldon, J.; Mellen, J.D.; Granger, D.A.; Walsh, M.T.; Miller, L.J. Environmental Enrichment Factors Associated with the Activity Level of Bottlenose Dolphins under Professional Care. *J. Zool. Bot. Gard.* **2022**, *3*, 44–55. [CrossRef]
34. Suyatno, A.P.; Afiati, N.; Muskananfola, M.R. Analysis of Enrichment Factors and Contamination Factors of Cd, Cr and Cu in Western Flood Canal Semarang, Central Java. *Berk. Perikan. Terubuk* **2021**, *49*, 1042–1052.
35. Qin, H.-B.; Yang, S.; Tanaka, M.; Sanematsu, K.; Arcilla, C.; Takahashi, Y. Scandium immobilization by goethite: Surface adsorption versus structural incorporation. *Geochim. et Cosmochim. Acta* **2021**, *294*, 255–272. [CrossRef]
36. Artemenkov, A. Content of Rare-Earth Elements in Soils of Technogeochemical Anomalies. *Russ. J. Gen. Chem.* **2020**, *90*, 2573–2581. [CrossRef]
37. Fahman, G.; Sianita, M.M. Calculation of enrichment factor in chloramphenicol analysis in shrimp with variation of concentration using molecularly imprinted polymer. *World J. Adv. Res. Rev.* **2022**, *15*, 362–368. [CrossRef]
38. Yasuda, K.; Hamada, K.; Asakura, Y. Enrichment of amino acids from its aqueous solution by ultrasonic atomization and ultrafine bubbles. *Jpn. J. Appl. Phys.* **2022**, *61*, SG1009. [CrossRef]

39. Abdullah, M.I.C.; Sah, A.S.R.M.; Haris, H. Geoaccumulation index and enrichment factor of arsenic in surface sediment of Bukit Merah Reservoir, Malaysia. *Trop. Life Sci. Res.* **2020**, *31*, 109. [CrossRef]
40. Wardani, N.; Prartono, T.; Sulistiono, S. Sediments quality based on Geo-Accumulation Index in heavy metals (Pb, Cu, and Cd) of Cengkok coastal waters, Banten Bay. *J. Pendidik. IPA Indones.* **2020**, *9*, 574–582. [CrossRef]
41. Senoro, D.B.; Monjardin, C.E.F.; Fetalvero, E.G.; Benjamin, Z.E.C.; Gorospe, A.F.B.; de Jesus, K.L.M.; Ical, M.L.G.; Wong, J.P. Quantitative Assessment and Spatial Analysis of Metals and Metalloids in Soil Using the Geo-Accumulation Index in the Capital Town of Romblon Province, Philippines. *Toxics* **2022**, *10*, 633. [CrossRef] [PubMed]
42. Wang, Y.; Zhao, S.; Sun, M.; Lv, X.; Cai, W.; Xu, X.; Ge, H.; Lei, K. Assessment of river ecosystem health in Tianjin City, China: Index of ecological integrity and water comprehensive pollution approach. *Front. Earth Sci.* **2021**, *15*, 936–946. [CrossRef]
43. Yang, H.; Jia, C.; Yang, F.; Yang, X.; Wei, R. Water quality assessment of deep learning-improved comprehensive pollution index: A case study of Dagu River, Jiaozhou Bay, China. *Environ. Sci. Pollut. Res.* **2023**, *30*, 66853–66866. [CrossRef]
44. Wang, X.-B.; Xiang, Y.; Li, X.-Y. Environmental risks for application of magnesium slag to soils in China. *J. Integr. Agric.* **2020**, *19*, 1671–1679. [CrossRef]
45. Nalan, K.; Dengiz, O. Assessment of potential ecological risk index based on heavy metal elements for organic farming in micro catchments under humid ecological condition. *Eurasian J. Soil. Sci.* **2020**, *9*, 194–201.
46. Ranjani, M.; Veerasingam, S.; Venkatachalapathy, R.; Mugilarasan, M.; Bagaev, A.; Mukhanov, V.; Vethamony, P. Assessment of potential ecological risk of microplastics in the coastal sediments of India: A meta-analysis. *Mar. Pollut. Bull.* **2021**, *163*, 111969. [CrossRef]
47. Saha, A.; Gupta, B.S.; Patidar, S.; Martínez-Villegas, N. Evaluation of potential ecological risk index of toxic metals contamination in the soils. *Chem. Proc.* **2022**, *10*, 59.
48. Khamraev, K.; Cheriyan, D.; Choi, J.-h. A review on health risk assessment of PM in the construction industry–Current situation and future directions. *Sci. Total Environ.* **2021**, *758*, 143716. [CrossRef]
49. Cabral-Pinto, M.M.; Inácio, M.; Neves, O.; Almeida, A.A.; Pinto, E.; Oliveiros, B.; Ferreira da Silva, E.A. Human health risk assessment due to agricultural activities and crop consumption in the surroundings of an industrial area. *Expo. Health* **2020**, *12*, 629–640. [CrossRef]
50. Arshad, H.; Mehmood, M.Z.; Shah, M.H.; Abbasi, A.M. Evaluation of heavy metals in cosmetic products and their health risk assessment. *Saudi Pharm. J.* **2020**, *28*, 779–790. [CrossRef]
51. Nag, R.; Cummins, E. Human health risk assessment of lead (Pb) through the environmental-food pathway. *Sci. Total Environ.* **2022**, *810*, 151168. [CrossRef]
52. Bazeli, J.; Ghalehaskar, S.; Morovati, M.; Soleimani, H.; Masoumi, S.; Rahmani Sani, A.; Saghi, M.H.; Rastegar, A. Health risk assessment techniques to evaluate non-carcinogenic human health risk due to fluoride, nitrite and nitrate using Monte Carlo simulation and sensitivity analysis in Groundwater of Khaf County, Iran. *Int. J. Environ. Anal. Chem.* **2022**, *102*, 1793–1813. [CrossRef]
53. Liu, X.; Gu, S.; Yang, S.; Deng, J.; Xu, J. Heavy metals in soil-vegetable system around E-waste site and the health risk assessment. *Sci. Total Environ.* **2021**, *779*, 146438. [CrossRef]
54. Yu, G.; Wang, J.; Liu, L.; Li, Y.; Zhang, Y.; Wang, S. The analysis of groundwater nitrate pollution and health risk assessment in rural areas of Yantai, China. *BMC Public Health* **2020**, *20*, 437. [CrossRef] [PubMed]
55. *GB 15618-2018*; Soil Environmental Quality Risk Control Standard for Soil Contamination of Agriculture Land. China Environmental Science Press: Beijing, China, 2018.
56. Junior, A.B.; Espinosa, D.C.R.; Vaughan, J.; Tenório, J.A.S. Recovery of scandium from various sources: A critical review of the state of the art and future prospects. *Miner. Eng.* **2021**, *172*, 107148. [CrossRef]
57. Gentzmann, M.C.; Schraut, K.; Vogel, C.; Gäbler, H.-E.; Huthwelker, T.; Adam, C. Investigation of scandium in bauxite residues of different origin. *Appl. Geochem.* **2021**, *126*, 104898. [CrossRef]

processes MDPI

Article

Integration of Slurry–Total Reflection X-ray Fluorescence and Machine Learning for Monitoring Arsenic and Lead Contamination: Case Study in Itata Valley Agricultural Soils, Chile

Guillermo Medina-González [1,*], Yelena Medina [2], Enrique Muñoz [3,4], Paola Andrade [5], Jordi Cruz [6], Yakdiel Rodriguez-Gallo [7] and Alison Matus-Bello [1]

1 Department of Environmental Chemistry, Faculty of Sciences, Universidad Católica de la Santísima Concepción, Concepción 4090541, Chile; amatus@qaciencias.ucsc.cl
2 EMOingenieros Ltda., Concepción 4090070, Chile; ymedina@emoingenieros.cl
3 Departament of Civil Engineering, Faculty of Engineering, Universidad Católica de la Santísima Concepción, Concepción 4090541, Chile; emunozo@ucsc.cl
4 Centro de Investigación en Biodiversidad y Ambientes Sustentables CIBAS, Universidad Católica de la Santísima Concepción, Concepción 4090541, Chile
5 Departament of Ecology, Faculty of Sciences, Universidad Católica de la Santísima Concepción, Concepción 4090541, Chile; pandrade@ucsc.cl
6 Escola Universitària Salesiana de Sarrià (EUSS School of Engineering, Barcelona), 08017 Barcelona, Spain; jcruz@euss.cat
7 Faculty of Engineering, Don Bosco University, Calle a Plan del Pino Km 1 1/2, Soyapango 1874, El Salvador; yakdiel.rodriguez@udb.edu.sv
* Correspondence: guillermo.medina@ucsc.cl

Abstract: The accuracy of determining arsenic and lead using the optical technique Slurry–Total Reflection X-ray Fluorescence (Slurry-TXRF) was significantly enhanced through the application of a machine learning method, aimed at improving the ecological risk assessment of agricultural soils. The overlapping of the arsenic Kα signal at 10.55 keV with the lead Lα signal at 10.54 keV due to the relatively low resolution of TXRF could compromise the determination of lead. However, by applying a Partial Least Squares (PLS) machine learning algorithm, we mitigated interference variations, resulting in improved selectivity and accuracy. Specifically, the average percentage error was reduced from 15.6% to 9.4% for arsenic (RMSEP improved from 5.6 mg kg^{-1} to 3.3 mg kg^{-1}) and from 18.9% to 6.8% for lead (RMSEP improved from 12.3 mg kg^{-1} to 5.03 mg kg^{-1}) compared to the previous univariable model. This enhanced predictive accuracy, within the set of samples concentration range, is attributable to the efficiency of the multivariate calibration first-order advantage in quantifying the presence of interferents. The evaluation of X-ray fluorescence emission signals for 26 different synthetic calibration mixtures confirmed these improvements, overcoming spectral interferences. Additionally, the application of these models enabled the quantification of arsenic and lead in soils from a viticultural subregion of Chile, facilitating the estimation of ecological risk indices in a fast and reliable manner. The results indicate that the contamination level of these soils with arsenic and lead ranges from moderate to considerable.

Keywords: ecological indices; contamination; TXRF; machine learning; arsenic; lead

1. Introduction

In environmental geochemistry, it is crucial to understand the concepts of geological baselines and critical levels. This understanding is important for differentiating between naturally occurring substance levels and pollution caused by human activities [1]. Several ecological indicators based on these geological baselines have been suggested to evaluate the extent and impact of soil contamination by trace elements, including the Geoaccumulation Index (GEOI), Enrichment Factor (EF), and Contamination Factor (CF) [2].

Establishing the baseline levels of trace elements and ecological indices in a particular location is important for detecting contamination. Still, it requires the systematic gathering and examination of a large number of samples [3].

Quantitative analyses in soil studies often utilize various optical analytical methods, including traditional approaches like inductively coupled plasma mass spectrometry (ICP-MS) [4], flame atomic absorption spectroscopy (FAAS) [5], inductively coupled plasma emission spectroscopy (ICP-OES) [6], graphite furnace atomic absorption spectroscopy (GF-AAS) [7], and cold vapor atomic absorption spectrometry (CV-AAS) [8].

Total X-ray fluorescence spectroscopy (TXRF) is a reliable and convenient alternative method for determining trace element concentrations. This versatile technique can be applied with or without sample digestion, making it a highly favored choice [2,9,10].

However, these methods often lead to significant spectral interferences. Studies have been carried out to explore the use of machine learning algorithms such as Partial Least Squares Regression (PLS-R) in liquid matrices [11], the Synchrotron Radiation X-ray Fluorescence technique [12], and Cd-K in soils by TXRF [13], where these models improve the accuracy of quantification in the presence of spectral interferences.

Several studies have examined the levels of heavy metals or potentially toxic elements in areas primarily used for agriculture and mining [2,14–21]. However, there is a lack of studies on effectively monitoring arsenic and lead contamination in agricultural soils, particularly in regions like Itata Valley. While there are standardized methods suggested by international quality standards for the determination of arsenic and lead [22], these methods are often slow, expensive, and environmentally unfriendly.

Therefore, it is crucial to continue developing and improving analytical methods for trace element analysis in soil samples, as well as exploring innovative techniques and approaches such as machine learning and X-ray fluorescence to overcome challenges and improve quantification accuracy in the context of low-resolution analytical methods. [23].

This study aimed to assess the effectiveness of the specific method of slurry-TXRF combined with machine learning models in analyzing arsenic and lead levels in viticultural soils. This method offers a rapid and efficient way to estimate and monitor contamination levels using various ecological indices, providing an innovative alternative for gaining valuable insights into the extent and potential risks of heavy-metal contamination in vineyard soils. These insights can help make informed decisions about soil management practices and ensure sustainable agriculture for the quality and safety of products.

2. Materials and Methods

This study on improving lead and arsenic measurement using slurry-TXRF and machine learning is part of a larger research effort focused on studying heavy-metal contamination in agricultural soils. Medina-Gonzalez et al.'s (2023) [2] study served as the basis for our methodology with certain adaptations.

2.1. Study Area and Samples

A well-known wine-producing region in southern Chile was chosen as the focus of this study. This region covers around 311,418 hectares, with approximately 13,030 hectares allocated to agricultural activities, specifically for wine production [24]. The Itata Valley is notable for its cultivation of grape varieties like País, Muscat of Alexandria, and Cinsault, with the latter being the most widely grown in terms of cultivation area [25].

Geographically, the Itata Valley is part of the Central Valley and Coastal Range, with elevations not exceeding 400 m above sea level. The area benefits from a Mediterranean climate, with distinct rainy and dry seasons, influenced by the Itata River, which flows from east to west across the valley. The presence of the Coastal Range creates climatic variations, with the western slopes receiving more rainfall compared to the drier eastern slopes [24].

The valley's soil originates primarily from three parent materials:

- Eroded Metamorphic Rock: Found in higher elevations with steep or complex slopes, this soil is characterized by shale, sandstone, phyllite, and slates. It has a clayey texture with slow water infiltration and is prone to forming Catena due to topography and drainage patterns.
- Granite Origin: Derived from granite and diorite rocks, this soil also has a clayey texture and low water infiltration, making it susceptible to erosion. It is typically found in hilly areas with complex slopes.
- Fine Alluvial Sediment: Formed from fluvio-glacial sediments transported by rivers, this soil has a loamy–clayey texture and poor drainage. The thickness of these sediments varies significantly across the valley.

In summary, the soils in the Itata Valley are primarily derived from eroded metamorphic rocks, granite and diorite, and fine alluvial sediments, each contributing to unique characteristics such as soil texture, water infiltration, and erosion susceptibility [25].

The soil sampling was carried out in two phases, namely June and October of 2023. These samples were obtained from five vineyards (A, B, C, D, E) at a total of 48 sampling points within the area corresponding to the Itata Valley, following the owners' authorization and using the combined method as recommended by Servicio Agrícola y Ganadero de Chile [26] (Figure 1).

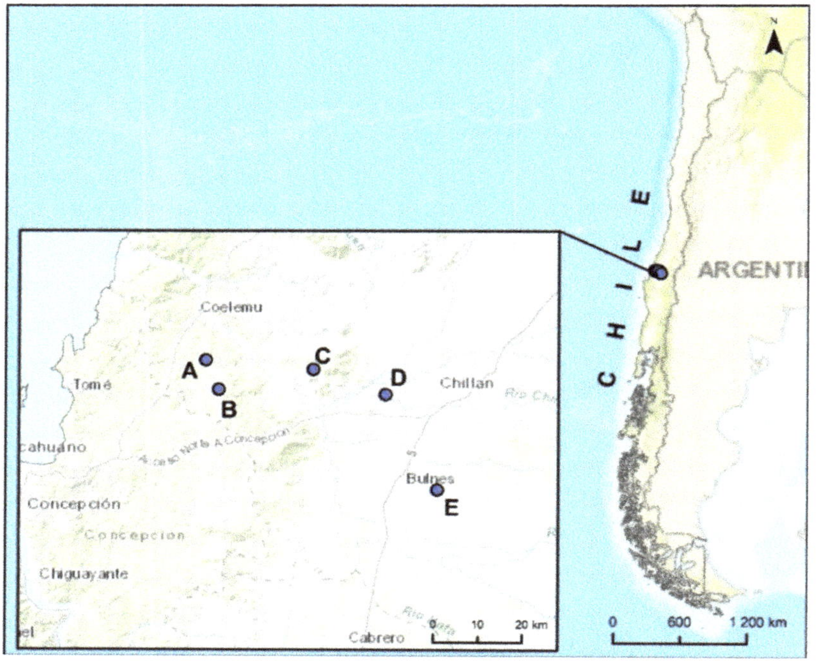

Figure 1. The study area shows the vineyards where samples were collected.

Approximately two kilograms of the sample were collected at each sampling location by taking five subsamples from the corners and center of a 1 × 1 m grid to create a composite sample. The topsoil samples (0–20 cm depth) were collected while removing any plant cover present and deep soil samples at a depth of 150 cm using a Dutch auger. Surface soils are most affected by human activities, whereas deep soils represent relatively undisturbed samples considered free from contamination [27].

The samples were packed in polyethylene bags for transport to the laboratory and then dried in an oven at 40 °C for 48 h. Next, each sample was sieved using a nylon sieve to remove coarse material and other residues, leaving only the fine material (<2 mm). The sieved samples were stored in tightly sealed plastic bags for further analysis [28].

2.2. Sample Preparation

2.2.1. PLS Model Samples

Standard calibration and external validation samples were prepared using 1000 mg L^{-1} solutions of As and Pb. Ultra-pure deionized water with a resistance of 18.2 MΩ cm^{-1} obtained from a Milli-Q quality purification system was used for dilution purposes. A silicone solution in isopropanol was applied to create a silicon film on reflective disks of quartz glass before the deposition of the sample. Working solutions with concentrations of 100 mg L^{-1} were utilized to generate a total of 26 synthetic calibration solutions (Table 1) and 8 external validation samples (Table 2) for these elements in 2 mL Eppendorf tubes.

Table 1. Composition of synthetic calibration mixtures.

Sample	As (mg kg^{-1})	Pb (mg kg^{-1})
C1	10	34
C2	12	72
C3	48	13
C4	42	20
C5	37	67
C6	33	15
C7	33	19
C8	32	49
C9	8	39
C10	43	65
C11	35	36
C12	49	5
C13	10	4
C14	45	78
C15	13	31
C16	33	69
C17	41	36
C18	24	22
C19	4	36
C20	7	72
C21	28	23
C22	33	36
C23	42	18
C24	36	47
C25	0	43
C26	12	27
Range	49	74
Minimum	0	4
Maximum	49	78

Table 2. Composition of synthetic mixtures for external validation.

Sample	As (mg kg^{-1})	Pb (mg kg^{-1})
V1	36	6
V2	19	54
V3	1	32
V4	12	51
V5	33	78
V6	16	42
V7	21	71
V8	21	7

All samples contained 10 ng of germanium as internal standard.

Each sample preparation included a random mixing of concentrations to avoid any potential correlations that could affect the calibration process. The concentration ranges were intentionally chosen to facilitate the measurement of arsenic and lead in environmental samples with varying levels of contamination. Sample preparation involved using the drop-drying method, which entails applying a small volume (such as 10 µL) of the mixture onto 30 mm diameter quartz sample disks and drying them with an infrared lamp.

The resulting solid residues on the sample holders were then analyzed in an S4TStar TXRF spectrometer after the evaporation of the liquid phase [2].

To create a model in milligrams per kilogram (mg kg^{-1}) concentration units from liquid calibration samples, enabling us to evaluate its suitability in solid matrices, conversions in concentrations were conducted using Equation (1):

$$\left[\mathrm{mgkg}^{-1}\right] = \frac{\left[\mathrm{mgL}^{-1}\right] \times V_{dep}}{S_{dep}} \tag{1}$$

where V_{dep} represents the volume of liquid calibration sample deposited in liters, while S_{dep} denotes the quantity of dry solid sample deposited in kilograms, based on the specific methodologies employed.

2.2.2. Slurry Sample Preparation

The soil sludge samples were prepared in sterilized 2.5 mL Eppendorf vials by combining 30 mg of dried and pulverized soil with 1500 µL of Triton-X as a surfactant using an agate mortar. The mixtures were homogenized for 300 s with an electronic shaker. After this, 10 µL of the thoroughly homogenized suspension was extracted and placed into the TXRF's 30 mm diameter quartz sample holder immediately to prevent potential sedimentation. A silicone solution in isopropanol was then applied to create a silicon film on the quartz glass disk before depositing the sample, which was followed by drying it with an infrared lamp. To evaluate accuracy and suitability, loamy clay certified reference material for trace metals from Sigma-Aldrich was used to test the model's performance in soil samples.

2.3. TXRF Spectra and Data Acquisition

This study used a TXRF S4TSTAR analytical system (Bruker® AXS Microanalysis GmbH, Berlin, Germany) for fluorescent emission measurements. The X-ray source operated at 50 W with an electron flow of 1 mA at 50 kV and was cooled by air. The system included a multi-layer carbon–nickel monochromator of 150 nm and a high-resolution silicon semiconductor detector XFlashR with an active area of 100 mm2 and energy resolution < 140 eV. The specific range of interest was defined from 10.27 keV to 12.78 keV, focused on the overlap of the arsenic Kα line (10.54 keV) and lead Lα line (10.55 keV). Counts were recorded using the SPECTRA 7.5.0.3 software for data analysis [29], generating a .txt format file processed through PIROUETTE 4.5 software [30].

2.4. Quantification Methods

For the aim of assessing the efficacy of the PLS model in quantifying arsenic and lead, in comparing its performance to classical deconvolution methods utilizing the SPECTRA 7.5.0.3 software, specifically the deconvolution routine (SuperBayes), the quantification was carried out employing internal standardization with germanium (10 ng).

A set of 26 synthetic mixtures served as the basis for PLS analysis. These mixtures were deliberately correlated with their known concentrations and employed as calibration samples. The fluorescence spectra within the calibration set underwent preprocessing. Specifically, mean centering was applied as the sole transformation method without additional alterations.

Latent variables were chosen through the cross-validation procedure employing the leaving-one-out method. This method systematically excluded each sample from the calibration set one at a time, utilizing it as a prediction object. With this, the root mean square error of cross-validation (RMSECV), calculated as per Equation (2), served as the criterion to determine the optimal number of latent variables essential for establishing the calibration model.

To assess the effectiveness of the Partial Least Squares model in quantifying arsenic and lead, its performance was compared to those of classical deconvolution methods using Spectra 7.5.0.3 software's SuperBayes routine with internal standardization using germanium (10 ng). A set of 26 synthetic mixtures correlated with known concentrations served as calibration samples for PLS analysis. These fluorescence spectra underwent mean-centering preprocessing without any further transformation. The optimal number of latent variables required for establishing the calibration model was determined through cross-validation and root mean square error of cross-validation [31], calculated according to Equation (2):

$$\text{RMSECV} = \sqrt{\sum (y_i - \hat{y}_i)^2 / n} \tag{2}$$

where $\hat{y}_i$ represents the estimated concentration values obtained through cross-validation, with each sample i being estimated using a model constructed from the sample set excluding that particular sample. Here, y_i denotes the known concentration variable value, and n denotes the total number of samples in the calibration set.

The same sample set used for the direct quantification of arsenic and lead through deconvolution (Table 2) using the SPECTRA software also played another important role: external validation of the model. External validation is a crucial practice in chemometrics that helps guard against overfitting in calibration models, ensuring their robustness and generalizability [32].

The root mean square error of prediction (RMSEP) was used to evaluate the predictive accuracy of the multivariate model [33]. It is calculated similarly to RMSECV, but in this case, each predicted sample (y_i) from the final model is compared with the corresponding reference value from an external validation set ($\hat{y}_{(i,ref)}$), where N represents the size of the external validation set. The calculation for the RMSEP follows Equation (3):

$$\text{RMSEP} = \sqrt{\sum (y_i - \hat{y}_{i,ref})^2 / N} \tag{3}$$

Linear regression can identify biases in the predictions of external samples, similar to its use in internal validation procedures. Equation (4) shows the linear regression approach used to assess systematic biases within the predictions for external samples.

$$\hat{y} = a + yb \tag{4}$$

The comparison between predicted values ($\hat{y}$) and reference values (y) is crucial for evaluation. Ideally, predictions made by the calibration model should exhibit a linear relationship with known values—a line with a slope (b) of 1 and an intercept (a) of 0. Any deviations from this linear behavior indicate issues arising during the analytical procedure [34]. The Elliptical Joint Confidence Region (EJCR) method was employed to assess the correlation of results and obtain a statistically significant confidence level of 95% [35].

Analytical Figures of Merit (AFOM's)

Figures of Merit for the PLS approach were calculated according to the NAS theory. This concept works by isolating relevant signals associated with a particular component of interest from the various interfering elements present in the spectra. The specific method utilized in this research follows the approach described by Short et al. [36], concentrating on deriving figures of merit using the NAS theory applied to spectral data.

2.5. Background Values and Ecological Indices

The assessment of natural or baseline values required the examination of data from deep soil samples following the procedures outlined in ISO 19258 [37]. There is no standardized method for setting threshold values to identify samples with abnormally high levels of element concentrations, as noted by Reimann et al. [38]. Nonetheless, this research applies the "MAD" technique, which makes use of the range from Median ± 2MAD (median absolute deviation) to pinpoint values suggestive of elevated element concentrations.

Various environmental indicators are frequently utilized to evaluate possible soil pollution, aiding in the differentiation between natural trace element accumulation and that caused by human actions [18,39]. One commonly used method is the Geoaccumulation Index, which is determined using a specific equation (Equation (5)):

$$\text{GEOI} = \log_2(C_i/1.5C_b) \quad (5)$$

where C_i is the measured content of the element in the soil, and C_b is the estimated background value of that element in the soil. According to Muller [40], the GEOI for a particular element is calculated and then classified as shown in Table 3.

Table 3. Contamination level according to the Geoaccumulation Index [18].

GEOI Group	Level of Contamination
GEOI ≤ 0	Uncontaminated
0 < GEOI ≤ 1	Slightly contaminated
1 < GEOI ≤ 2	Moderately contaminated
2 < GEOI ≤ 3	Moderately to heavily contaminated
3 < GEOI ≤ 4	Heavily contaminated
4 < GEOI ≤ 5	Heavily to extremely contaminated
GEOI > 5	Extremely contaminated

The Enrichment Factor (EF) is determined by standardizing the concentration of a measured element with respect to a reference element. The selection of the reference element relies on its minimal variability in occurrence within the specific environment [38] and its independence from anthropogenic influences [41]. Commonly used reference elements comprise Sc, Mn, Ti, Al, and Fe. This research chose Fe as the reference element due to its significantly low variation in viticultural soils in the Itata Valley [1]. Its abundance in the Earth's crust and its connection with the soil matrix validate its choice as a stable reference element [42]. The EF is calculated using Equation (6):

$$\text{EF} = (C_i/CFe_i)/(C_b/CFe_b) \quad (6)$$

where C_i represents the concentration of the particular element, CFe_i denotes the quantity of Fe as a reference constituent, C_b is the reference content derived via the MAD method, while CFe_b signifies background content for Fe. Enrichment Factor values are divided into five categories based on Table 4:

Table 4. Contamination level according to the Enrichment Factor (EF) [18].

EF Group	Level of Contamination
EF < 2	Deficiency to minimal enrichment
2 < EF < 5	Moderate enrichment
5 < EF < 20	Significant enrichment
20 < EF < 40	Very high enrichment
EF > 40	Extremely high enrichment

The Contamination Factor (CF) is calculated using Equation (7):

$$\text{CF} = C_i/C_b \quad (7)$$

where C_i represents the concentration of the examined element in the soil, and C_b is the content of background values obtained using the MAD method. Hakanson [43] defines the categories for CF as presented in Table 5.

Table 5. Contamination level according to Contamination Factor (CF) [18].

CF Group	Level of Contamination
CF < 1	Low contamination
1 < CF < 3	Moderate contamination
3 < CF < 6	Considerable contamination
CF > 6	Very high contamination

3. Results

3.1. Classical Regression: Deconvolution

Table 6 shows the results obtained for the quantification of arsenic and lead in conditions where there is overlap, using the classical method of internal standardization followed by signal deconvolution.

Table 6. Comparison of As and Pb quantification results by Internal Standardization vs. PLS.

Sample	Concentration (mg kg^{-1})		Internal Standardization (mg kg^{-1})		Relative Bias (%)		PLS (mg kg^{-1})		Relative Bias (%)	
	As	Pb	As	Pb	As	Pb	As	Pb	As	Pb
A	36	6	45.3	8.1	25.8	35.0	39.8	7.2	10.6	20.0
B	19	54	24.2	68.2	27.4	26.3	15.2	62.4	−20.0	15.6
C	1	32	2.5	33.9	150.0	5.9	1.9	31.4	90.0	−1.9
D	12	51	16.2	60.3	35.0	18.2	9.3	56.6	−22.5	11.0
E	33	80	40.3	87.6	22.1	9.5	38.5	83.4	16.7	4.3
F	16	42	18.5	52.2	15.6	24.3	18.2	37.4	13.8	−11.0
G	21	71	28.3	98.1	34.8	38.2	17.3	64.2	−17.6	−9.6
H	21	7	23.5	12.2	11.9	74.3	22.3	10.8	6.2	54.3
RMSEP (mg kg^{-1})			5.6	12.3			3.3	5.0		
RMSEP (%)			15.6	18.9			9.4	6.8		

The results indicate significant positive biases for both arsenic and lead, with relative biases ranging from 11.9% to 150% for arsenic and from 5.9% to 74.3% for lead. These values, respectively, in relation to the reference values of the external validation set, reveal a noticeable systematic error. The calculated RMSEP for this quantification method is 5.6 mg kg^{-1} for arsenic and 12.3 mg kg^{-1} for lead, resulting in an average percentage error of 15.6% and 18.9%, respectively, within the concentration range of the sample set.

3.2. Machine Learning Regression: Partial Least Squares

The X-ray fluorescent emission spectra of the 26 samples, which were utilized to develop the PLS models, cover a range of energies, including the Kα (10.54 keV) and Kβ (11.73 keV) lines for arsenic, as well as Lα (10.55 keV) and Lβ (12.61 keV) lines for lead, as shown in Figure 2.

When considering two latent variables, the model shows RMSECVs of 2.95 mg kg^{-1} and 3.92 mg kg^{-1}. For external validation, there is an RMSEP of 3.31 mg kg^{-1} and 5.03 mg kg^{-1} for As and Pb, respectively. The discrepancy between both parameters is expected due to the tendency of cross-validation to be overly optimistic in comparison to external sample validation.

Additionally, it is observed that both cross-validation and external validation predictions for these elements exhibit a high level of correlation with the real concentrations of the samples, achieving correlation coefficients rVal = 0.981 and rPred = 0.972 for As and rVal = 0.986 and rPred = 0.982 for Pb (Figure 3).

The quantification biases for arsenic and lead, observed in both multivariate models and classical quantification methods (internal standardization and deconvolution), have significantly reduced. There is no longer a pronounced positive trend, and the percentage of biases has notably decreased. The summary of all these parameters for the PLS model is presented in Table 7.

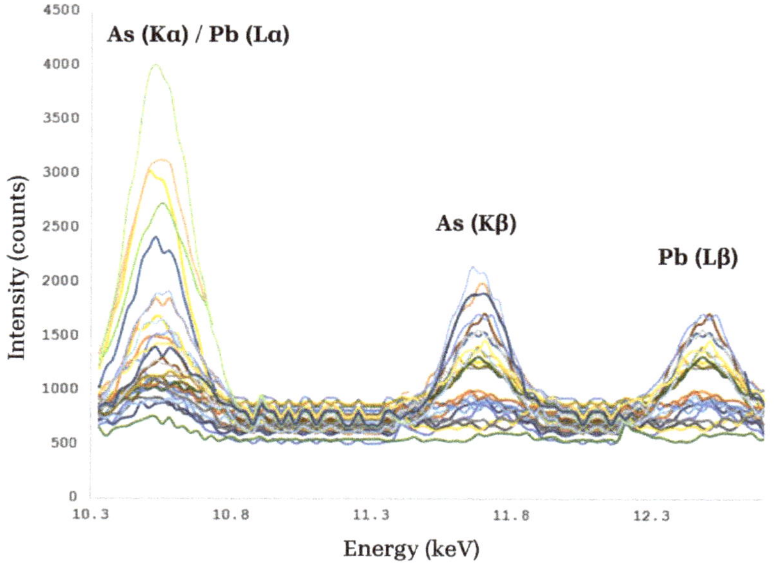

Figure 2. X-ray fluorescence spectra of the 26 calibration samples. The spectra of each sample are represented in different colors.

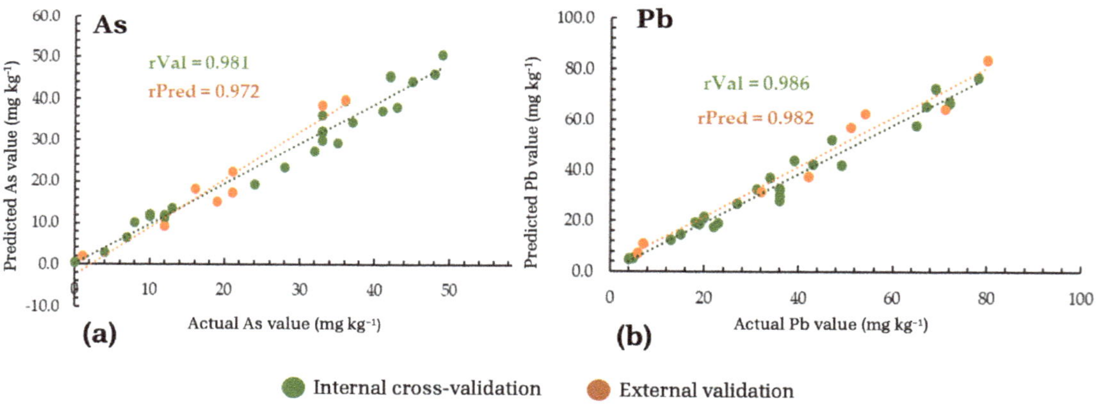

Figure 3. Actual vs. predicted values of arsenic (**a**) and lead (**b**) in cross-validation and external validation procedure.

Table 7. Characteristics and parameters of the PLS model for As and Pb.

	As Model	Pb Model
RMSECV (mg kg^{-1})	2.95	3.92
rVal	0.981	0.986
RMSEP (mg kg^{-1})	3.31	5.03
rPred	0.972	0.982
Calibration Samples	26	
Number of Variables	257	
Pretreatment	Mean-centering	
Latent Variables	2	
Range [As] (mg kg^{-1})	0 to 49	
Range [Pb] (mg kg^{-1})	4 to 78	

When examining the regression vectors depicted in Figure 4, it can be observed that the multivariate model's predictions for arsenic concentration predominantly rely on signals produced within the energy range corresponding to its primary emission line (Kα). Signals from the secondary emission line (Kβ) also have a lesser impact. This finding highlights how incorporating multiple variables and assessing their contributions improves the accuracy of predicting arsenic levels, effectively managing interference from lead's fluorescent signal.

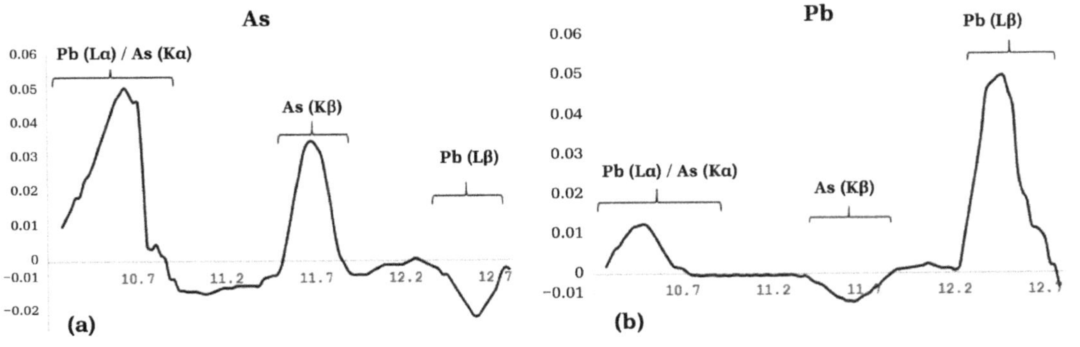

Figure 4. Regression coefficients for the PLS models for the quantification of arsenic (**a**) and lead (**b**), including lines Kα (10.54 keV) and Kβ (11.73 keV) of arsenic and Lα (10.55 keV) and Lβ (12.61 keV) of lead.

In contrast, there is a more evident overlap of signals, particularly from lead's most intense emission line (Lα). Consequently, signals from its weaker line (Lβ) become more significant within the model. This indicates that reliance on signals from the Lβ zone is higher due to interference with the stronger Lα line.

3.3. Elliptical Joint Confidence Region (EJCR)

In employing the EJCR correlation method at a significance level (α) of 0.05, it has been statistically shown that determinations made using univariate methods like internal standardization and deconvolution lack accuracy. This conclusion is drawn from the failure to meet the criteria of a slope of 1 and an intercept of 0, indicating a deviation from the ideal correlation.

Figure 5 visually illustrates this difference. The ellipse in the figure demonstrates the disparity between the ideal point representing slope vs. intercept and the observed correlation, highlighting that there is a significant distance between the observed data points and their ideal position.

The deconvolution via software prior internal standardization fails to satisfy the EJCR criteria, as evidenced by significant deviations in slope and intercept from their ideal values ($\alpha = 0.05$).

In contrast, multivariate calibration using Partial Least Squares demonstrates accuracy in quantification. Adhering to the EJCR criteria with an ideality point within the ellipse confirms that both slope and intercept statistically approach 1 and 0, respectively ($\alpha = 0.05$).

This comparison emphasizes how effective multivariate calibration through PLS is in addressing overlapping X-ray fluorescence signals of arsenic and lead. Consequently, this method produces more precise and accurate results than deconvolution processes.

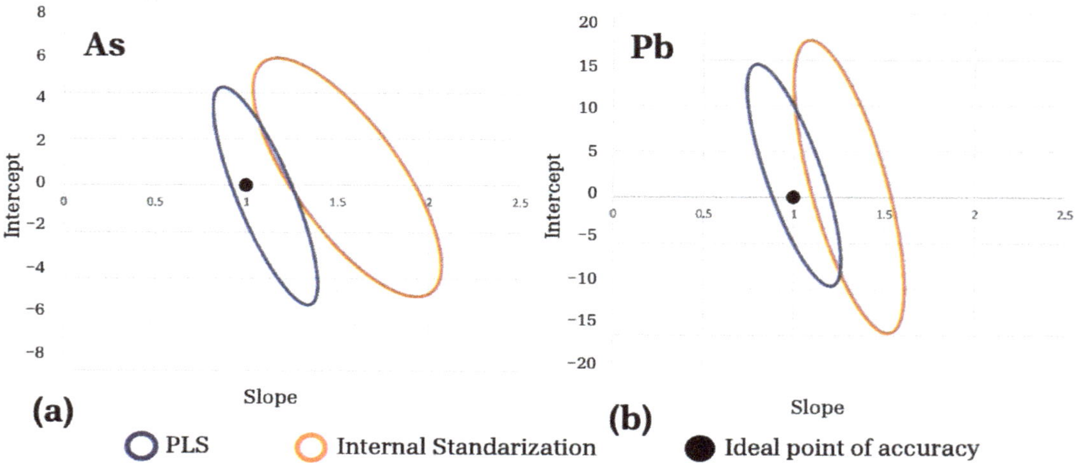

Figure 5. EJCR for the quantification of arsenic (**a**) and lead (**b**) by means of internal standardization (orange ellipse) and PLS (blue ellipse). The black dot indicates the ideal: a slope of 1 and an intercept of 0.

3.4. AFOM

Table 8 displays the analytical performance metrics for each quantification approach. It is evident from the table that the PLS method has resulted in an enhanced limit of detection by two orders of magnitude and limit of quantification by three orders and two orders of magnitude for As and Pb, respectively, compared to internal standard calibration. The analytical sensitivities obtained through PLS models have also seen important improvements.

Table 8. Analytical figures of merit using internal standardization and PLS methods are presented, including limit of detection (LOD), limit of quantification (LOQ), sensitivity, and analytical sensitivity.

Figure of Merit	Internal Standard		PLS	
	As	Pb	As	Pb
LOD (mg kg^{-1})	5.42×10^{-2}	6.28×10^{-2}	2.56×10^{-4}	3.48×10^{-4}
LOQ (mg kg^{-1})	1.64×10^{-1}	1.90×10^{-1}	7.69×10^{-4}	1.05×10^{-3}
Sensitivity (counts (mg kg^{-1})$^{-1}$)	432	298	474	402
Analytical Sensitivity ((mg kg^{-1})$^{-1}$)	18.3	12.7	59.6	63.3

3.5. Application in Soil Samples

Due to the lack of noticeable differences in the spectra between liquid and well-prepared solid samples within the relevant range, a comparison of techniques was conducted using a validated soil sample (Clay 2—CRM051). The assessment employing a certified soil sample, based on the t-test criteria ($\alpha = 0.05$, $n = 3$), reveals that both techniques yield results that are not statistically different from the certified value (Table 9). However, in terms of bias and accuracy, the PLS model surpasses the quantification achieved through internal standardization and signal deconvolution. The PLS model demonstrates lower bias and higher accuracy compared to other methods. Subsequently, the developed PLS models for arsenic and lead quantification were utilized to evaluate these metals in the viticultural soils of Itata Valley.

Table 9. Comparison of accuracy for arsenic (As) and lead (Pb) quantification in Certified Soil Sample Clay 2—CRM051 using internal standardization and PLS methods.

	Certified Value		Internal Standard (n = 3)		PLS Model (n = 3)	
	As	Pb	As	Pb	As	Pb
Mean (mg kg^{-1})	45.5	68.1	55.3	79.2	51.2	63.2
SD [1]	4.45	7.6	6.23	11.2	7.23	8.23
RSD [2] (%)	9.78	11.16	11.27	14.14	14.12	13.02
p (α = 0.05)	-	-	0.077	0.215	0.31	0.483

[1] Standard deviation; [2] relative standard deviation.

3.6. Ecological Assessment of Viticultural Soils

Table 10 provides a statistical overview of the contents of As and Pb in topsoil samples. The difference between the highest and lowest concentrations of As and Pb indicates spatial diversity within the topsoil samples. This variance suggests natural disparities in soil composition among the locations sampled, potentially influenced by variations in mineral content, organic matter, geological origins, or human activities [2].

Table 10. Statistical summary of As and Pb concentrations (mg kg^{-1}) in soils from Itata Valley.

	Median	Min	Percentile					Max	Background Upper Limit	Outliers (%)
			5	25	50	75	95			
As	2.34	0.56	0.92	1.24	2.34	3.45	4.72	5.01	0.83	95.8
Pb	22.5	6.02	7.45	13.0	22.5	31.1	42.7	45.5	6.90	95.6

After establishing the upper limits based on the background levels, an evaluation was conducted to determine how many samples exceed these limits for both arsenic and lead, as detailed in Table 10. The results indicate that a majority of samples—approximately 95% for both elements—have concentrations exceeding these established upper limits.

Figure 6 graphically illustrates the results of the GEOI analysis, suggesting that the soils exhibit relatively low to moderate levels of contamination. This evaluation likely considers specific element concentrations in relation to their natural or background levels. EF assessments indicate a significant number of sampling points showing moderate to substantial enrichment, particularly when compared to iron concentrations, which show minimal vertical variation in the soil. The more rigorous CF index, not considering natural variations, indicates that most sampling points demonstrate moderate to considerable contamination levels for both elements. These conclusions align with the areas identified by the GEOI as moderately contaminated and reinforce their importance in terms of contamination severity.

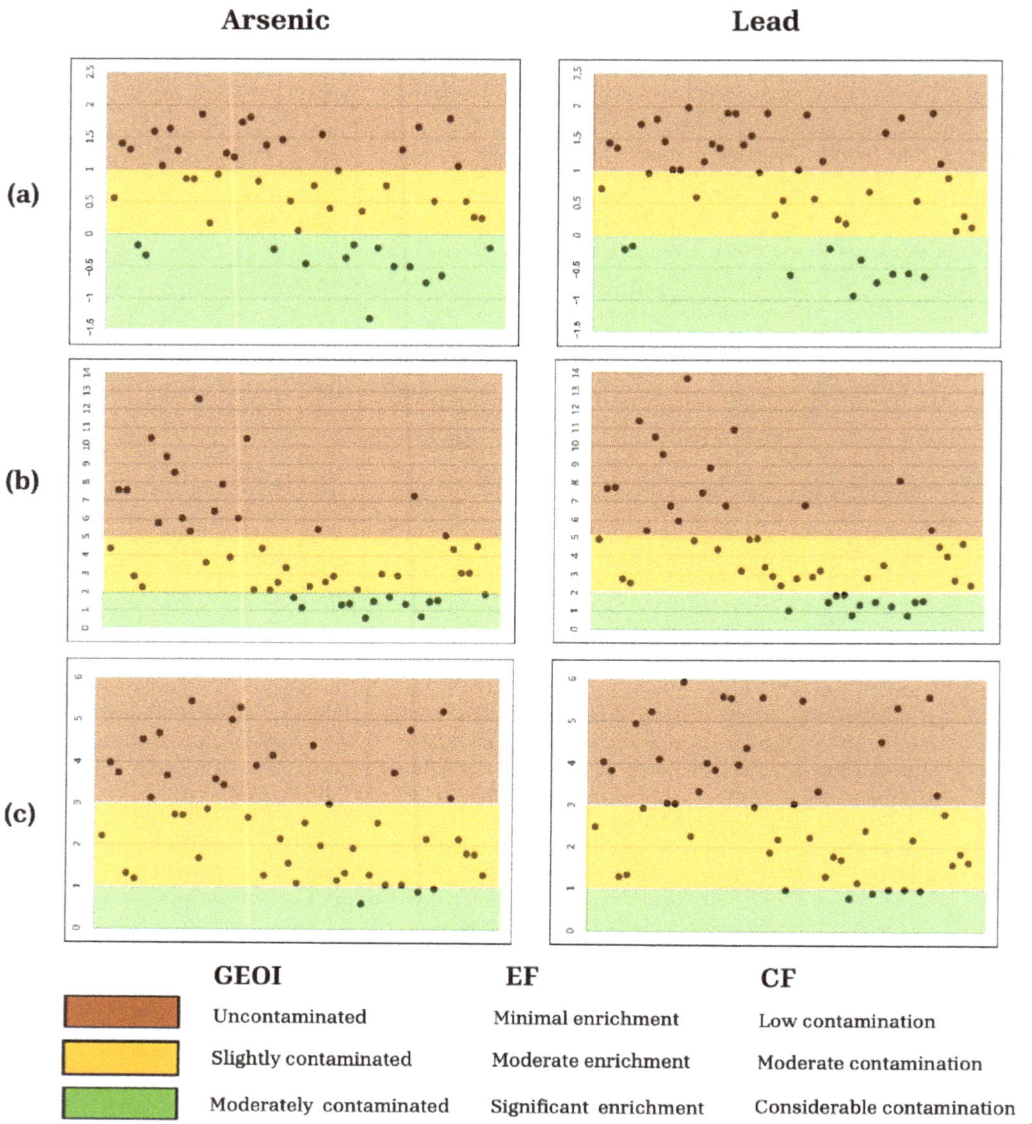

Figure 6. Ecological indices of every viticultural soil sample (black dots) from the Itata Valley for As (arsenic) and Pb (lead). Geoaccumulation Index (**a**), Enrichment Factor (**b**), and Contamination Factor (**c**).

4. Discussion

While lower RMSE values are preferable, the average percentage errors of 15.6% for arsenic and 18.9% for lead quantification using classical regression and deconvolution are considered reasonably acceptable, particularly given the analytical technique and concentration range involved. Such error magnitudes are common in this field, as evident from various studies on solid matrices [2,44–46].

Conversely, the utilization of the proposed PLS machine learning algorithm demonstrates improved performance, with the average percentage error notably reduced from

15.6 to 9.4% for arsenic (from an RMSEP of 5.6 mg kg^{-1} to 3.3 mg kg^{-1}) and from 18.9% to 6.8% for lead (from an RMSEP of 12.3 mg kg^{-1} to 5.03 mg kg^{-1}) compared to the previous univariable model. This enhanced predictive accuracy within the sample set's concentration range is attributable to the efficiency of the multivariate calibration's first-order advantage in quantifying in the presence of interferents, which enables the identification and use of lower-intensity signals to mitigate the errors generated by signal overlaps [46]. The RMSEP results undoubtedly demonstrate improvement; however, they do not surpass the performance achieved for liquid samples in TXRF using PLS models. Notably, the multivariate model developed by Nagata et al. (2006) [12] facilitated the determination of lead and arsenic with RMSEPs of 0.03 and 0.24 mg L^{-1}, respectively. However, the Slurry-TXRF with PLS improvements are limited in comparison to liquid sample analysis, primarily due to two factors. Firstly, liquid samples present fewer physical interferences for total reflection [10]. Achieving a comparable state from solid samples would necessitate significant resources, including time and reagents for digestion. Secondly, the TXRF equipment utilized in the aforementioned research employs synchrotron radiation TXRF (SR-TXRF), which enhances the instrument performance [12].

According to the EJCR criteria, the deconvolution method provided by the univariate calibration software for mitigating the overlap of fluorescent arsenic and lead signals appears inadequate in generating reliable quantifications, in contrast with the superior performance of the PLS method. However, as mentioned earlier, the quality of results obtained through classical calibration is commonly encountered within this analytical technique and concentration range.

Alongside the improved quality of results observed with PLS quantification, all figures of merit are also enhanced, potentially due to the multivariate calibration's ability to exploit modeled interferents, thereby improving selectivity and reducing background noise [36].

The proposed methodology also reduces costs and analysis time compared to traditional methods [11]. This novel approach, based on Slurry-TXRF and PLS quantification models, facilitates soil monitoring for the measurement of arsenic and lead contents, and it is presented as an alternative with a positive technical and environmental impact. This is a preliminary advancement in an era where automation and unmanned systems, known for their reliability and cost-effectiveness, are becoming increasingly popular. [47].

The enhanced methodology for quantifying arsenic and lead indicates that approximately 95% of the agricultural soil samples from the Itata Valley exceed the background limit, which is expected for a soil type influenced by productive anthropogenic activity where the addition of these elements is possible through the use of fertilizers [48] and fossil fuels from machinery [49]. Additionally, the geomorphological fluvial–glacial–volcanic plain zone that the Itata Valley covers [24] could significantly contribute to the presence of these two elements in the environment, especially arsenic [50].

In general, the ecological indices suggest that the agricultural soils in the studied area are moderately contaminated by these elements, consistent with previous findings on other potentially toxic substances. However, this level of contamination does not necessarily imply immediate adverse biological effects on the resident communities or pose significant health risks to humans.

5. Conclusions

The methodology proposed in our study, which integrates Slurry-TXRF with a Partial Least Squares (PLS) machine learning algorithm, offers a significant advancement in the field of environmental monitoring, particularly for arsenic and lead contamination in soils. While the accuracy of our results does not surpass that of traditional methods, such as hydride generation for arsenic (which typically achieves errors of less than 5% with quantification limits of 0.2 mg/kg) [51] and flame atomic absorption for lead (with expected errors around 2%) [52], the application of machine learning techniques has notably improved the performance of Slurry-TXRF. Specifically, our study achieved a reduction in the average percentage error from 15.6% to 9.4% for arsenic (an RMSEP from 5.6 mg kg^{-1} to 3.3 mg kg^{-1}) and from

18.9% to 6.8% for lead (an RMSEP from 12.3 mg kg^{-1} to 5.03 mg kg^{-1}) just through using multivariate calibration.

These improvements, while not surpassing the precision and accuracy of the most established methods, provide significant practical advantages for environmental monitoring. The proposed methodology is faster, simpler, less costly, and more environmentally friendly, making it a valuable alternative in situations where these factors are critical.

Additionally, the PLS calibration effectively corrects the mutual influence of overlapping signals between arsenic and lead, enhancing the reliability of quantitative analyses. This is particularly advantageous in complex soil matrices where traditional methods may struggle with signal interference.

Overall, while the traditional methods remain superior in numerical accuracy, the combined use of Slurry-TXRF and machine learning represents a promising alternative that balances reasonable accuracy with substantial operational benefits. This approach is well suited for efficient environmental and ecological evaluations, particularly in regions where rapid and cost-effective monitoring is necessary.

Our assessment indicates moderate to significant contamination levels with arsenic and lead in the soils of the Itata Valley, which is typical for agriculturally used soils. However, further studies would be required to assess the potential risks to human health, including an evaluation of factors such as elemental speciation and migration from soil to plant parts, which are outside the scope of the present study.

Author Contributions: Conceptualization, G.M.-G.; data curation, Y.M. and P.A.; formal analysis, G.M.-G., Y.M., and A.M.-B.; investigation, G.M.-G., E.M., and P.A.; methodology, G.M.-G. and J.C.; project administration, Y.M. and P.A.; resources, G.M.-G.; supervision, G.M.-G.; founding, E.M.; validation, G.M.-G., E.M., and J.C.; visualization, Y.R.-G. and A.M.-B.; writing—original draft, P.A. and A.M.-B.; writing—review and editing, G.M.-G., Y.M., E.M., J.C., and Y.R.-G. All authors have read and agreed to the published version of the manuscript.

Funding: This research was funded by the Agencia Nacional de Investigación, grant number 21170516. The APC was funded by the Universidad Católica de la Santísima Concepción by means of projects Ingeniería 2030 (ING222010004) and InES Ciencia Abierta (INCA210005).

Data Availability Statement: The data supporting this article will be made available by the authors upon request.

Acknowledgments: The authors are deeply grateful for the support of project REDTPI4.0-320RT0006 CYTED program.

Conflicts of Interest: Author Yelena Medina was employed by the company EMOingenieros Ltda. The remaining authors declare that the research was conducted in the absence of any commercial or financial relationships that could be construed as a potential conflict of interest.

References

1. Guo, J.; Liu, L.; Feng, Q.; Okonkwo, C.U.; Zhang, X.; Chen, J.; Yang, S.; Yang, C. Heavy Metal Pollution near an Abandoned Mercury-Bearing Waste Recovery Enterprise in Southwestern China: Spatial Distribution and Its Sources in Soil and Plants. *Environ. Res.* **2024**, *252*, 118694. [CrossRef]
2. Medina-González, G.; Medina, Y.; Muñoz, E.; Fuentes, P. Rapid and Convenient Assessment of Trace Element Contamination in Agricultural Soils through Slurry-TXRF and Ecological Indices: The Ñuble Region, Chile as a Case Study. *Sustainability* **2023**, *15*, 9190. [CrossRef]
3. Saha, J.K.; Selladurai, R.; Coumar, M.V.; Dotaniya, M.L.; Kundu, S.; Patra, A.K. Assessment of Heavy Metals Contamination in Soil. In *Soil Pollution—An Emerging Threat to Agriculture*; Environmental Chemistry for a Sustainable World; Springer Singapore: Singapore, 2017; Volume 10, pp. 155–191. ISBN 978-981-10-4273-7.
4. Mao, J.; Liu, X.; Chen, B.; Luo, F.; Wu, X.; Jiang, D.; Luo, Z. Determination of Heavy Metals in Soil by Inductively Coupled Plasma Mass Spectrometry (ICP-MS) with Internal Standard Method. *Electron. Sci. Technol. Appl.* **2017**, *4*. [CrossRef]
5. Yahaya, S.M.; Abubakar, F.; Abdu, N. Ecological Risk Assessment of Heavy Metal-Contaminated Soils of Selected Villages in Zamfara State, Nigeria. *SN Appl. Sci.* **2021**, *3*, 168. [CrossRef]
6. Al-Momani, I.F.; Al-Massaedh, A.A. Assessment of Heavy Metal Contamination in Roadside Soils along Irbid-Amman Highway, Jordan by ICP-OES. *Jordan J. Chem.* **2020**, *15*, 1–12. [CrossRef]

7. Ahmad, I.; Ansari, T.M. An Assessment of Toxic Heavy Metals in Soil and Plants (Allium Cepa and Daucus Carota) by GFAAS. *Int. J. Environ. Anal. Chem.* **2022**, *102*, 1029–1048. [CrossRef]
8. Ismičić-Tanjo, D.; Huremović, J.; Selović, A.; Šehović, E. Human Health Risk Assessment of Mercury in Soil–Plants System. *Int. J. Environ. Res.* **2021**, *15*, 797–805. [CrossRef]
9. Carvalho, P.M.; Marguí, E.; Kubala-Kukuś, A.; Banaś, D.; Machado, J.; Casal, D.; Pais, D.; Santos, J.P.; Pessanha, S. Evaluation of Different Analytical Approaches Using Total Reflection X-Ray Fluorescence Systems for Multielemental Analysis of Human Tissues with Different Adipose Content. *Spectrochim. Acta Part B At. Spectrosc.* **2022**, *198*, 106548. [CrossRef]
10. Klockenkämper, R.; Bohlen, A. *Von Total-Reflection X-Ray Fluorescence Analysis and Related Methods*; John Wiley & Sons: Hoboken, NJ, USA, 2014; ISBN 978-1-118-98587-8.
11. Liu, C.; Zhang, F.; Wang, X.; Chan, N.W.; Rahman, H.A.; Yang, S.; Tan, M.L. Assessing the Factors Influencing Water Quality Using Environment Water Quality Index and Partial Least Squares Structural Equation Model in the Ebinur Lake Watershed, Xinjiang, China. *Environ. Sci. Pollut. Res.* **2022**, *29*, 29033–29048. [CrossRef]
12. Nagata, N.; Peralta-Zamora, P.G.; Poppi, R.J.; Perez, C.A.; Bueno, M.I.M.S. Multivariate Calibrations for the SR-TXRF Determination of Trace Concentrations of Lead and Arsenic in the Presence of Bromine. *X-Ray Spectrom.* **2006**, *35*, 79–84. [CrossRef]
13. González, G.M.; Castillo, R.d.P.; Neira, J.Y. Multivariate Calibration for the Improvement of the Quantification of Cadmium in the Presence of Potassium as Interferent by Total Reflection X-Ray Fluorescence. *X-Ray Spectrom.* **2019**, *48*, 700–707. [CrossRef]
14. Carvajal, M.; Jeldres, P.; Vergara, A.; Lobaina, E.; Olivares, M.; Meza, D.; Velásquez, A.; Dorta, F.; Jorquera, F.; Seeger, M.; et al. Bioremoval of Copper by Filamentous Fungi Isolated from Contaminated Soils of Puchuncaví-Ventanas Central Chile. *Environ. Geochem. Health* **2023**, *45*, 4275–4293. [CrossRef] [PubMed]
15. Corradini, F.; Meza, F.; Calderón, R. Trace Element Content in Soil after a Sediment-Laden Flood in Northern Chile. *J. Soils Sediments* **2017**, *17*, 2500–2515. [CrossRef]
16. Madrid, E.; Pino, J.A.; Muñoz, S.; Cardemil, F.; Martinez, F.; Berasaluce, M.; San Martín, S.; Reyes, J.; González-Miranda, I. Oxidative Damage Associated with Exposure to Heavy Metals Present in Topsoils in Central Chile. *Environ. Geochem. Health* **2023**, *45*, 9891–9901. [CrossRef] [PubMed]
17. Rubinos, D.A.; Jerez, Ó.; Forghani, G.; Edraki, M.; Kelm, U. Geochemical Stability of Potentially Toxic Elements in Porphyry Copper-Mine Tailings from Chile as Linked to Ecological and Human Health Risks Assessment. *Environ. Sci. Pollut. Res.* **2021**, *28*, 57499–57529. [CrossRef]
18. Tume, P.; González, E.; Reyes, F.; Fuentes, J.P.; Roca, N.; Bech, J.; Medina, G. Sources Analysis and Health Risk Assessment of Trace Elements in Urban Soils of Hualpen, Chile. *Catena* **2019**, *175*, 304–316. [CrossRef]
19. Tume, P.; Melipichún, T.; Ferraro, F.; Sepúlveda, B.; Roca, N.; Bech, J. Contamination of As, Cd, Cr, Hg and Pb in Soils in Arica Commune (Chile). *Environ. Geochem. Health* **2023**, *45*, 9199–9213. [CrossRef]
20. Li, Z.; He, M.-Y.; Li, B.; Wen, X.; Zhou, J.; Cheng, Y.; Zhang, N.; Deng, L. Multi-Isotopic Composition (Li and B Isotopes) and Hydrochemistry Characterization of the Lakko Co Li-Rich Salt Lake in Tibet, China: Origin and Hydrological Processes. *J. Hydrol.* **2024**, *630*, 130714. [CrossRef]
21. Liu, J.; Wang, Y.; Li, Y.; Peñuelas, J.; Zhao, Y.; Sardans, J.; Tetzlaff, D.; Liu, J.; Liu, X.; Yuan, H.; et al. Soil Ecological Stoichiometry Synchronously Regulates Stream Nitrogen and Phosphorus Concentrations and Ratios. *Catena* **2023**, *231*, 107357. [CrossRef]
22. US EPA, O. SW-846 Test Method 6010D: Inductively Coupled Plasma-Optical Emission Spectrometry (ICP-OES). Available online: https://www.epa.gov/hw-sw846/sw-846-test-method-6010d-inductively-coupled-plasma-optical-emission-spectrometry-icp-oes (accessed on 16 August 2024).
23. Yang, Z.; Lu, W.; Long, Y.; Bao, X.; Yang, Q. Assessment of Heavy Metals Contamination in Urban Topsoil from Changchun City, China. *J. Geochem. Explor.* **2011**, *108*, 27–38. [CrossRef]
24. Serra, I.; Calderón-Orellana, A.; Hidalgo, M. Heroic Viticulture in Itata Valley, Chile: Characteristics and Challenges for the Development of Unique Wines in Southern Chilean Vineyards. In *Latin American Viticulture Adaptation to Climate Change: Perspectives and Challenges of Viticulture Facing up Global Warming*; Gutiérrez Gamboa, G., Fourment, M., Eds.; Springer International Publishing: Cham, Switzerland, 2024; pp. 215–228. ISBN 978-3-031-51325-1.
25. Stolpe, B.N. *Descripciones de Los Principales Suelos de La VIII Región de Chile*; Universidad de Concepción: Concepción, Chile, 2006.
26. Protocolo de Toma de Muestras de Suelos | SAG. Available online: https://www.sag.gob.cl/content/protocolo-de-toma-de-muestras-de-suelos (accessed on 21 June 2024).
27. Kang, Z.; Wang, S.; Qin, J.; Wu, R.; Li, H. Pollution Characteristics and Ecological Risk Assessment of Heavy Metals in Paddy Fields of Fujian Province, China. *Sci. Rep.* **2020**, *10*, 12244. [CrossRef]
28. Feng, F.; Jiang, Y.; Jia, Y.; Lian, X.; Shang, C.; Zhao, M. Exogenous-Organic-Matter-Driven Mobilization of Groundwater Arsenic. *Environ. Sci. Ecotechnol.* **2023**, *15*, 100243. [CrossRef] [PubMed]
29. XRF Software. Available online: https://www.bruker.com/en/products-and-solutions/elemental-analyzers/xrf-spectrometers/xrf-software.html (accessed on 21 June 2024).
30. *Pirouette Software*, Version 4.5; Infometrix Inc.: Bothell, WA, USA, 2014.
31. Kusumiyati; Mubarok, S.; Sutari, W.; Hadiwijaya, Y. Application of Spectra Pre-Treatments on Firmness Assessment of Intact Sapodilla Using Vis-Nir Spectroscopy. *IOP Conf. Ser. Earth Environ. Sci.* **2021**, *644*, 012001. [CrossRef]

32. Montesinos López, O.A.; Montesinos López, A.; Crossa, J. Overfitting, Model Tuning, and Evaluation of Prediction Performance. In *Multivariate Statistical Machine Learning Methods for Genomic Prediction*; Montesinos López, O.A., Montesinos López, A., Crossa, J., Eds.; Springer International Publishing: Cham, Switzerland, 2022; pp. 109–139. ISBN 978-3-030-89010-0.
33. Westad, F.; Marini, F. Variable Selection and Redundancy in Multivariate Regression Models. *Front. Anal. Sci.* **2022**, *2*, 897605. [CrossRef]
34. Monago-Maraña, O.; Domínguez-Manzano, J.; Muñoz de la Peña, A.; Durán-Merás, I. Second-Order Calibration in Combination with Fluorescence Fibre-Optic Data Modelling as a Novel Approach for Monitoring the Maturation Stage of Plums. *Chemom. Intell. Lab. Syst.* **2020**, *199*, 103980. [CrossRef]
35. Beh, E.J. Elliptical Confidence Regions for Simple Correspondence Analysis. *J. Stat. Plan. Inference* **2010**, *140*, 2582–2588. [CrossRef]
36. Short, S.M.; Cogdill, R.P.; Anderson, C.A. Determination of Figures of Merit for Near-Infrared and Raman Spectrometry by Net Analyte Signal Analysis for a 4-Component Solid Dosage System. *AAPS PharmSciTech* **2007**, *8*, 96. [CrossRef] [PubMed]
37. *ISO 19258:2018(En)*; Soil Quality—Guidance on the Determination of Background Values. Available online: https://www.iso.org/obp/ui/en/#iso:std:iso:19258:ed-2:v1:en (accessed on 21 June 2024).
38. Reimann, C.; de Caritat, P. Distinguishing between Natural and Anthropogenic Sources for Elements in the Environment: Regional Geochemical Surveys versus Enrichment Factors. *Sci. Total Environ.* **2005**, *337*, 91–107. [CrossRef] [PubMed]
39. Kowalska, J.B.; Mazurek, R.; Gąsiorek, M.; Zaleski, T. Pollution Indices as Useful Tools for the Comprehensive Evaluation of the Degree of Soil Contamination—A Review. *Environ. Geochem. Health* **2018**, *40*, 2395–2420. [CrossRef]
40. Muller, G. Index of Geoaccumulation in Sediments of the Rhine River. *GeoJournal* **1969**, *2*, 108–118.
41. Samuel, B.; Solomon, S.; Daniel, F.; Zinabu, G.M.; Riise, G. Heavy Metals Contamination of Soil in the Vicinity of Hawassa Industrial Zone, Ethiopia. *J. Appl. Sci. Environ. Manag.* **2020**, *24*, 1447–1454. [CrossRef]
42. Zhu, Y.; Wang, L.; Zhao, X.; Lian, J.; Zhang, Z. Accumulation and Potential Sources of Heavy Metals in Soils of the Hetao Area, Inner Mongolia, China. *Pedosphere* **2020**, *30*, 244–252. [CrossRef]
43. Hakanson, L. An Ecological Risk Index for Aquatic Pollution Control.a Sedimentological Approach. *Water Res.* **1980**, *14*, 975–1001. [CrossRef]
44. Towett, E.K.; Shepherd, K.D.; Cadisch, G. Quantification of Total Element Concentrations in Soils Using Total X-Ray Fluorescence Spectroscopy (TXRF). *Sci. Total Environ.* **2013**, *463–464*, 374–388. [CrossRef] [PubMed]
45. Bilo, F.; Borgese, L.; Cazzago, D.; Zacco, A.; Bontempi, E.; Guarneri, R.; Bernardello, M.; Attuati, S.; Lazo, P.; Depero, L.E. TXRF Analysis of Soils and Sediments to Assess Environmental Contamination. *Environ. Sci. Pollut. Res.* **2014**, *21*, 13208–13214. [CrossRef] [PubMed]
46. Espinoza-Quiñones, F.R.; Módenes, A.N.; dos Santos, J.; Obregón, P.L.; de Pauli, A.R. Insights on Limits of Detection, Precision and Accuracy in TXRF Analysis of Trace and Major Elements in Environmental Solid Suspensions. *Appl. Radiat. Isot.* **2018**, *137*, 80–90. [CrossRef] [PubMed]
47. Liu, L.; Gong, Y.; Miao, Y.; Guo, J.; Long, H.; Feng, Q.; Chen, Y. New Trends in Pollution Prevention and Control Technology for Healthcare and Medical Waste Disposal in China. *Processes* **2024**, *12*, 7. [CrossRef]
48. Khatun, J.; Intekhab, A.; Dhak, D. Effect of Uncontrolled Fertilization and Heavy Metal Toxicity Associated with Arsenic(As), Lead(Pb) and Cadmium (Cd), and Possible Remediation. *Toxicology* **2022**, *477*, 153274. [CrossRef] [PubMed]
49. Inumaru, J.; Hasegawa, T.; Shirai, H.; Nishida, H.; Noda, N.; Ohyama, S. 1—Fossil Fuels Combustion and Environmental Issues. In *Advances in Power Boilers*; Ozawa, M., Asano, H., Eds.; JSME Series in Thermal and Nuclear Power Generation; Elsevier: Amsterdam, The Netherlands, 2021; Volume 2, pp. 1–56. ISBN 978-0-12-820360-6.
50. Bundschuh, J.; Schneider, J.; Alam, M.A.; Niazi, N.K.; Herath, I.; Parvez, F.; Tomaszewska, B.; Guilherme, L.R.G.; Maity, J.P.; López, D.L.; et al. Seven Potential Sources of Arsenic Pollution in Latin America and Their Environmental and Health Impacts. *Sci. Total Environ.* **2021**, *780*, 146274. [CrossRef]
51. Nakagoe, S.; Olivares, J.C.; García, L.; Yubero, F.; Monteiro, M.; Mereles, L.; Diez-Pérez-Núñez, D. Validación de un método analítico para la determinación de arsénico total en muestras de suelo por espectrofotometría de absorción atómica con generación de hidruros. *Innotec* **2023**, *25*, e635. [CrossRef]
52. Sabol, J. Major Analytical Methods for Determining Lead in Environmental and Biological Samples. In *Lead in Plants and the Environment*; Gupta, D.K., Chatterjee, S., Walther, C., Eds.; Springer International Publishing: Cham, Switzerland, 2020; pp. 1–13. ISBN 978-3-030-21638-2.

MDPI AG
Grosspeteranlage 5
4052 Basel
Switzerland
Tel.: +41 61 683 77 34

Processes Editorial Office
E-mail: processes@mdpi.com
www.mdpi.com/journal/processes

www.ingramcontent.com/pod-product-compliance
Lightning Source LLC
Chambersburg PA
CBHW061602210926
51850CB00009B/2189
9783725850471